D1538758

St. Olaf College

SEP 1 1 1989

Science Library

765

# TECTONIC EVOLUTION
# OF THE HIMALAYAS AND TIBET

# TECTONIC EVOLUTION OF THE HIMALAYAS AND TIBET

PROCEEDINGS OF
A ROYAL SOCIETY DISCUSSION MEETING
HELD ON 11 AND 12 NOVEMBER 1987

ORGANIZED AND EDITED BY
R. M. SHACKLETON, F.R.S., J. F. DEWEY, F.R.S.,
AND B. F. WINDLEY

LONDON
THE ROYAL SOCIETY
1988

Printed in Great Britain for the Royal Society
by the
University Press, Cambridge
ISBN 0 85403 359 9

First published in *Philosophical Transactions of the Royal Society of London*,
series A, volume 326 (no. 1589), pages 1–325

∞ The text paper used in this publication meets the minimum requirements of American National Standard for Information Sciences—Permanence of Paper for Printed Library Materials, ANSI Z39.48-1984.

## Copyright

© 1988 The Royal Society and the authors of individual papers.

It is the policy of the Royal Society not to charge any royalty for the production of a single copy of any one article made for private study or research. Requests for the copying or reprinting of any article for any other purpose should be sent to the Royal Society.

**British Library Cataloguing in Publication Data**

Tectonic evolution of the Himalayas and Tibet.
1. Himalayas. Tectonic features
I. Shackleton, R. M.   II. Dewey, J. F.
(John Frederick),   III. Windley, B. F.
(Brian Frederick)   IV. Royal Society
V. Philosophical transactions of the Royal
Society of London. Series A
555.4

ISBN 0-85403-359-9

Published by the Royal Society
6 Carlton House Terrace, London SW1Y 5AG

# PREFACE ✦

Two factors led us to arrange a Discussion Meeting on the tectonic evolution of the Himalayas: first, the development of plate tectonic theory, and the problems of applying it to the continental crust, have focused attention on the Himalayas and Tibet as the prime examples of the result of the recent collision of two continents; and second, a very large amount of work has just been completed in both the Himalayas and Tibet. Besides presenting the results of this new work, including geological mapping, geochemistry, structure, palaeontology, seismic studies and palaeomagnetism, we attempted to cover the whole span of the Himalayan belt, extending some 3000 km from Afghanistan to Burma, by inviting review papers on regions not otherwise discussed. The results of the recent Royal Society–Academia Sinica Tibet Geotraverse are presented separately in a volume shortly to be published by the Society.

The Tibetan Plateau, to the north of the Himalayas, evolved by the successive accretion to Eurasia, during the Palaeozoic and Mesozoic, of terranes derived from the south. The Karakoram Range is an island arc terrane accreted to Eurasia in the Cretaceous. Northward subduction of the Tethyan oceanic crust under an Andean-type margin (the Gangdise Shan or Transhimalaya) culminated in collision and the indentation of Eurasia by the Indian Plate. This doubled the thickness of the Tibetan crust and led to the formation of the huge Himalayan Range. Thus the Tibetan–Himalayan region shows a progressive southward migration of magmatism and tectonism from late Palaeozoic onwards. Activity now is concentrated in the southern foothills of the Himalayas.

In many ways, the Himalayas represent the classic collisional orogenic belt. They contain every significant stage in the Wilson cycle from Permian intracontinental rifting to the Quaternary neotectonics and rifting of Tibet. Precollisional ophiolite obduction and thrusting of shelf sediments, syncollisional suturing and thrusting, and postcollisional indentation of Eurasia by India, giving rise to late thrusts, inverted isograds and Miocene leucogranites, are all easily separable. Thus the Himalayas provide a model of collisional orogenesis that may be applicable to many more complex and more obscure older orogenic belts.

*April 1988*

R. M. Shackleton
J. F. Dewey
B. F. Windley

## Dedication

This book is dedicated to the memory of Rein Tirrul who died tragically in a motor accident in Ottawa in May 1987. Rein was to have presented the talk on the Karakoram at this Royal Society Meeting following our two expeditions to the Baltoro glacier system in 1985 and the Hushe Valley along the southern flanks of Masherbrum in 1986. His meticulous mapping of the remote mountain country south of the Baltoro glacier must rank among the most accurate and detailed mapping accomplished anywhere in the Himalaya–Karakoram Ranges.

As a person, Rein was one of the very best: good natured, even in the most desperate of Karakoram blizzards, good humoured, generous and honest. As a geologist, Rein was a true professional, dedicated to total accuracy, who taught us all a tremendous amount.

M.P.S.

# CONTENTS

[Eight plates]

# CONTENTS

Phil. Trans. R. Soc. Lond. A **326**, 3–16 (1988)

Printed in Great Britain

[ 3 ]

# Tectonic framework of the Himalaya, Karakoram and Tibet, and problems of their evolution

By B. F. Windley

Department of Geology, University of Leicester, Leicester LE1 7RH, U.K.

The Himalaya, the Karakoram and Tibet were assembled by the successive accretion to Asia of continental and arc terranes during the Mesozoic and early Tertiary. The Jinsha and Banggong Sutures in Tibet join continental terranes separated from Gondwana. Ophiolites were obducted onto the shelf of southern Tibet in the Jurassic before the formation of the Banggong Suture. The Kohistan–Ladakh Terrane contains an island arc that was accreted in the late Cretaceous on the Shyok Suture and consequently evolved into an Andean-type batholith. Further east this Trans-Himalayan batholith developed on the southern active margin of Tibet without the prior development of an island arc. Ophiolites were obducted onto the shelf of India in the late Cretaceous to Lower Palaeocene before the closing of Tethys and the formation of the Indus–Yarlung Zangbo Suture at about 50 Ma. Post-collisional northward indentation of India at *ca.* 5 cm a$^{-1}$ since the Eocene has redeformed this accreted terrane collage; palaeomagnetic evidence suggests this indentation has given rise to some 2000 km of intracontinental shortening. Expressions of this shortening are the uplift of mid-crustal gneisses in the Karakoram on a late-Tertiary breakback thrust, folding of Palaeogene redbeds in Tibet, south-directed thrust imbrication of the foreland and shelf of the Indian Plate, north-directed back-thrusts along the Indus Suture Zone, post-Miocene spreading and uplift of thickened Tibet, giving rise to N–S extensional faults, and strike-slip faults, which allowed eastward escape of Tibetan fault blocks.

## 1. Introduction

Central Asia is dominated by the Tibetan Plateau and the mountain ranges of the Karakoram and the Himalaya. The tectonic evolution of these regions took place in three stages. (1) Following northward drift of several plates separated from Gondwanaland, growth of magmatic arcs during closure of Tethys in the Mesozoic–Lower Tertiary. (2) Accretion of these plates and arcs terminated in the collision between India and the amalgamated Asian Block to the north, giving rise to the Indus–Yarlung Zangbo Suture. (3) Post-collisional northward indentation of India since about 40 Ma has given rise to some 2000 km of crustal shortening (on palaeomagnetic evidence) and has caused redeformation of major segments of this accreted collage of plates.

The aim of this paper is to provide a synoptic account of the main tectonic zones and their mutual relations. A variety of tectonic problems are then discussed, because considerable disagreement has arisen concerning the interpretation of some key geological relationships.

## 2. Tectonic zones

This region of Central Asia is divisible into the following tectonic zones, which retain their coherence and continuity for considerable distances along strike (see figure 1).

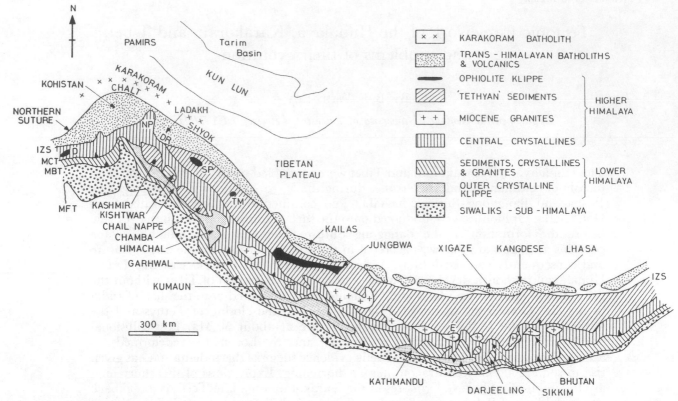

FIGURE 1. Map of the Himalaya showing the main tectonic zones and key localities. D = Dargai ophiolite, Dr = Dras, E = Everest, NP = Nanga Parbat, SP = Spongtang ophiolite, TM = Tso Morari, IZS = Indus–Yarlung Zangbo Suture, MCT = Main Central Thrust, MBT = Main Boundary Thrust, MFT = Main Frontal Thrust. After Windley (1983).

### (a) Tibet

For the purpose of this description, the Tibetan Plateau is taken to extend from the south side of the Kun Lun Range to the north side of the Trans-Himalayan magmatic belt. This region is divisible into three micro-continental fragments, the Kun Lun, Changtang and Lhasa Terranes separated by the Jinsha and Banggong Sutures (Chang et al. 1986). Faunal data suggest that the Kun Lun was already part of Laurasia by the Carboniferous, and that the Changtang and Lhasa Terranes were separated from Gondwana in the pre-Permian and Triassic respectively. These terranes were accreted successively northwards to the southern margin of Asia. The Jinsha Suture formed in the late Triassic–early Jurassic and the Banggong Suture in the late Jurassic–early Cretaceous (Lin & Watts, this symposium). Palaeomagnetic data by Patriat & Achache (1984) and biostratigraphic data by Wang & Sun (1985) indicate that the Lhasa Terrane remained stationary through most of the Upper Cretaceous and Eocene, but that it has moved 20° north since the suturing of India to it in the Eocene. The folding and thrusting of Palaeogene red beds across a wide extent of the Tibetan Plateau allows a minimum estimate of overall shortening of 12% (Chang et al. 1986). Remnants of dismembered ophiolites occur along the Banggong Suture Zone and in isolated, gently dipping outcrops over the plateau for at least 200 km south of it (Girardeau et al. 1985a). They were obducted probably in a single flat thrust sheet over the continental margin of the Lhasa Terrane (Chang et al. 1986). Granites of 140–120 Ma within the Lhasa Terrane were probably formed by lower crustal anatexis during intracrustal thrusting and thickening following the collision between the Changtang and Lhasa Terranes (Xu *et al.* 1982).

Neotectonic structures are widespread across the Tibetan Plateau. These are extensional structures which are significantly different in south and north Tibet (Chang *et al.* 1986; Mercier *et al.* 1987). South Tibet (extending to the northern part of the Tibetan-Tethys Zone) has extended in an E–W direction on major N–S-aligned rifts (Tapponnier *et al.* 1981*a*). The rate of Quaternary extension is about 1% Ma$^{-1}$, corresponding to a 'spreading' rate of $1.0 \pm 0.6$ cm a$^{-1}$ (Armijo *et al.* 1986). In contrast, north Tibet has been able to escape more easily eastwards and thus has extended along a conjugate set of NW–SE and NE–SW strike-slip faults (Mercier *et al.* 1987).

### (b) The Trans-Himalayan magmatic belt

This Andean-type magmatic belt (Le Fort, this symposium) is situated along the northern margin of the Indus–Yarlung Zangbo Suture. In Tibet, the Gangdese batholith of plutonic bodies is divisible into four simultaneously emplaced units, the predominant rocks of which are adamellites, quartz monzodiorites and granodiorites (Debon *et al.* 1986). U–Pb dates of the intrusives range from 94 to 41 Ma in Tibet (Scharer *et al.* 1984*a*) and from 103 to 60 Ma in Ladakh (Scharer *et al.* 1984*b*). Considering that the India–Eurasia collision took place in the period 50–40 Ma, the Trans-Himalayan plutons were emplaced before and during the period of collision. This may not be surprising in view of the expected time lapse between subduction of oceanic lithosphere and the intrusion of calc-alkaline melts (Zhou 1985).

North of the Gangdese batholith there is a belt of ignimbrites, andesites and alkaline rhyolites belonging to the Lingzigong Formation, which is at least 1500 m thick and which has $^{39}$Ar/$^{40}$Ar ages of 60 and 48 Ma (Maluski *et al.* 1982; Coulon *et al.* 1986) and Rb–Sr ages of 60 and 56 Ma (Xu *et al.* 1982). These volcanics are widely regarded as the extrusive and comagmatic equivalents of the Gangdese plutonic batholith.

In the western Himalaya the Kohistan–Ladakh Terrane is situated between the Shyok Suture and the Indus Suture. Recent detailed field and geochronological studies of rocks in Pakistan and NW India demonstrate that the magmatic belt in this terrane underwent at least three stages of crustal growth, each separated by phases of deformation, caused by collision tectonics.

1. An island arc (Tahirkheli & Jan 1979; Bard *et al.* 1980), which developed in Tethys from the Jurassic(?) to the mid-Cretaceous, has the following units from top to bottom: (*a*) The Yasin Group of slates, turbidites, volcaniclastics and limestones has an Aptian–Albian microfauna and formed in intra-arc basins (Pudsey *et al.* 1985*b*); (*b*) The Chalt (Pakistan) and Dras (India) volcanics of basaltic tholeiites, andesites, rhyolites, tuffs and volcaniclastic sediments (Dietrich *et al.* 1983; Radhakrishna *et al.* 1984); (*c*) Plutons of tonalite, granodiorite and diorite, which have ages of 103 Ma – U/Pb (Honegger *et al.* 1982), $101 \pm 2$ Ma – U/Pb (Scharer *et al.* 1984*b*), and $102 \pm 12$ Ma – Rb/Sr (Petterson & Windley 1985); (*d*) The Chilas Complex (in Kohistan) of layered gabbros and norites (now two pyroxene granulites), which formed in the sub-arc magma chamber (Khan *et al.* 1988); there are comparable rocks in the Kargil Complex in Ladakh (Rai & Pande 1978). All these rocks were deformed and metamorphosed when the arc was accreted to the Karakoram Terrane on the southern margin of Eurasia (Scharer *et al.* 1984*a, b*; Coward *et al.* 1986; Debon *et al.* 1987).

2. Subsequent northward subduction of Tethys gave rise to the Andean-type Trans-

Himalayan batholith, which was intruded into the steeply dipping arc volcanics. Dioritic dykes, representing the earliest phase of the calc-alkaline magmatism, have a $^{39}$Ar/$^{40}$Ar hornblende age of 75 Ma (D. C. Rex, personal communication). Plutons of granodiorite and granite have ages of $60.7 \pm 0.4$ Ma – U/Pb (Scharer *et al.* 1984 *a, b*), $59 \pm 2$ Ma – Rb/Sr (Debon *et al.* 1987), $54 \pm 3$ Ma and $40 \pm 6$ Ma – Rb/Sr (Petterson & Windley 1985), Rb/Sr and K/Ar mineral ages between 79 and 45 Ma (Honegger *et al.* 1982), $^{39}$Ar/$^{40}$Ar ages of $44 \pm 0.5$ Ma and $39.7 \pm 0.1$ Ma (Reynolds *et al.* 1983) and a Rb/Sr age of 40 Ma (Brookfield & Reynolds 1981). These calc-alkaline plutons (Petterson & Windley 1986) are overlain further west by the Dir-Kalam Group of calc-alkaline volcanics and Eocene sediments that are relics of the roof of the Andean-type batholith. Deformation of these volcanics and sediments took place when the Kohistan–Ladakh Terrane on the leading edge of the Eurasian Plate collided with the Indian Plate in the Eocene.

3. Abundant sheets of garnet/tourmaline-bearing aplite and pegmatite have poorly defined Rb/Sr whole-rock ages of $34 \pm 14$ Ma and $29 \pm 8$ Ma (Petterson & Windley 1985). They were intruded in post-collisional times; they formed by partial melting of continental crust thickened by tectonic and magmatic processes.

On the south side of the Kohistan–Ladakh arc-batholith lies the Kamila amphibolite belt, which was interpreted by Bard *et al.* (1980) as a relic of the oceanic crust on which the arc was built. However, the presence of granitic gneisses in the amphibolites led Coward *et al.* (1986) to suggest this is a belt of highly deformed arc-type plutonics and volcanics. It is possible that this is a relic of a second arc within the Kohistan–Ladakh Terrane.

### (c)  The Indus–Yarlung Zangbo Suture Zone

This suture zone formed as a result of the collision between India and the Kohistan–Ladakh arc-batholith in the west and Tibet in the central–eastern Himalaya. There are significant differences in the nature of the zone along strike.

1. In Pakistan, west of the Nanga Parbat syntaxis the suture (or Main Mantle Thrust; Bard *et al.* 1980) separates high-grade amphibolites to the north from high-grade gneisses to the south. Key rock units along the suture are (*a*) a 3 km wide thrust belt of blueschists (Shams *et al.* 1980; Majid & Shah 1985) and (*b*) a 200 km$^2$ tectonic wedge, the Jijal Complex of high-pressure garnet granulites (Jan & Howie 1981), which may have been recrystallized during sub-arc metamorphism, before collision (Coward *et al.* 1986).

2. In India, the suture zone contains the following units. (*a*) The Lamayuru Complex of allochthonous, pre-orogenic basin sediments (Thakur 1981; Searle 1983). They comprise shales, turbidites and deep-water radiolarian cherts and they range in age from Triassic to late Cretaceous. (*b*) Ophiolitic mélanges, which occur in shear zones up to 150 m thick and which contain serpentinized harzburgites and dunites, gabbros, rodingites, sheared volcanics, blueschists, shales and cherts (Frank *et al.* 1977). They commonly border and cut through the Lamayuru sediments and the Dras volcanic rocks. Some granitic rocks of the Ladakh batholith intrude the mélanges, demonstrating that the emplacement of these dismembered ophiolites (Ramana *et al.* 1986) in the Indus Suture Zone occurred in the Cretaceous. (*c*) The Jurassic–Cretaceous Dras volcanics (Radhakrishna *et al.* 1984), which are equivalent to the Chalt volcanics in Pakistan. (*d*) The Indus Group of Eocene to (?)Miocene molasse sediments, which are dominantly fluviatile and lacustrine sandstones and conglomerates, the pebbles of

which contain granitic and volcanic rocks derived from the Ladakh batholith, over which parts of the Indus Group still overlie unconformably.

3. In Tibet, the suture zone (Burg & Chen 1984; Burg *et al.* 1987) contains the remains of three rock units.

The Xigaze Group consists of Aptian–Albian flysch and conglomerates, which contain pebbles of plutonics and volcanics derived from the Gangdese Belt to the north. It probably formed in a fore-arc on the southern margin of the Lhasa Terrane. It was deformed by the collision of the Indian Plate (Shackleton 1981).

The Xigaze ophiolite occurs in several thrust slices along the suture zone. It displays a complete ophiolitic sequence from marine radiolarian cherts and pillow-bearing basalts to 5 km of harzburgites and minor lherzolites. It has few cumulate gabbros and a sill–dyke complex (Nicholas *et al.* 1981; Girardeau *et al.* 1985 *b*). It has a U–Pb whole-rock age of $120 \pm 10$ Ma and overlying radiolarian cherts have Albian–Aptian microfossils (Gopal *et al.* 1984). Dolerite dykes have a constant trend of N080. By assuming that the dykes were injected in tensional fractures parallel to the oceanic spreading axis and by taking into account the 90° anticlockwise rotation of the whole ophiolite determined by Pozzi *et al.* (1984), Girardeau *et al.* (1985 *c*) concluded that the Xigaze ophiolite formed in a spreading centre oriented at N160, which is close to the N175 determined by Pozzi *et al.* (1984) from palaeomagnetic data. Low-temperature fabrics suggest a very low heat flow at the spreading centre, which may have been a discontinuous, slowly accreting type located in a small basin (Girardeau *et al.* 1985 *d*). The palaeolatitude of the accreting centre was at 10–20° N (Pozzi *et al.* 1984), which was close to the southern margin of Eurasia. Intraoceanic thrusting began at about 110 Ma soon after accretion, and final thrusting was at 50 Ma during terminal collision.

South of the ophiolite is a thrust slice containing turbidites which are equivalent to the Lamayuru complex in Ladakh and which were deposited in a sedimentary apron on the northern continental margin of the Indian Plate (Burg *et al.* 1987), and the Yamdrock tectonic mélange, which probably formed in the subduction zone trench on the footwall of the Xigaze ophiolite (Searle *et al.* 1987).

4. In Nagaland in NE India there are ophiolites (Agrawal & Kacker 1980) and blueschists (Ghose & Singh 1980) in the suture zone.

### (d) Tibetan-Tethys Zone

Extending from Zanskar along the southern edge of Tibet to the NE corner of India is a continuous zone of essentially conformable Palaeozoic and Mesozoic sediments, which are 6 km thick and which were deposited on the northern passive continental margin of the Indian Plate (Gansser 1964; Gupta & Kumar 1975; Fuchs 1979; Thakur 1981; Tapponnier *et al.* 1981 *b*). There is an almost continuous succession from the Cambrian to the Eocene (Baud *et al.* 1984; Gaetani *et al.* 1986). The Panjal Group of continental tholeiitic to mildly alkaline basalts in Kashmir and the equivalent Abor volcanics in NE India (Bhat 1984) formed in Carboniferous–Permian rifts which evolved into the passive continental margin of Neo-Tethys. In general, the Mesozoic sediments pass northwards from Triassic to middle-Jurassic shelf carbonates to late-Jurassic shales deposited in a widespread transgression, to early Cretaceous shelf carbonates, and to a deeper water Campanian–Maastrichtian flysch facies. Shelf limestones continued to form locally until the Lower Eocene. The sediments have been thrust southwards towards the foreland of the Indian Plate. From balanced cross sections, Searle (1986) calculated a shortening of 126 km across this thrust shelf sequence.

The Tibetan-Tethys Zone also contains Lower Ordovician gneissic porphyritic granites as in the Kangmar dome (Burg *et al.* 1984*a*), that have a Rb–Sr isochron of about 485 Ma. They are similar in type and age to the Central Crystalline gneisses of the Higher Himalaya to the south; their aluminous character and high initial Sr isotope ratios suggest they developed in the continental crust, which was part of Gondwana at the time (Debon *et al.* 1986*a*).

There are two ophiolite complexes that have been thrust onto the Tethyan shelf sediments: Spongtang in Zanskar and Jungbwa in SW Tibet (Gansser 1979). The Spongtang klippe is composed of ultramafic and gabbroic rocks with poorly developed cumulates and dykes in a thrust nappe overlying a mélange unit with volcanics, limestone and chert blocks. The ophiolite rests tectonically on a thrust plane over sediments ranging in age from Eocene to Jurassic (Srikantia & Razdan 1981). Relations between mylonitic and ductile shear zones in peridotites led Reuber (1986) to suggest that the ophiolite formed at a transform boundary. The age of both ophiolites is unknown, but presumed to be Cretaceous. The age of tectonic emplacement of the ophiolites over the shelf sediments of the Indian continental margin is controversial and will be discussed later.

At the base of the Tibetan-Tethys Zone there is a northward-dipping normal fault, along which there has been at least several tens of kilometres movement. It formed by gravity collapse during late-Tertiary spreading of the Himalayan crust, which had been thickened during post-collisional convergence of India and Tibet (Burg *et al.* 1984*b*; Burchfiel & Royden 1985).

### (e) The Higher Himalaya

This 3.5–10 km thick 'Tibetan Slab' of high-grade metamorphic rocks (the Central Crystallines) is bounded on the north by the normal fault referred to above and to the south by the Main Central Thrust (MCT). The rocks have undergone amphibolite facies medium pressure-type metamorphism during Himalayan southward-thrusting, giving rise to wide-spread pelitic schists, marbles, paragneisses, orthogneisses, amphibolites and migmatites (Burg *et al.* 1987).

Two-mica leucogranites have intruded the Central Crystalline Complex and the northern part of the Tibetan-Tethys Zone sediments. The well-studied Monaslu granite has very inhomogeneous Rb–Sr isotopic compositions; magmatic ages range from 25 Ma (U–Pb) and 18 Ma (Rb–Sr isochron), suggesting that the magmatic activity lasted at least 7 Ma (Scharer *et al.* 1984*a*; Deniel *et al.* 1987). In general, the High Himalayan leucogranites are peraluminous in composition and often contain tourmaline, sillimanite, and garnet, and have very high initial Sr isotope ratios of 0.7800–0.7550 (Le Fort 1981; Dietrich & Gansser 1981; Vidal *et al.* 1982; Searle & Fryer 1986). They were most likely generated by anatexis of lower crustal Precambrian paragneisses comparable to those of the Central Crystalline Complex.

### (f) The Lesser Himalaya

The Lesser Himalayan tectonic zone is bounded to the north by the Main Central Thrust (MCT) and to the south by the Main Boundary Thrust (MBT). It consists of weakly metamorphosed, late Proterozoic (Riphean–Vendian) and Palaeozoic sediments, which have been overridden by thrust nappes of high-grade gneiss derived from the Central Crystallines

(Stocklin 1980; Valdiya 1981; Sinha 1981). Predominant amongst the low-grade sediments are carbonates with Riphean stromatolites, slates, phyllites, quartzites, flysch and Permo-Carboniferous tillites. The rocks have been transported southwards in several thrust slices (e.g. the Krol and Chail nappes).

The MCT is a thrust that has problems of definition (Sinha Roy 1982). It has a displacement of more than 100 km (Andrieux *et al.* 1981). In Nepal it is represented by a 10 km wide ductile shear zone (Bouchez & Pecher 1981). Inverted, 'Barrovian-type' isograds are parallel and genetically related to the northward-dipping MCT (Le Fort 1975). Problems associated with the origin of this reversed metamorphism are discussed later.

The MBT underwent movement in the Pliocene–Pleistocene that folded the overlying nappes.

### (g) The Karakoram Terrane

This tectonic zone is situated between the Pamir–Kun Lun Ranges and the Kohistan–Ladakh Terrane. Early reconnaissance on the geological evolution of the Karakoram was reviewed by Desio (1979). Recent detailed work has only been undertaken in the southern and central sectors of this very high mountain range (Rex *et al.*, this symposium). The central Karakoram is dominated by the composite granitic batholith, which evolved in two major stages. (1) Subalkaline to calc-alkaline hornblende-bearing granitic rocks have a U–Pb age of 95 Ma (Le Fort *et al.* 1983b) and a Rb–Sr whole-rock age of $97 \pm 17$ Ma (Debon *et al.* 1987). These subduction-related granites were intruded as an Andean-type batholith into the southern margin of the Karakoram Terrane. (2) Mildly peraluminous granites were intruded at *ca.* 20 Ma; they formed as a result of crustal thickening and partial melting of crystalline basement (Searle *et al.* 1988; Rex *et al.*, this symposium).

South of the batholith there is a high-grade metamorphic series consisting of sillimanite gneisses, kyanite–staurolite–garnet mica schists, marbles and amphibolites (Searle *et al.* 1988). These represent shales, carbonates and minor volcanics that reached peak metamorphic conditions at 550 °C and 5.5 kbar† at a depth of about 17.5 km. Uplift of these mid-crustal rocks with older components of the Karakoram batholith took place along the hangingwall of the Main Karakoram Thrust, which has been interpreted as a late-Tertiary breakback thrust (Rex *et al.*, this symposium).

### (h) The Shyok (Northern) Suture Zone

This suture zone is commonly but inappropriately termed the 'Northern Suture' in Pakistan (Pudsey *et al.* 1985a; Coward *et al.* 1986; Pudsey 1986), and the Shyok Suture in NW India (Rai 1982, 1983; Thakur & Misra 1984). It joins the late-Palaeozoic shelf sediments in the western part of the Karakoram Terrane with the Cretaceous Kohistan–Ladakh Terrane. In Pakistan the suture is a zone of mélange 150 m–4 km wide which contains blocks of volcanic greenstone, limestone, red shale, conglomerate, quartzite and serpentinite in a slate matrix; the mélange is interpreted by Pudsey (1986) as an olistostrome, largely derived from the arc to the south. In the Shyok Suture in NW India (Rai 1983; Thakur & Misra 1983) there is a dismembered ophiolite with peridotite, pyroxenite, gabbro, intermediate volcanics and chert. The ophiolite is overlain discordantly by a 500 m thick molasse containing pebbles of granite and rhyolite derived from the Ladakh batholith.

† 1 kbar = $10^8$ Pa.

### (i) Foredeep and intermontane basins

South of the MBT there is an apron of fluvial molasse sediments constituting the Siwalik Group with conglomerates, arkoses, siltstones and shales, which were derived from the uplift and erosion of the Himalayas and which were deposited in a foredeep along the southern flank of the mountain range. Most modern studies of these sediments have been in the western Himalaya. Sedimentation began at least 15 Ma ago (Johnson et al. 1981, 1985). Sedimentation rates ranged from 15 to 52 cm per 1000 years in the last 13 Ma; this is comparable to the uplift rates of the Himalayan source areas. Southward migration of thrusts across the foredeep has folded and thrust the Siwaliks. However, although the rate of convergence of the Indian subcontinent with Eurasia appears to have been steady in the time concerned, detailed studies of these sediments show that prolonged periods of tectonic quiescence and uniform molasse sedimentation were punctuated by brief, intense intervals of deformation as the locus of thrust faults encroached in a stepwise fashion across the foredeep (Burbank & Raynolds 1984).

The Peshawar and Kashmir Intermontane Basins are situated just north of the MBT and thus are embedded in the still-developing thrust belt (Burbank & Johnson 1983). They contain up to 1300 m of Plio-Pleistocene synorogenic sediments. Basin sedimentation began by 4 Ma in the Kashmir and 3 Ma in the Peshawar Basin at rates of 16–64 cm per 1000 years (Burbank & Johnson 1983; Burbank & Tahirkheli 1985). During the Pliocene sedimentation patterns were related to the uplift of the Pir Panjal and Attock Mountain Ranges respectively.

## 3. TECTONIC PROBLEMS

There is considerable controversy about a variety of key tectonic problems, which will now be briefly discussed.

### (a) Models of Himalayan structure

There are several models for the structure and tectonic evolution of the Himalaya that are based on the concept of the drift and collision of India into Eurasia. These all recognize the Indus–Zangbo Suture as the main collisional boundary formed in the early Tertiary, the progressive southward migration of the locus of thrusting through the Tertiary, and the MCT as a major intracontinental thrust formed in the mid-Tertiary as a result of the thrusting of the mid-crustal Tibetan slab over the upper crust to the south. However, these models differ considerably in their interpretation of the structures formed in the late Tertiary.

An important development in the ideas on Himalayan structure came with the recognition of Seeber et al. (1981) (see also Seeber & Armbruster 1984), that in the western Himalaya, earthquake foci indicate the presence of a major detachment surface that dips shallowly to the north. Before this discovery, the MCT had to join the MBT at depth (Powell & Conaghan 1973; Powell 1979). The importance of the detachment surface is enhanced in the model of Ni & Barazangi (1984) in which it extends under the Tibetan slab and under Tibet north of the Indus–Zangbo Suture.

### (b) The deep structure of Tibet

The Tibetan crust is 60–70 km thick (Hirn et al. 1984; Hirn, this symposium). Several models have been proposed to explain this anomalous structure. (a) Following Argand (1924), Powell (1979, 1986), Powell & Conaghan (1975), Barazangi & Ni (1982) and Knopoff (1983) suggested that the Indian crust was thrust under Tibet, so doubling the crustal thickness.

According to this model there should be geophysical evidence under northern Tibet of a remnant northward-dipping subducting slab. From palaeomagnetic data, Lin & Watts (this symposium) conclude that since collision at 40 Ma the Lhasa Terrane has moved northwards $2000 \pm 600$ km; this challenges the underthrusting model for Tibet. (b) Dewey & Burke (1973) and Sengör & Kidd (1979) argued that the thick Tibetan crust was caused by intracrustal N–S shortening. The recent geophysical data of Hirn et al. (1984) and Hirn (this symposium) and the geological information of Chang et al. (1986) support this idea. According to Chang et al. (1986) the Tibetan Plateau has been shortened by at least 40% in post-Eocene times, and this allows all the crustal thickening to be explained by internal deformation. (c) Mattauer (1986) proposed that the Tibetan crust is of multiple age and origin. It was thickened during collisional tectonics in the Hercynian, the Triassic, the Jurassic and the Cretaceous, before the post-Eocene intracontinental thrusting and thickening. This idea is increasingly appealing as more information is produced about the Kun Lun, Jinsha and Banggong Sutures. Geophysical evidence for the deep structure of Tibet is reviewed by Molnar (this symposium).

### (c) Inverted metamorphic isograds

Along the length of the Himalaya there are major inverted metamorphic zones related to crustal-scale thrusting associated with the MCT (Bordet et al. 1981; Lal et al. 1981; Sinha Roy 1981; Windley 1983). Historically, problems have arisen concerning the positioning of the MCT in the different parts of the Himalaya, and this in turn has created confusion regarding which isograds lie above or below this thrust (Sinha Roy 1982). The inverted metamorphism passes upwards from the chlorite–biotite zone through a Barrovian-type sequence of garnet, staurolite, kyanite and sillimanite zones. Migmatites and leucogranites are characteristically developed in the sillimanite zone.

The thermal model of Le Fort (1975), according to which the thrusting caused the inversion, formed the influential basis of most later attempts to explain this metamorphic problem. According to these ideas the metamorphic zones are related to the post-collisional intracontinental thrusts which were responsible for crustal thickening and which reached a peak some 20 Ma after collision with formation of the MCT in the Miocene. An alternative model concerns the thrusting of a hot slab of Central Crystalline gneisses over a cold slab of Lower Himalayan sediments; cooling of the former and conductive heating of the latter would give rise to the inverted metamorphism. Isograds formed at such a major tectonic boundary could be inverted by syn-metamorphic or post-metamorphic folding (Harte & Dempster 1987; Searle et al., this symposium). England & Thompson (1984) demonstrated that thrusting was the most efficient means of producing high-grade metamorphism and melting in a thickened crust, and Jaupart & Provost (1985) showed that thrusting variably affects the thermal conductivity of low- and high-grade rocks, and so gives rise to temperature maxima at shallow depths. The position of the rocks within a thrust slice, and the number of thrust units in the nappe pile both affect the generation of $P$–$T$–time paths during growth and uplift of thickened crust (Davy & Gillet 1986). Hodges et al. (this symposium) make the first attempt to relate segments of the Himalayan $P$–$T$ trajectory to specific tectonic events.

### (d) Timing of suture formation

There is serious disagreement about the age of formation of the Shyok (Northern) and Indus–Yarlung Zangbo Sutures, and this raises the question: which formed first? Let us start with the Shyok Suture. Did it form in the Eocene–Oligocene (Brookfield & Reynolds 1981; Andrews-Speed & Brookfield 1982; Reynolds *et al.* 1983; Thakur 1987*a*), or in the late Cretaceous–Palaeocene (Petterson & Windley 1985; Coward *et al.* 1986; Pudsey 1986; Sharma 1986; Debon *et al.* 1987; Srimal *et al.* 1987; Searle *et al.* 1987, 1988; Rex *et al.*, this symposium)?

The following are constraints on this problem. From palaeomagnetic data on a granodiorite from Ladakh, Klootwijk *et al.* (1979) and Klootwijk & Radhakrishnamurty (1981) concluded that after the accretion of the Kohistan–Ladakh island arc to India this combined mass moved northwards over 25° of latitude before colliding with Asia along the Shyok Suture at the Eocene–Oligocene boundary. However, the isotopic age of the palaeomagnetically determined samples was not known, and therefore this conclusion is doubtful, especially in view of subsequent isotopic results, which show that the Kohistan–Ladakh arc-batholith underwent an evolution from at least 100 to 40 Ma (Scharer *et al.* 1984*b*; Petterson & Windley 1985).

Andrews-Speed & Brookfield (1982) concluded that the Shyok Suture did not form until the Oligocene or Miocene, and Brookfield & Reynolds (1981) and Reynolds *et al.* (1983) until the Miocene, because they considered that the Karakoram batholith formed in the Palaeocene and Oligocene–Miocene as a result of northward subduction of oceanic lithosphere sited on the eventual Shyok Suture. However, isotopic data by Le Fort *et al.* (1983), Debon *et al.* (1987) and Searle *et al.* (1988) showed that the calc-alkaline part of the batholith was intruded in the period 110–95 Ma. (Miocene isotopic dates are probably from cooling ages related to uplift–erosion, etc. (Searle *et al.* 1988).)

The most cogent evidence for a relatively early age for the Shyok Suture comes from the Naz Bar pluton, which is an undeformed granite, which transects the suture (figure 6 in Pudsey 1986). Nine $^{39}Ar/^{40}Ar$ ages on biotites from this granite all lie in the range of late Cretaceous–Palaeocene (D. C. Rex, unpublished data), and preclude the possibility that the suture formed in the mid-Tertiary. Furthermore, dioritic dykes near Gilgit, that cross-cut structures correlated with the formation of the suture (Coward *et al.* 1986), have a $^{39}Ar/^{40}Ar$ hornblende age of 75 Ma (D. C. Rex in Petterson & Windley 1985). These results indicate that the Shyok Suture must have formed before the end of the Cretaceous (Coward *et al.* 1986), and at the latest before the end of the Palaeocene (Debon *et al.* 1987).

Let us now consider the Indus–Yarlung Zangbo Suture, whose postulated age ranges from 55 to 40 Ma. According to Thakur (1987*a, b*) the initial collision of India with Tibet and the Kohistan–Ladakh Terrane was in the middle Eocene, and according to Debon *et al.* (1987) it was close to the end of the Eocene. This conclusion was largely based on the fact that the calc-alkaline plutons of the batholith continued to form until the mid–late Eocene. However, a time lapse of at least 10 Ma may be expected between the closure of an ocean and the final calc-alkaline magmatism (Zhou 1985). More reliable information is provided by a combination of continental palaeomagnetic data and Indian Ocean magnetic anomaly data, which suggest that initial contact between India and Asia occurred at about 55 Ma (Besse *et al.* 1984; Klootwijk 1984; Patriat & Achache 1984). This age is supported by the change from marine to continental (molasse-type) sedimentation in the suture zone after the Lower Palaeocene (Searle *et al.* 1987).

The question arises: did collision take place at different times along the Himalayan belt? From palaeomagnetic data, Klootwijk *et al.* (1985) concluded that anticlockwise rotation of India allowed progressively eastward suturing from about 62–60 Ma in the NW Himalaya to 55 Ma in Tibet; suturing was completed by the end of the Palaeocene. U–Pb radiometric dating of plutons in the Trans-Himalayan batholith led Xu *et al.* (1982) to suggest that collision of India with the Tibetan block occurred at 60 Ma in Ladakh and at 40 Ma in Tibet. However, more recent Rb–Sr isochron dates on plutons in Ladakh (Reynolds *et al.* 1983) and Kohistan (Petterson & Windley 1985) indicate that the plutonic activity continued until 40 Ma, and in turn, this suggests a similar age for the terminal intrusion of calc-alkaline plutons along the Himalayan range.

In conclusion, the best evidence currently available indicates that the Shyok (Northern) Suture formed in the mid–late Cretaceous, and the Indus–Yarlung Zangbo Suture in the Lower–Middle Eocene.

### (e) Timing of ophiolite obduction

Baud *et al.* (1984) and Thakur (1987 *a*, *b*) argued that, because the Spongtang ophiolite rests on fossiliferous early Eocene sediments, it must have been thrust southwards onto shelf sediments in post-early Eocene times. Because such a well-preserved slab of ocean floor can only have been emplaced onto the continental margin before the continent–continent collision, this age would imply that Tethys could not have closed until the mid-Eocene. However, such an age contradicts the biostratigraphic and palaeomagnetic data. In contrast, Brookfield & Reynolds (1981) and Brookfield (1981) proposed that the ophiolite was thrust over the shelf sediments in the late Cretaceous and that the formation of the Indus Suture was complete by the Maastrichtian. The problem lies in the interpretation of the thrust history and geometry. Searle (1983, 1986) and Searle *et al.* (1987, this symposium) emphasized the complexity of the three-stage sequence of thrust events (during ophiolite emplacement, during continental collision, and in post-collisional times). Many of the thrusts currently observed formed in the second and third stages, such as breakback thrusts that reversed the earlier stacking order of the thrust sheets. Only the most detailed structural studies will reveal the position and geometry of the stage-one thrusts responsible for the emplacement of the ophiolites.

In summary, a combination of palaeomagnetic, structural and stratigraphic studies indicate that ophiolite obduction could not have been in the post-early Eocene, because Tethys was closed by that time, but rather in the late Cretaceous – early Palaeocene during collapse of the continental shelf and formation of a foredeep, as documented by Brookfield & Andrews-Speed (1984).

## 4. Conclusions

The Himalaya, the Karakoram and Tibet were assembled by the accretion of continental and arc terranes to Asia. Following the terminal collision of India with the accreted collage to the north, the continued convergence and indentation of India gave rise to some 2000 km of intracontinental shortening. The structural history can be divided into pre-collisional thrusts and obduction of ophiolites, syn-collisional thrusts and suture formation, and post-collisional thrusts and extensional faults. Neotectonics is dominated by extensional structures and related intrusions and extrusions, as a result of thrust-controlled crustal thickening, consequent weakening of the quartz-dominated lower crust, and spreading and stretching of the upper crust.

I thank NERC for grant GR3/4242 for a decade's research in the Himalaya and the Karakoram, and especially for all the post-doctoral research fellows and Ph.D. students at Leicester who made a multidisciplinary project possible. Also I am grateful to Mike Searle for constructive comments on this paper.

REFERENCES

Agrawal, O. P. & Kacker, R. N.  1980  In *Ophiolites* (ed. A. Panayioutou), pp. 454–461. Cyprus Geological Survey.

Andrews-Speed, C. P. & Brookfield, M. E.  1982  *Tectonophysics* **82**, 253–275.

Andrieux, J., Arthaud, F., Brunel, M. & Sauniac, S.  1981  *Bull. Soc. géol. Fr.* **23**, 651–661.

Argand, E.  1924  *Rep. Int. Geol. Congr. 13th*, **1**, 170–372.

Armijo, R., Tapponnier, P., Mercier, J. L. & Han, T.-L.  1986  *J. geophys. Res.* **91**, 803–811.

Barazangi, M. & Ni, J.  1982  *Geology* **10**, 179–185.

Bard, J. P., Maluski, H., Matte, P. & Proust, F.  1980  *Geol. Bull. Univ. Peshawar* **13**, 87–94.

Baud, A., Gaetani, M., Garzanti, E., Fois, E., Nicora, A. & Tintori, A.  1984  *Eclog. geol. Helv.* **77**, 171–197.

Besse, J., Courtillot, V., Pozzi, J. P., Westphal, M. & Zhou, Y. X.  1984  *Nature, Lond.* **311**, 621–626.

Bhat, M. I.  1984  *J. geol. Soc. Lond.* **141**, 763–775.

Bordet, P., Colchen, M., Le Fort, P. & Pecher, A.  1981  In *Zagros, Hindu Kush, Himalaya: geodynamic evolution* (ed. H. K. Gupta & F. M. Delaney), pp. 149–168. Washington, D.C.: American Geophysics Union.

Bouchez, J. L. & Pecher, A.  1981  *Tectonophysics* **78**, 23–50.

Brookfield, M. E.  1981  In *Metamorphic tectonites of the Himalaya* (ed. P. S. Saklani), pp. 1–14. New Delhi: Today & Tomorrow Publishers.

Brookfield, M. E. & Andrews-Speed, C. P.  1984  *Sed. Geol.* **40**, 249–286.

Brookfield, M. E. & Reynolds, P. H.  1981  *Earth planet. Sci. Lett.* **55**, 157–162.

Burbank, D. W. & Johnson, G. D.  1983  *Palaeogeog. Palaeoclim. Palaeoecol.* **43**, 205–235.

Burbank, D. W. & Raynolds, R. G. W.  1984  *Nature, Lond.* **311**, 114–118.

Burbank, D. W. & Tahirkheli, R. A. K.  1985  *Bull. geol. Soc. Am.* **96**, 539–552.

Burchfiel, B. C. & Royden, L. H.  1985  *Geology* **13**, 679–682.

Burg, J. P., Brunel, M., Gapais, D., Chen, G. M. & Liu, G. H.  1984*b*  *J. struct. Geol.* **6**, 535–542.

Burg, J. P. & Chen, G.-M.  1984  *Nature, Lond.* **311**, 219–223.

Burg, J. P., Guiraud, M., Chen, G. M. & Li, G. C.  1984*a*  *Earth planet Sci. Lett.* **69**, 391–400.

Burg, J. P., Leyreloup, A., Girardeau, J. & Chen, G.-M.  1987  *Phil. Trans. R. Soc. Lond.* A **321**, 67–86.

Chang, C. *et al.*  1986  *Nature, Lond.* **323**, 501–507.

Coulon, C., Maluski, H., Bollinger, C. & Wang, S.  1986  *Earth planet. Sci. Lett.* **79**, 281–302.

Coward, M. P., Windley, B. F., Broughton, R. D., Luff, I. W., Petterson, M. G., Pudsey, C. J., Rex, D. C. & Khan, M. A.  1986  In *Collision Tectonics* (ed. M. P. Coward & A. C. Ries) (Geol. Soc. Lond. Spec. Publ. no. 19), pp. 203–219. London: Blackwell.

Davy, P. & Gillet, P.  1986  *Tectonics* **5**, 913–929.

Debon, F., Le Fort, P., Sheppard, S. M. F. & Sonet, J.  1986  *J. Petr.* **27**, 219–250.

Debon, F. B., Le Fort, P., Dautel, D., Sonet, J. & Zimmermann, J. L.  1987  *Lithos* **20**, 19–40.

Deniel, C., Vidal, P., Fernandez, A., Le Fort, P. & Peucat, J. J.  1987  *Contr. Miner. Petr.* **96**, 78–97.

Desio, A.  1979  In *Geodynamics of Pakistan* (ed. A. Farah & K. A. DeJong), pp. 111–124. Quetta: Geol. Surv. Pakistan.

Dewey, J. F. & Burke, K.  1973  *J. Geol.* **81**, 683–692.

Dietrich, V. J. & Gansser, A.  1981  *Schweiz. miner. petrogr. Mitt.* **61**, 177–202.

Dietrich, V. J., Frank, W. & Honegger, K.  1983  *J. volc. geotherm. Res.* **18**, 405–433.

England, P. C. & Thompson, A. B.  1984  *J. Petr.* **25**, 894–928.

Frank, W., Gansser, A. & Trommsdorff, V.  1977  *Schweiz miner. petrogr. Mitt.* **57**, 89–113.

Fuchs, G.  1979  *Geol. Rdsch.* **122**, 513–540.

Gaetani, M., Casnedi, R., Fois, E., Garzanti, E., Jadoul, F., Nicora, A. & Tintori, A.  1985  *Riv. ital. Paleont. Strat.* **91**, 443–478.

Gansser, A.  1964  *Geology of the Himalayas.* (289 pages.) London: Wiley.

Gansser, A.  1979  *J. geol. Soc. India* **20**, 277–281.

Ghose, N. C. & Singh, R. N.  1980  *Geol. Rdsch.* **69**, 41–48.

Girardeau, J., Marcoux, J., Fourcade, E., Bassouillet, J. P. & Tang, Y.  1985*a*  *Geology* **13**, 330–333.

Girardeau, J., Mercier, J. C. C. & Wang, X.  1985*b*  *Contr. Miner. Petr.* **90**, 309–321.

Girardeau, J., Mercier, J. C. C. & Zao, Y.  1985*c*  *Tectonics* **4**, 267–288.

Girardeau, J., Mercier, J. C. C. & Zao, Y.  1985*d*  *Tectonophysics* **119**, 407–433.

Gopal, C., Allègre, C. J. & Xu, R.-H.  1984  *Earth planet. Sci. Lett.* **69**, 301–310.

Gupta, V. J. & Kumar, S.  1975  *Geol. Rdsch.* **64**, 540–563.

Harte, B. & Dempster, T. J.  1987  *Phil. Trans. R. Soc. Lond.* A **321**, 105–127.

Hirn, A. *et al.*  1984  *Nature, Lond.* **307**, 25–27.

Honegger, K., Dietrich, V., Frank, W., Gansser, A., Thoni, M. & Trommsdorff, V.  1982  *Earth planet. Sci. Lett.* **60**, 253–292.

Jan, M. Q. & Howie, R. A.  1981  *J. Petr.* **22**, 85–126.

Jaupart, C. & Provost, A.  1985  *Earth planet. Sci. Lett.* **73**, 385–397.

Johnson, N. M., Opdyke, N. D., Johnson, G. D., Lindsay, E. H. & Tahirkheli, R. A. K. 1982 *Palaeogeog. Palaeoclim. Palaeoecol.* **37**, 17–42.

Johnson, N. M., Stix, J., Tauxe, L., Cerveny, P. F. & Tahirkheli, R. A. K. 1985 *J. Geol.* **93**, 27–40.

Khan, M. Asif, Jan, M. Q., Windley, B. F. & Tarney, J. 1988 Geol. Soc. Am. Spec. Paper. (In the press.)

Klootwijk, C. T. 1984 *Tectonophysics* **105**, 331–353.

Klootwijk, C. T., Conaghan, P. J. & Powell, C. McA. 1985 *Earth planet. Sci. Lett.* **75**, 167–183.

Klootwijk, C. T. & Radhakrishnamurty, C. 1981 In *Paleoreconstruction of the continents*, Geodyn. Ser. vol. 2, pp. 93–105. Washington, D.C.: American Geophysics Union.

Klootwijk, C., Sharma, M. L., Gergan, J., Tirkey, B., Shah, S. K. & Agarwal, V. 1979 *Earth planet. Sci. Lett.* **44**, 47–64.

Knopoff, L. 1983 *Geophys. Jl R. astr. Soc.* **74**, 55–81.

Lal, R. K., Mukerji, S. & Ackermand, D. 1981 In *Metamorphic tectonites of the Himalaya* (ed. P. S. Saklani), pp. 231–278. Delhi: Today & Tomorrow Publishers.

Le Fort, P. 1975 *Am. J. Sci.* **75**, 1–44.

Le Fort, P. 1981 *J. geophys. Res.* **86**, 10545–10568.

Le Fort, P., Michard, A., Sonet, J. & Zimmermann, J. L. 1983 In *Granites of Himalayas, Karakorum and Hindu Kush* (ed. F. A. Shams), pp. 377–387. Lahore: Punjab University.

Majid, M. & Shah, M. T. 1985 *Geol. Bull. Univ. Peshawar* **18**, 41–52.

Maluski, H., Proust, F. & Xiao, X.-C. 1982 *Nature, Lond.* **298**, 152–154.

Mattauer, M. 1986 *Bull. Soc. géol. Fr.* **8**, 142–157.

Mercier, J. L., Armijo, R., Tapponnier, P., Carey-Gailhardis, E. & Han, T.-L. 1987 *Tectonics* **6**, 275–304.

Molnar, P. 1984 *A. Rev. Earth planet. Sci.* **12**, 489–518.

Molnar, P., Burchfiel, B. C., Liang, K. & Zhao, Z. 1987 *Geology* **15**, 249–253.

Ni, J. & Barazangi, M. 1984 *J. geophys. Res.* **89**, 1147–1163.

Nicholas, A., Girardeau, J., Marcoux, J., Dupré, B., Wang, X., Cao, Y., Zheng, H. & Xiao, X. 1981 *Nature, Lond.* **294**, 414–417.

Patriat, P. & Achache, J. 1984 *Nature, Lond.* **311**, 615–621.

Petterson, M. P. & Windley, B. F. 1985 *Earth planet. Sci. Lett.* **74**, 45–57.

Petterson, M. P. & Windley, B. F. 1986 *Geol. Bull. Univ. Peshawar* **19**, 121–149.

Powell, C. McA. 1979 In *Geodynamics of Pakistan* (ed. A. Farah & K. A. de Jong), pp. 5–24. Quetta: Geological Survey of Pakistan.

Powell, C. McA. 1986 *Earth planet. Sci. Lett.* **81**, 79–84.

Powell, C. McA. & Conaghan, P. J. 1973 *Earth planet. Sci. Lett.* **20**, 1–12.

Powell, C. McA. & Conaghan, P. J. 1975 *Geology* **3**, 727–731.

Pozzi, J. P., Westphal, M., Girardeau, J., Besse, J., Yao, X.-Z., Xian, Y.-C. & Li, S.-X. 1984 *Earth planet. Sci. Lett.* **70**, 383–394.

Pudsey, C. J. 1986 *Geol. Mag.* **123**, 405–423.

Pudsey, C. J., Coward, M. P., Luff, I. W., Shackleton, R. M., Windley, B. F. & Jan, M. Q. 1985a *Trans. R. Soc. Edinb.* **76**, 463–479.

Pudsey, C. J., Schroeder, R. & Skelton, P. W. 1985b In *Geology of Western Himalayas* (ed. V. J. Gupta *et al.*), Contr. to Himal. Geol. no. 3, pp. 150–168. Delhi: Hindustan Publishing Corporation.

Radhakrishna, T., Divakara Rao, V. & Murali, A. V. 1984 *Tectonophysics* **108**, 135–153.

Rai, H. 1982 *Nature, Lond.* **297**, 142–144.

Rai, H. 1983 Publ. Centre Adv. Study in Geol. Panjab Univ. Chandigarh, India, no. 3, pp. 170–179.

Rai, H. & Pande, I. C. 1978 *Rec. Res. Geol.* **5**, 219–228.

Ramana, Y. V., Gogte, B. S. & Sarma, K. V. L. N. S. 1986 *Phys. Earth planet. Int.* **43**, 104–122.

Reuber, I. 1986 *Nature, Lond.* **321**, 592–596.

Reynolds, P. H., Brookfield, M. E. & McNutt, R. H. 1983 *Geol. Rdsch.* **72**, 981–1004.

Scharer, U., Xu, R.-H. & Allègre, C. J. 1984a *Earth planet. Sci. Lett.* **69**, 311–320.

Scharer, U., Hamet, J. & Allègre, C. J. 1984b *Earth planet. Sci. Lett.* **67**, 327–339.

Searle, M. P. 1983 *Trans. R. Soc. Edinb.* **73**, 205–219.

Searle, M. P. 1986 *J. struct. Geol.* **8**, 923–936.

Searle, M. P. & Fryer, B. J. 1986 In *Collision tectonics* (ed. M. P. Coward & A. C. Ries), publ. no. 19, pp. 185–201. Geological Society of London.

Searle, M. P. *et al.* 1987 *Bull. geol. Soc. Am.* **98**, 678–701.

Searle, M. P., Rex, A. J., Tirrul, R., Rex, D. C. & Barnicoat, A. 1988 Geol. Soc. Am. Symp. Vol. (In the press.)

Seeber, L. & Armbruster, J. G. 1984 *Tectonophysics* **105**, 263–278.

Seeber, L., Armbruster, J. G. & Quittmeyer, R. C. 1981 In *Zagros, Hindu Kush, Himalaya, geodynamic evolution* (ed. H. K. Gupta & F. M. Delaney), Geodyn. Ser. vol. 3, pp. 215–241. Washington, D.C.: American Geophysics Union.

Sengör, A. M. C. & Kidd, W. S. F. 1979 *Tectonophysics* **55**, 361–376.

Shackleton, R. M. 1981 *J. struct. Geol.* **3**, 97–105.

Shams, F. A., Jones, G. C. & Kempe, D. R. C. 1981 *Mineralog. Mag.* **43**, 941–942.

Sharma, K. K. 1986 In *Proc. Int. Symp. Neotectonics in south Asia*, pp. 25–50. Dehradun: Geological Survey of India.

Sinha, A. K. 1981 In *Zagros, Hindu Kush, Himalaya: geodynamic evolution* (ed. H. K. Gupta & F. Delaney), Geodyn. Ser. vol. 3, pp. 122–148. Washington, D.C.: American Geophysics Union.

Sinha Roy, S. 1981 In *Metamorphic tectonites of the Himalaya* (ed. P. S. Saklani), pp. 279–302. Delhi: Today & Tomorrow Publishers.

Sinha Roy, S. 1982 *Tectonophysics* **84**, 197–224.

Srikantia, S. V. & Razdan, M. L. 1981 *J. geol. Soc. India* **22**, 227–234.

Srimal, N., Basu, A. R. & Kyser, T. K. 1987 *Tectonics* **6**, 261–273.

Stocklin, J. 1980 *J. geol. Soc. Lond.* **137**, 1–34.

Tahirkheli, R. A. K. & Jan, M. Q. 1979 *Geol. Bull. Univ. Peshawar, Spec. Issue*, **11**, 1–30.

Tapponnier, P., Mercier, J. L., Armijo, R., Han, T.-L. & Zhou, J. 1981*a* *Nature, Lond.* **294**, 410–414.

Tapponnier, P. *et al.* 1981*b* *Nature, Lond.* **294**, 405–410.

Thakur, V. C. 1981 *Trans. R. Soc. Edinb.* **72**, 890–897.

Thakur, V. C. 1987*a* *Tectonophysics* **134**, 91–102.

Thakur, V. C. 1987*b* *Tectonophysics* **135**, 1–13.

Thakur, V. C. & Misra, D. K. 1983 In *Geology of Indus Suture Zone in Ladakh* (ed. V. C. Thakur & K. K. Sharma), pp. 33–40. Dehra Dun: Wadia Institute of Himalayan Geology.

Thakur, V. C. & Misra, D. K. 1984 *Tectonophysics* **101**, 207–220.

Valdiya, K. S. 1981 In *Zagros, Hindu Kush, Himalaya: geodynamic evolution* (ed. H. K. Gupta & F. Delaney), Geodyn. Ser. vol. 3, pp. 87–110. Washington, D.C.: American Geophysics Union.

Vidal, Ph., Cocherie, A. & Le Fort, P. 1982 *Geochim. cosmochim. Acta* **46**, 2279–2292.

Wang, Y.-G. & Sun, D.-L. 1985 *Can. J. Earth Sci.* **22**, 195–204.

Windley, B. F. 1983 *J. geol. Soc. Lond.* **140**, 849–865.

Xu, R.-H., Schzarer, U. & Allègre, C. J. 1982 *Eos, Wash.* **62**, 1094.

Zhou, J. 1985 *J. geol. Soc. Lond.* **142**, 309–318.

*Phil. Trans. R. Soc. Lond.* A **326**, 17–32 (1988)     [ 17 ]

*Printed in Great Britain*

# Features of the crust–mantle structure of Himalayas–Tibet: a comparison with seismic traverses of Alpine, Pyrenean and Variscan orogenic belts

By A. Hirn

*Laboratoire de Sismologie and Project ECORS, Institut de Physique du Globe, 4 pl. Jussieu, 75252 Paris, France*

Seismic data able to resolve the crustal structure are limited in quantity and quality with respect to the size and complexity of Tibet–Himalayas. They may be interpreted as indicating a strong heterogeneity: lack of continuity of even major interfaces across strike, defining different crustal blocks, but also lack of continuity of surface tectonic features down through the whole lithosphere. A thickening by imbrication of both the upper crustal and the lower crust–upper mantle levels is suggested. Indications from recent high-resolution surveys in other domains of thickened crust are also of a less smooth geometry of structures and depth than intuitively considered.

## 1. Introduction

Evidence on the seismic structure of Tibet–Himalayas has traditionally only been obtained with sensors outside the region. In 1977 explosion seismology, deep seismic sounding by so-called refraction profiles, was initiated by the Institute of Geophysics, Chinese Academy of Sciences (1981). The use of this technique was developed in 1981–1982 by a Sino–French cooperative programme that also deployed temporarily long-period seismographs for surface-wave dispersion studies (Jobert *et al.* 1985), did gravity (Van de Meulebrouck *et al.* 1983) and heat-flow (Francheteau *et al.* 1984) measurements and other geophysical and geological studies (Allègre *et al.* 1984). In other regions like the Alps, similar studies initiated over 30 years ago are still currently supplemented almost every year without the picture of the deep structure having reached a sharpness and precision such that a consensus about its nature and the evolutionary models it supports being reached. For Tibet–Himalayas, that single experimental effort has unfortunately remained quite isolated and further sampling of the medium to control the significance of data and suggested interpretations is not forthcoming.

We review here particular aspects of the presentation of the seismic data acquired and of suggested interpretations, which still remain of a speculative nature as there has been no chance to test them by new experiments. Recently we gained a new insight into the deep structure of other regions of abnormally thickened crust in the Alps and the Pyrenees that shall be used to shed nevertheless some light on interpretations or speculations we suggested of the deep structure of Tibet–Himalayas. First, in the frame of our ECORS programme the particular type of gross explosion-seismology single-trace oblique-incidence profiling as used in Tibet could be tested against the much more resolving multiple-coverage vertical-reflection profiling derived from expensive industrial prospecting, now currently used (Matthews & Smith 1987). Second, the particular structures we detected (ECORS Pyrenees Team 1988; Daignières *et al.* 1988; Bayer *et al.* 1987; ECORS–CROP Deep Seismic Sounding Group 1988) provide interesting examples of structural styles unsuspected from other data or at odds with currently admitted interpretations, a situation that might bear some resemblance to that in Tibet where the simplest explanation of some seismic observations would yield pictures unexpected by commonly considered models.

## 2. Seismic sounding for the lithospheric structure

Seismic waves artificially generated at the surface of the Earth penetrate the lithosphere and can be sensed and studied if turned back to the surface either by refraction in a medium of strong positive velocity gradient with depth or by reflection on interfaces of strong local velocity contrast or by propagation as head-waves along such interfaces. The main, unavoidable limitation of the seismic-prospecting method comes from both the sources and the sensors being at the surface, which causes non-unicity of the interpretations of data in terms of the underlying structure. Nevertheless, variations in experimental designs are possible, mainly in the relative geometry of the sources and sensors that control the incidence of waves detected, the spectral content and amplitude level of the signal, and in the number of data, which is controlled by financial aspects. The sensitivity of an experiment may be adapted, for instance, rather to the study of the velocity–depth function of a presumably horizontally stratified structure by following the wave field as a function of range from the source and hence as a function of penetration into the structure. Commonly called a refraction profile, this arrangement where source and sensors are indeed on a straight profile line provides, in fact, a seismic sounding and the clearest waves detected are most often the wide-angle reflections of the sharpest changes in the velocity–depth function. In contrast to this reconnaissance method for the average layering of the structure, the vertical reflection method used by the prospection industry and now also in lithospheric studies (see, for example, Brown & Barazangi 1986) profiles the topography of subsurface interfaces with an extremely fine resolution, because of the density of source and sensor arrays. This succeeds if signals can be made strong enough by adding many individual observations in the multiple coverage, and provided the interface responds to this particular vertical incidence and its response and the attenuation in the medium of propagation are fitted to the high-frequency character of sources and sensors tuned for fine resolution.

For the investigation of the lithospheric structure of Tibet, deep seismic sounding along two 400 km long bases with shots at both ends and in the middle was performed along the strike, just north of the High Himalayas and 400 km further north. To gain furthermore an impression on the variation of the deep crust across the strike, we used a wide-angle reflection method that we designed and used some years earlier through the Pyrenees (Hirn *et al.* 1980) and more recently through the Variscan Belt of northern and western France (Hirn *et al.* 1987; Matte & Hirn 1988) and the Alps (ECORS–CROP Deep Seismic Sounding Group 1988) as will be shown later for comparison. Sensors placed on fan profiles with respect to the source, i.e. at constant offset of it, record waves for which this range corresponds to a critical or wide-angle reflection of maximum amplitude. Differences of arrival times of a given wave on neighbouring traces mirror differences in reflector depth or average velocity above it at the different turning points of waves. From one single source, wide-angle reflection profiling provides as many data points as there are sensors and with a separation adjustable by that of the sensors, e.g. with 30 sensors spaced at 5 km intervals along the fan a 75 km segment of reflector can be followed with a 2.5 km sampling, mid-way between the shotpoint and the profile of receivers. The usual vertical reflection method of industrial prospection may be viewed as an extreme case of this scheme in which the constant offset is set to zero. In this layout each shot is recorded by a multichannel array of nearby geophones but data are summed to form about only one single zero-offset trace for each shot, hence the sampling of the medium along the profile line depends on source separation. However, the energy of each source may then be reduced as only the vertical distance to the reflector has to be penetrated and stacking and signal-processing techniques can be used on a pattern of close-by sensors. In the first approximation interesting for the Tibetan expedition, the use of wide-angle reflection profiling is a cheap reconnaissance ersatz to vertical reflection resolution having thus to be subordinated to the length of cross section to be obtained; this section is, however, only for the depth range to which the offset is tuned. Another important reason for using it was that adequate response at wide angle to a low-frequency signal is much more probable than at steep angle to a high-frequency one in the case possible in Tibet of a transition zone at crust–mantle boundary and of long paths of propagation through a possibly attenuating crust.

## 3. Average lithospheric layering

### 3.1. *Average layering of the crust between High Himalayas and Yarlung Zangbo, depth to Moho and a comparison with the western Alps*

*Seismic nature and depth of the crust–mantle boundary*

Probably the most firmly established result of explosion seismology in Himalaya–Tibet is the unequivocal identification of the crust–mantle boundary and determination of its depth just north of the High Himalayas (Hirn *et al.* 1984*a, c*). It is probably one of the most clear and precise examples of such determination worldwide as both the upper mantle is well identified by Pn head waves propagating just beneath this limit with a 8.7 km s$^{-1}$ reversed apparent velocity, and the lower crust and transition zone are well expressed in, and can be modelled by, amplitudes and waveforms of wide-angle reflections (figure 1).

The uppermost mantle has a high velocity: even correcting the measured value for Earth curvature and pressure due to abnormal depth leaves us with a very high value for the material that might call for reduced temperatures and anisotropy or only anisotropy with an east–west orientation of the fast direction. There is no straightforward interpretation for this were it not that it is the direction of tectonic escape of the lithosphere or certain levels of it in the general shortening between points respectively in undeformed Asia and India and that high velocities in the uppermost mantle have been proposed elsewhere to be related to anisotropy induced by stress or differential motion (see, for example, Hirn 1977; Fuchs 1983).

The mantle is identified as lying about 80 km beneath the surface, or 75 km beneath sea level; crustal material reaches 65 km thickness, i.e. extends over 60 km below sea level, and a 10–15 km transition layer in between allows us to model the low-frequency seismograms. This passage from crust to mantle may look differently if it could be resolved with shorter wavelengths signals.

The region where this large crustal thickness is measured is situated only 30–50 km north of the highest Himalayan peaks, nearer to them than to their border with Tibetan terranes at the Yarlung Zangbo Suture. We are here still on the slope of the Bouguer anomaly, with values of the order of $-300$ to $-350$ mGal,† far from the minimum of the order of $-500$ mGal. This constraint from seismic sounding, the maximum thickness of the crust extending far south

---

† 1 mGal = $10^{-3}$ cm s$^{-2}$.

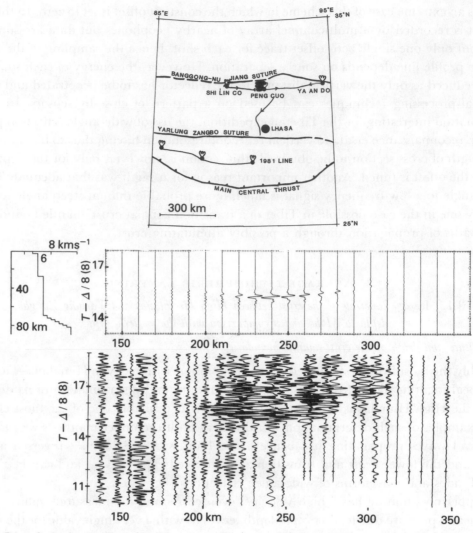

FIGURE 1. Part of seismic sounding along the reversed line recorded along strike around 40 km north of the High Himalayas. Constant gain on all traces reveals strong energy at offsets around 200 km, modelled as a reflection on the Moho, a transition zone about 10 km thick centred around 70 km beneath the surface. Mantle refractions with 8.7 km s⁻¹ high apparent velocity are identifiable on amplified section only (figure 7 in Hirn *et al.* 1984*c*). Insert shows position of main seismic lines in Tibet.

towards the High Himalayas, is difficult to reconcile with the considerations developed by Karner & Watts (1983) and Lyon-Caen & Molnar (1983, 1985) when trying to account for the support of the topography of the Himalayas and Ganga Basin and for the gravity anomaly by the sole flexure of the Indian Plate without introduction of an anomalous repartition of masses. A smooth slope of the Moho of an Indian Plate continuous to the Yarlung Zangbo would have this Moho situated still shallower just north of the Himalayas than we find. Furthermore, we establish that as the crust is already very thick at this latitude it does not have a marked further increase towards the Yarlung Zangbo, whereas this is the region where Lyon-Caen & Molnar (1983) would need the strongest flexure, i.e. gradient in Moho depth (from 50 to 115 km), after having already been obliged to allow a strong variation in elastic parameters of the Indian crust, which in fact does weaken the simplicity of the model of its continuity towards north.

*Comparison with the response of the Alpine thickened crust*

Although the topography of the Moho across the Himalayas that we propose later as the simplest interpretation of seismic data may be debatable in its complexity, it would appear that maximum crustal thickness is reached earlier across the mountain range than if its topography were only supported by a flexed plate. A possibly similar feature was revealed by our wide-angle reflection profiling of the Moho across the western Alps (ECORS–CROP Deep Seismic Sounding Group 1988), the time section of which is shown in figure 2, converted to depth after correcting for normal moveout. A strong change in the rate of increase of crustal thickness is seen to occur at the edge of a relatively horizontal Moho that reaches to the east of the External Crystalline Massif of Belledonne. If the Moho reflection was strictly continuous and straight to the east of it, the reflector should be obtained by migrating the reflection from its position in

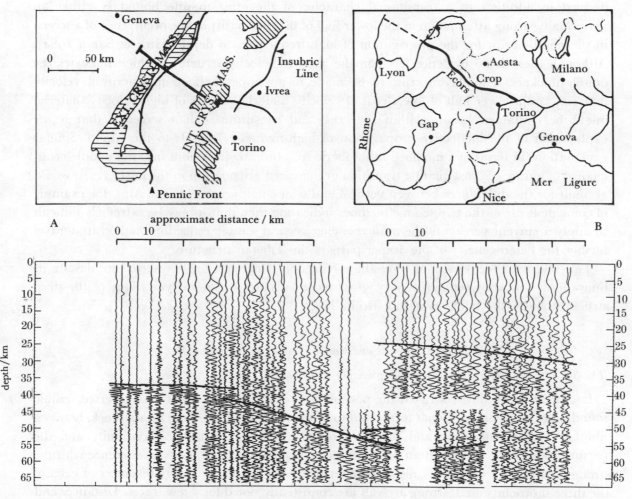

FIGURE 2. Western Alps, wide-angle reflection composite cross section of Moho topography, after ECORS–CROP Deep Seismic Sounding Group (1988). Data corrected for normal moveout with an average crustal velocity of 6.25 km s$^{-1}$ from their offset between 100 and 150 km (insert map shows position of section, mid-way between shotpoints and recorders). Note change in rate of variation in crustal thickness east of Belledonne, Moho would be even steeper there if migrated in the assumption that its reflection is a piecewise continuous and straight segment. Note also the change in signal character and frequency content, the reflection signal losing higher frequency when Mono gets deeper (recording of the same shots outside the region of thick crust allows to exclude a source effect).

figure 2 and would be even steeper than it appears. The change in rate of increase of crustal thickness to the east would then be even more important with respect to that considered for simple support of topography by flexure, which model has hence to be complicated by introduction of abnormal repartition of mass (Karner & Watts 1983).

The Alpine example illustrates another feature for which hints exist in Tibet: propagation of seismic waves to the Moho is not just like twice the propagation through a normal crust, but indicates that a change of internal or Moho structure occurred upon thickening. In the Alpine wide-angle Moho cross section (figure 2) the reflection signal loses its high frequencies from the foreland to the zone of thick crust. Also the coincident vertical reflection section, for which the prospection method derived from industry was used, so that signals cannot contain frequencies lower than 10 Hz and incidence is restricted to near vertical, completely loses track of the Moho in this internal part (Bayer *et al.* 1987). In Himalaya–Tibet, we modelled the contrasting broad-spectrum intracrustal and low-frequency Moho wide-angle reflections obtained and derived in addition to a transitional character of the crust–mantle boundary either an abnormally strong attenuation in the lower half of the thick crust or the occurrence of a screen in the middle crust for the propagation of high frequencies to depth (Hirn & Sapin 1984). Although there is no evidence for it in the average velocity structure, which indicates the double-thickness Himalayan crust to be made of an upper half of upper-crustal velocity material and a lower half of lower-crustal velocity material, slivers of high-velocity material might be contained locally within the crust and constitute such a screen if that is the explanation of the ineffective propagation of high-frequency signals to the Moho. Similar explanations in terms of a modified lithospheric structure, by inclusions into the middle crust, change of nature of crust–mantle transition or increased attenuation in the lower crust, would account for the differences between vertical and wide-angle seismics in the Alps, the example of coincident use of the two seismic methods indicating then that it may be extremely difficult to apply a current version of the most resolving vertical seismic reflection method to unravel further the heterogeneity of the deeper parts of the Tibetan structure.

This relative effectiveness of wide-angle reflection hence has to be resorted to, despite its limitations, for what remains a very gross reconnaissance of lateral variations of the deep structure across the Himalayas and within Tibet.

### 3.2. *Hints at crust and mantle layering in the Lhasa Block*

*The Shi Lin Lake to Ya An Do sounding* (Sapin et al. 1985)

Logistical constraints restricted the possibility of achieving a 400 km base reversed seismic sounding to the vicinity of what turned out to be the northern limit of the Lhasa Block, between Shi Lin Lake, Peng Lake and the town of Ya An Do. The lack of continuity and the perturbation of the wave pattern with distance is probably caused by its interference with the strong north–south variability discussed later. Nevertheless at very large distances of each of the three shotpoints used, strong arrivals are consistently noted for a few traces. Distances and times of arrivals are almost comparable to the Moho reflection beneath the line south of the Yarlung Zangbo and may hence be attributed to a Moho around 65 km deep; it seems difficult to imagine an alternative way of bringing such seismic energy to these offsets and times of propagation.

The lack of the complete wavefield prevents us from estimating precisely both the depth to Moho and the average crustal velocity, between which there is a trade-off. If the average velocity was smaller than that south of the Yarlung Zangbo, Moho depth would be smaller than proposed above, and conversely.

The upper 30 km of the crust, as their sampling occurs by shorter sounding lines, are better defined and exhibit fine structural layering. The particular situation of the profile probably prevents us, however, from generalizing to the whole-block details of this layering, although a line from the central shotpoint of the line, Peng Lake, recorded to the south all through the block is strikingly similar as far as the reflectivity around 20 km depth is concerned.

### The upper mantle from surface-wave dispersion (Jobert et al. 1985)

No penetration below the crust could be achieved by explosion seismology. To cope with this, the acquisition of a complementary data set by a different method had been prepared. Four long-period seismic stations could hence be positioned and maintained for several weeks by taking advantage of personnel and logistical support of the explosion seismology experiment. The geometry of the array, 400 km distance between stations, and the occurrence of a few distant earthquakes of adequate magnitude and azimuth allow us to constrain the shear velocity–depth function by the inversion of Rayleigh wave phase-velocity dispersion curves between pairs of stations. The two southern stations are between the Himalayas and Yarlung Zangbo, the two northern ones at the northern limit of the Lhasa Block, which is thus the region principally sampled by this study, in the region of the explosion seismology surveys. The velocity–depth function (for shear waves) below 65 km depth shows a similarity with that for the Western Europe platform or mobile region rather than with the Eastern Europe shield region. Besides the need for an increase of velocity between the upper half and the lower half of the crust, the inversion of dispersion data mainly tends to show a reduced thickness of a mantle high-velocity lid for these different starting models. Above a region of velocity decrease in the mantle that may correspond to the base of the lithosphere and is situated at similar depths of 100–150 km in the models for Tibet and Westen Europe, the crustal part of the lithosphere in Tibet is thicker by a factor of two whereas its mantle part is not, or is even thinner. If originating from a lithosphere originally having a thicker mantle than crustal part, this would mean that while the crust thickened the mantle lithosphere returned partly to the asthenosphere, by delamination or thermal modification.

This relatively low average velocity in the mantle under Tibet (Molnar & Chen 1984; Romanowicz 1982) has also been more recently confirmed by Lyon-Caen (1986) with the independent study of direct and surface-reflected shear waves to be confined to the uppermost mantle, i.e. to a reduced thickness of the mantle lithosphere.

### 3.3. Evidence for internal structure of the lithosphere south of the High Himalayas

The seismic determination of the crustal structure of the Lesser Himalayas would well deserve a reversed sounding along the strike; unfortunately such data of a quality similar to southern Tibet are still missing. This leaves much room for speculation as to the evolution of this structure across the Himalayas as it is not even known where the Indian Plate approaches this complex domain. Having only shotpoints north of the Himalayas we could do no better than at least place recording stations in Nepal, unfortunately across strike and without shot in the south to reverse the profile. Several relatively clear waves are recorded showing a heterogeneous crust but obviously there is a trade-off between velocities and dips of interfaces. A later section will present the composite picture derived across the Himalayas (Lépine et al. 1984; Hirn & Sapin 1984).

## 4. MAJOR LATERAL VARIATIONS: THE SUTURES OR WRENCH FAULTS LIMITING THE LHASA BLOCK

The Mohorovičić discontinuity between crust and upper mantle is the major seismic marker in the lithosphere. Most recently results of vertical reflection profiles of COCORP in the Western United States and BIRPS around Britain would suggest the Moho be less immutable than previously thought, perhaps able to change its identity and migrate in depth in the lithosphere in geodynamical contexts where extension and high temperatures prevail (see, for example, Barton 1986). Nevertheless in compressive contexts its depth situation and its shape may be remnants of the major deformational history.

In the north–south cross section of the Tibetan Moho based on wide-angle reflection fan-shooting we proposed (Hirn *et al.* 1984 *b*), the two limits of the Lhasa Block inferred from surface geology appear to be associated with clear perturbation to the topography of the Moho. This whole section was obtained as a composite from several shotpoints but the part of data across each of these limits comes each time from only one shot situated near its surface trace. Also the stations either side of the limit are really at the same offset from shotpoints so that the uncertainty that may be introduced elsewhere in normal moveout correction by the lack of precise knowledge of the velocity–depth function does not disturb the picture of the Moho topography obtained here. The coincidence of changes of structure at Moho depth with the vertical of limits between terranes at the surface would in general favour the idea of a strike-slip enhanced limit of the sutured blocks. However, as the structure between surface and the Moho marker is unknown there may be an unknown amount of imbrication (see figure 3).

### 4.1. *The southern limit of the Lhasa Block: the Yarlung Zangbo Fault Zone*

The Angren drillhole shotpoint on the Yarlung Zangbo provides a picture of deep reflections across this feature. With respect to a situation at around the maximum depth of 70 km on either side at 40 km distance, the Moho stays at about this position in the south, possibly sloping very slightly northward, whereas a clear reflection gets shallower from 70 to 50 km depth when approaching the suture from the north. These figures suppose the same average crustal velocities on either side, which seems to hold regionally but is not precisely established at this particular location. There is no clear observation of a prolongation of the southern Moho towards the north under the shallower one. It might, however, exist; rays that might sample it could be screened off deeper propagation by the shallow Moho. It is also clear that, at this scale, we are detecting lateral heterogeneities of spatial extent along the section that are on the order of, or smaller than, their depth and that the very precise way seismic wavefronts have travelled would need more data to be effectively modelled in detail. Fortunately, Van de Meulebrouck *et al.* (1983) were able to carry out a dense gravity profile across the suture zone, some tens of kilometres east of the zone sampled by seismology. After the general steep slope through the Himalayas that brings the Bouguer values to −470 mGal some 50 km south of the suture, a local maximum of −430 mGal is reached just on the suture before the anomaly decreases again to reach the minimum value of −530 mGal at the northernmost observation point, 70 km further north. The wavelength of this relative Bouguer maximum confirms that early reflection times on the Moho just north of the suture are caused by the shallow position of that reflector rather than by the influence on propagation of a local high average velocity in the crust above, as the mass heterogeneity of this would be situated higher and cause a

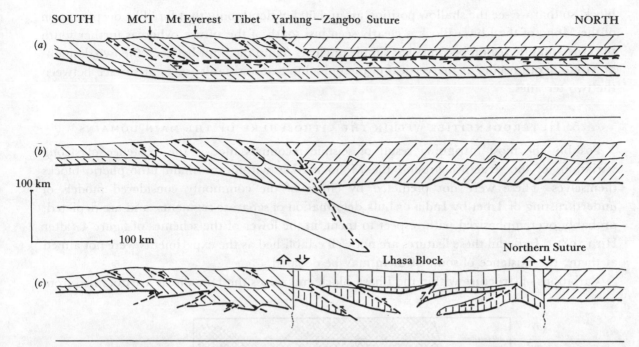

SOUTH    MCT  Mt Everest  Tibet  Yarlung–Zangbo Suture    NORTH

100 km

100 km

Northern Suture

Lhasa Block

FIGURE 3. Schematic cross sections through Tibet–Himalayas, after Hirn (1984). Indian and Asian crusts are indicated by shading in opposite directions, heavier for the lower than upper crustal material, and vertical for the Lhasa Block introduced in (c). The lower section (c) illustrates what explosion seismic data would suggest for the heterogeneity within the lithosphere. Difference in depths to the Moho are documented under two ophiolite belts and fault zones, which define a Lhasa Block between these sutures with India to the south and terranes accreted to the north to Eurasia. Wrench motions along these verticalized sutures and within the block allow shortening by escape of parts of the lithosphere out of the plane of the figure. In this block, if the crust has been thickened, it does not appear to have been homogeneously as the Moho and lower crustal marker are neither horizontal nor continuous. Tectonic deformation near the crust–mantle boundary and a zone of decoupling within the crust would allow the lower and upper parts of the crust to thicken separately. A similar behaviour would be indicated south of the Yarlung Zangbo Suture if the thickening of the crust had not occurred progressively from south to north through the Himalayas but rather stepwise as there would be an indication for it in the seismic data.

The upper section (a) is inspired from the Argand model of the crustal part of the Asian lithosphere being completely through the section underlain by the Indian lithosphere to provide the abnormal crustal thickness and north–south shortening (Powell & Conaghan 1973; Ni & Barazangi 1983). The middle section (b) inspired from Dewey & Burke (1973) limits the extent of Indian lithosphere to the north and deforms the Asian lithosphere.

narrower anomaly. The gross test of the gravity response of the seismic model containing the local topography of the Moho across the Yarlung Zangbo in Hirn & Sapin (1984) is consistent with observations.

What cannot be decided from seismic data is whether the signature on the deep structure has been left by the converging tectonics at the suture or by a later strike-slip motion along this limit during the tectonic escape of Tibet. The apparent limitation of the northern Moho just under the Yarlung Zangbo Fault might favour the second interpretation.

### 4.2. *The northern limit of the Lhasa Block: the Bang Gong–Nu Jiang line*

Across the Bang Gong–Nu Jiang ophiolite marking the northern suture of the Lhasa Block about 300 km further to the north, a similar picture emerges from seismic data, this time without the gravity control. Data could not be extended more than 30 km into the northern block, so that we see the shallow position around 50 km depth only that far. The deep position of the Moho of the Lhasa Block is clearly seen just south of the suture, whereas further south the crust–mantle region becomes less simple or clear. Again the limit between the two domains at Moho depth occurs at the vertical of the ophiolite belt and fault traces at surface between the two terranes.

## 5. Heterogeneities within the lithosphere of the main domains

Possibly the most intriguing result of seismic reconnaissance is the existence of strong variations of the structure around the Moho marker within the main lithospheric blocks themselves. These were not predicted by either of the commonly considered models of underthrusting of Tibet by India or bulk deformation of separate segments, and are depicted, probably overemphasized with respect to them, in the lower of the schemes of figure 4 (after Hirn 1984). In detail these features are not well established as the experiments were not aimed at them; the existence of some of them may be debated.

Variations of structure within the Lhasa Block and across the Himalayas are obtained from

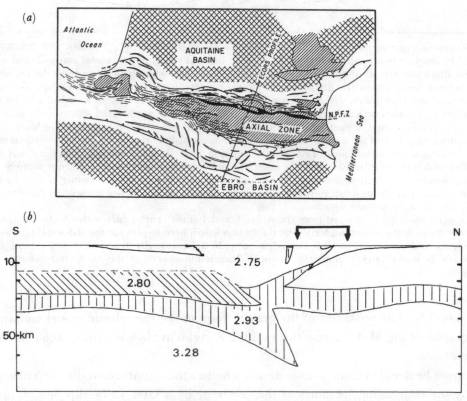

FIGURE 4. ECORS seismic profile through the Pyrenees (after ECORS Pyrenees Team 1988) and Daignieres *et al.* (1988). Along the line in (*a*), the cross section (*b*) derived from seismic reflection is compatible with gravity control. It shows an example of strong variation of Moho depth, lower limit of the black lower crust discovered by wide-angle and confirmed by vertical reflection seismics. South-verging tectonics in the upper part of the section are well documented north of the NPF which limits the Axial Zone to the north and even crossing its surface trace, whereas at lower crustal depth the vergence is towards north, resulting in an imbrication of the crusts. Vertical reflection seismic data displayed as (*c*) correspond to the part just north of the NPF between the arrows. South-dipping reflections are obvious around 3–6 s, TWT; the shallower, northern Moho is seen around 12 s. The significantly later very conspicuous straight reflection, strongly north-dipping from 15 to 21 s TWT, migrates to the south to give the (tectonic) limit of the wedge of southern thickened crust under the northern mantle.

wide-angle reflection fan-shooting. General limitations of this have been commented on earlier but here in addition, as the geometry was not exactly aimed at them, some additional problems and uncertainties arise from the necessity of resorting to composite sections with only limited knowledge of how different shotpoints should be tied in and also from the fact that the offset distances may vary. This, however, only limits the value of a precise correlation of absolute Moho depths between distant parts of the general, 500 km long traverse of Moho topography (figure 2 in Hirn *et al.* 1984*b*) but the existence of local lateral changes in the features of the seismograms is obvious and cannot be reconciled with a horizontally stratified lithosphere.

### 5.1. *Lateral variations across the Lhasa Block of Tibet*

What is established beyond doubt is that the Moho is not flat, horizontal and continuous across the Lhasa Block. There are places where branches of strong reflections, which at these very large offsets must be attributed to interfaces of strong, about crust–mantle, velocity

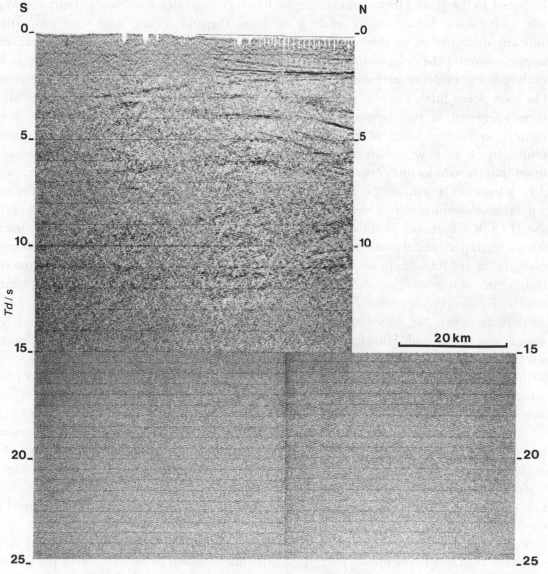

FIGURE 4*c*. For description see opposite.

contrasts, seem to relay each other at different times, i.e. corresponding to different distances of the reflector from the source–receiver line. There is no control with these data alone on the possible importance of echoes out of the vertical plane from source to receiver. Such a possibility is not taken into account by the way seismic time-sections are plotted, that is attributing all energy to reflection at half source–receiver distance, vertically. In the most similar case available, that of the Pyrenees (discussed later), the hypothesis attributing strong variation in aspect of wide-angle seismograms only to the abrupt change in structure at the vertical without any noticeable side echoes (Hirn *et al.* 1980) could be verified by coincident high resolution vertical seismic reflection (ECORS Pyrenees Team 1988). Hence it is reasonable to consider that strongly heterogeneous structure exists around the Moho marker in the Lhasa lithosphere.

We proposed that the imbrication of crust and mantle suggested by Moho profiling may mark an average thickening of the lower crust. Correlative, but possibly separated by an intracrustal ductile layer (Meissner 1974), thickening of the upper crust could have contributed to the total Tibetan thickening by heterogeneous deformation, a combination of brittle and ductile behaviour, depending on local thermal régime and rock properties. Significant deformation at the surface was implied by this inferred deformation with the lithosphere around the crust–mantle boundary. This was then poorly known but seems to be recently gaining evidence and support (Tertiary deformations, J. F. Dewey, unpublished).

The case needs further study, but the possibility of the existence of the most provocative features suggested in the simplest interpretation of gross seismic data at that time, non-continuity and imbrication of Moho and separation of deep crustal from upper-crustal tectonics, has been now shown elsewhere. Although not a common occurrence this shows *a posteriori* that the odd features once suggested from scarce deep seismic data may exit in Nature. Eighty kilometres north of the Pyrenees, high-resolution multichannel vertical reflection seismics shows unequivocally a loss of continuity accompanied by brutal change in depth of the Moho (ECORS Pyrenees Team 1988). Also on this line, the middle crustal domain clearly separates an upper-crustal domain of north-verging reflections from a lower one where all dips, intracrustal or at Moho depth are south verging. The presently favoured interpretation, on the considerations of regional geology, of different ages for the two levels of dipping reflectors, may somewhat weaken the argument, if correct. That opposite vergences are found on the same vertical is, however, the strongest case in favour of an intracrustal decoupling, which we proposed for Tibet and Himalayas having allowed upper and lower crusts to thicken separately. Indications of this decoupling may also be seen, as in the case of the Caledonides (Matthews & Hirn 1984), already in the lack of continuity through the middle or lower crust of structural reflectors even if these have the same general vergence on either side. The occurrence of a second brittle layer in addition to the upper crustal layer is attributed to the compositional change at the top of the upper mantle. Brittle tectonics at the Moho decoupled, as proposed here, from the surface by an intracrustal ductile layer should also be decoupled from deeper layers by a weak zone in the uppermost mantle. In Europe clear evidence for a sub-Moho low-velocity layer in the lithosphere was firmly documented by several very-long-range refraction profiles first in the Variscan part (Hirn *et al.* 1973), and this layer seemed to possess adequate rheology to allow accommodation of differential motions of small segments of crust in the continental part of the larger-scale lithospheric plate (Bottinga *et al.* 1973). In Tibet the reduced thickness of the mantle part of the lithosphere, which seems to underlie parts of the plateau itself, would allow the accommodation of the tectonics at Moho level to reach the asthenosphere. At the southern margin, near the Himalayas, the east–west seismic sounding gives indication of coherent energy after the Moho head-wave that bears some similarity to the situations encountered in Europe indicating seismic boundaries within the mantle.

### 5.2. *Variations across the Himalayas*

The variation of the structure of the lithosphere across the Himalayas has been proposed by fitting together three types of observation (Hirn *et al.* 1984*a*; Lépine *et al.* 1984; Hirn & Sapin 1984): that of a fan, but with increasing offset to the south of the High Himalayas; that of an unreversed profile across the range, from North to South; and isolated observations on short profiles at very large offsets. Other interpretations may be possible, but the question of the deep structure beneath the Himalayas and of the different evolutionary models suggested cannot be definitely settled without new data. We originally decided to publish the seismic data with the indication of what kind of structure and model they would suggest to us if we were not prejudiced by other knowledge (see figure 5 of Hirn *et al.* (1984*a*) or the southern extremity of the lower scheme of the present figure 3). Basically the seismic data suggested that the Moho may not be continuous, crustal thickening may occur by steps and deep tectonics may exist with a different vergence from that at the surface, which led to an image of imbrication of the lithosphere or a kind of indentation in the vertical section. Recently documented examples of lateral variations in the lithosphere elsewhere were given in the previous section. Admittedly further tests were needed to favour this picture or that of a postulated and more commonly admitted smooth and continuous dipping Moho suggestive of flexuration of the Indian Plate underriding Tibet. In the absence of new data to discuss the points where these approaches most diverge, it seems, however, possible to consider that there is not formal contradiction between that part of the seismic model that is best constrained, that is a thick crust extending to the south, and those features of the flexural model for Moho topography that are consistent with its initial aim at a simple explanation.

The extensive tests of the flexural model by Karner & Watts (1983) and Lyon-Caen & Molnar (1983, 1985) have not established that, in its simplest form, this type of model can account for the topography and gravity across the Himalayas, which limits the attractiveness of this approach. To achieve the fit by flexural models alone, the flexural rigidity of the Indian lithosphere had to be decreased by two to three orders of magnitude, i.e. its equivalent elastic thickness divided by seven north of the High Himalayas and the Moho postulated to increase continuously in depth from 50 to 115 km from the High Himalayas to the Yarlung Zangbo. The first of these complications introduces such a change along the plate that one might as well choose to consider that it is the negation of the continuity of the same plate north of the High Himalayas; the second, position and dip of the Moho, is contradicted by unequivocal high-quality seismic data on the reversed east–west sounding line situated 30–50 km north of the High Himalayas (§3.1).

The simplest, hence attractive, model considered by Lyon-Caen & Molnar (1983, 1985) that accounts for the shape of the Ganga Basin and the shape of the Bouguer anomaly is that where the Indian Plate is continuous, supports the topographic load and is flexed by it only as far north as the high peaks of the Himalayas. This limitation of its continuity does not allow it to support the topography north of the high peaks. There the observed Bouguer anomaly then demands that the crust be already of great thickness and uniform to the Yarlung Zangbo, which is just what was not foreseen but established by explosion seismology. This combination over the Himalayas of a flexural model where it works simply, and of a mass heterogeneity where it is well documented by seismic methods might be envisioned independently and before a further step where the much more debatable suggestions by wide-angle profiles of mass heterogeneity further south, would then be considered. It seems unavoidable that high-precision, adequate-penetration seismic data would be required on a traverse from the Ganga to the High Himalayas to proceed further with this discussion.

## 6. Case history of a limit between lithospheric domains suggested by wide-angle reflection fan profiling: the Pyrenees

In 1978–1980 we designed a wide-angle fan-profiling experiment to follow the Moho topography across the Eastern Pyrenean mountain range as an addition to the reversed seismic sounding lines recorded along strike that indicated that the crustal thickness differed by over 10 km between the axial Palaeozoic zone and the north Pyrenean zone (Daignières et al. 1982). Our simplest interpretation of that fan was that the change in Moho depth occurred in less than 10 km north–south horizontal distance (Hirn et al. 1980). In the western part of the Pyrenees, as the Moho did not respond adequately to reflection, we had to resort to a transmission method with waves from distant earthquakes, which in spite of its nominal low resolution yielded the same result (Hirn et al. 1984d).

One fashionable model for the structure and evolution of the Pyrenees advocated by Boillot & Capdevilla (1977) was then that of a southward-directed subduction of the northern, European Plate under the southern, Iberian Plate along the North Pyrenean Frontal Thrust. Such an underriding before collision and blocking would call for a smoothly dipping Moho. Our wide-angle seismic data could not completely disprove such a smoothness but we resisted pressure to erase a strong and sharp step in crustal thickness. We resisted also the temptation to consider that we had only measured artefacts even when the possibility of this structure resulting from our simplest seismic interpretation was completely negated by a later, all the more fashionable but completely different model of Williams & Fisher (1984). Since then, high-resolution vertical reflection seismic data confirmed that the picture suggested by some limited seismic data was not wrong but that the proposed models, although attractive could be rejected.

More interestingly, vertical reflection seismics, if it does not contradict that previous seismic evidence, shows that its simplest interpretation (if it was less wrong than the models cited) must be completed by taking into account structure between the surface and the Moho. The imbricated internal structure of the Pyrenean crust, which only vertical reflection seismics could reveal, is of particular importance in our present resistance to give up the complicated, imbricated image of the Himalayan crust with its opposite vergences at upper-crustal and upper-mantle levels. We do not pretend that the Pyrenees and Himalayas are similar, but only that some speculative features proposed for the Himalayas cannot be excluded from existing in Nature as we see them in the Pyrenees: the upper part of the southern, Iberian crust is thrust onto the European crust, whereas its lower part is overridden by it. We have here a thickening of crust related to an imbrication of lithospheres. We could only see the brutal change in Moho depth with wide-angle data, but vertical reflection sees also the south-verging contact of the upper-crustal part of the imbrication. It also sees the change of this vergence with depth when approaching the deeper, southern Moho and we contend that the upper, very long, continuous, straight and steeply north-dipping reflection, which appears very prominently at two-way times between 18 and 26 s on a 40 km long segment of the line further north, migrates into a position that allows it to be interpreted as the north-verging limit along which the northern lithosphere is imbricated into the southern crust (figure 4) (Daignières *et al.* 1988). That this feature cannot be seen by wide-angle seismics in the Pyrenees nor in the Himalayas if the crust is really thickened there by a succession of imbrications as we suggested from the unsmooth Moho topography, may be because of its seismic nature of a decrease in velocity with depth, from the European mantle wedge to the overridden Iberian lower crustal wedge, a situation that cannot generate critical, hence strong-amplitude reflections at large offset, whereas the impedance contrast is adequate for strong vertical reflection.

I acknowledge the action of Claude Allègre and Guy Aubert who imposed on us the challenge to gather significant seismic data in a single experimental effort in an obviously particular geographical, technical and geodynamical context. Georges Jobert shared enthusiasm and anxiety in the field and was instrumental in tying together the Sino–French crews. Logistical and technical expertise of my Chinese, French and Nepalese colleagues allowed us to secure fine data. More of the same are obviously needed for further progress. Main support in personnel came from Laboratoire de Sismologie, IPG Paris, the Changchun Institute of Geophysics and Geophysical Team 562, Beijing, of the Chinese Ministry of Geology, the Geophysical Institute of Academia Sinica and the Nepalese Department of Mines and Geology. Funding was under an agreement of the Chinese Ministry of Geology and the French CNRS-INSU.

Peter Molnar, while writing his review on geophysics in the Himalayas–Tibet for this symposium, kindly communicated his critical comments and reinterpretations of explosion seismology data. New interpretations cannot be rejected, but what is badly needed is new experimental evidence. Brian Windley forced me *in extremis* to the present discussion of both the points I consider well established and those I admit are very loosely constrained or controversial. For most of the latter, in the absence of additional data, I consider that the simplest interpretation from the strictly seismological point of view should still not be rejected in favour of others more generally or intuitively coherent with usually accepted models. After all, examples may be presented here of higher quality, more reliable seismic data in Europe that document structures that also turn out to be often at odds with reasonable or fashionable models based on a much better geological knowledge.

I thank my academic colleagues who have helped maintain active crustal research by explosion seismology within Institut National des Sciences de l'Univers and my colleagues in the different ECORS profile teams for having made reflection seismics available for crustal research. ECORS is a joint project of INSU, IFP, SNEA(P) and IFREMER.

## References

Allègre, C. J. *et al.* 1984 *Nature, Lond.* **307**, 17–22.
Barton, P. 1986 *Nature, Lond.* **323**, 392–393.
Bayer, R. *et al.* 1987 *C. r. Acad. Sci., Paris* (II) **305**, 1461–1470.
Boillot, G. & Capdevilla, R. 1977 *Earth planet. Sci. Lett.* **35**, 151–160.
Bottinga, Y., Hirn, A. S. & Steinmetz, L. 1973 *Bull. Soc. géol. Fr.* **7**, 500–505.
Brown, L. & Barazangi, M. (eds) 1986 *Reflection seismology*, vols 13 and 14, Geodynamics Series, American Geophysical Union. Washington, D.C.: American Geophysical Union.
Daignières, M., Gallart, J., Banda, E. & Hirn, A. 1982 *Earth planet. Sci. Lett.* **57**, 88–100.
Daignières, M., de Cabissole, B., Gallart, J., Hirn, A. & Surinach, E. 1988 *Tectonics*. (Submitted.)
Dewey, J. F. & Burke, K. C. A. 1973 *J. Geol.* **81**, 683–692.

ECORS Pyrenees Team 1988 *Nature, Lond.* **331**, 508–511.

ECORS–CROP Deep Seismic Sounding Group 1988 (Submitted.)

Francheteau, J., Jaupart, C., Shen, X. J., Kang, W. H., Lee, D. L., Bai, J. C., Wei, H. P. & Deng, H. Y. 1984 *Nature, Lond.* **307**, 32–36.

Fuchs, K. 1983 *Phys. Earth planet. Inter.* **31**, 93–118.

Hirn, A. 1977 *Geophys. Jl R. astr. Soc.* **49**, 49–58.

Hirn, A. 1984 *La Recherche* **156**, 878–881.

Hirn, A. & Sapin, M. 1984 *Ann. Geophys.* **2**, 123–130.

Hirn, A., Daignières, M., Gallart, J. & Vadell, M. 1980 *Geophys. Res. Lett.* **7**, 263–266.

Hirn, A., Steinmetz, L., Kind, R. & Fuchs, K. 1973 *Z. Geophys.* **39**, 363–384.

Hirn, A., Lépine, J. C., Jobert, G., Sapin, M., Wittlinger, G., Xu, Z. X., Gao, E. Y., Wang, X. J., Temg, J. W., Xiong, S. B., Pandey, M. R. & Tater, J. M. 1984*a Nature, Lond.* **307**, 23–25.

Hirn, A., Nercessian, A., Sapin, M., Jobert, G., Xu, Z. X., Gao, E. Y., Lu, D. Y. & Teng, J. W. 1984*b Nature, Lond.* **307**, 25–28.

Hirn, A., Jobert, G., Wittlinger, G., Xu, Z. X. & Gao, E. Y. 1984*c Ann. Geophys.* **2**, 113–118.

Hirn, A., Poupinet, G., Wittlinger, G., Gallart, J. & Thouvenot, F. 1984*d Nature, Lond.* **308**, 531–533.

Hirn, A., Damotte, B., Torreilles, G. & ECORS Scientific Party 1987 *Geophys. Jl R. astr. Soc.* **89**, 287–296.

Institute of Geophysics, Chinese Academy of Sciences 1981 *Acta geophys. sinica* **24**, 155–170.

Jobert, N., Journet, B., Jobert, G., Hirn, A. & Sun, K. Z. 1985 *Nature, Lond.* **313**, 386–388.

Karner, G. D. & Watts, A. B. 1983 *J. geophys. Res.* **88**, 10449–10477.

Lépine, J. C., Hirn, A., Pandey, M. R. & Tater, J. M. 1984 *Ann. Geophys.* **2**, 119–121.

Lyon-Caen, H. & Molnar, P. 1983 *J. geophys. Res.* **88**, 8171–8191.

Lyon-Caen, H. & Molnar, P. 1985 *Tectonics* **4**, 513–538.

Lyon-Caen, H. 1986 *Geophys. Jl R. astr. Soc.* **86**, 727–749.

Matte, Ph. & Hirn, A. 1988 *Tectonics*. (In the press.)

Matthews, D. & Hirn, A. 1984 *Nature, Lond.* **308**, 497–498.

Matthews, D. & Smith, C. 1987 *Geophys Jl R. astr. Soc.* **89**, 1–447.

Meissner, R. 1974 *J. Geophys.* **40**, 57–73.

Molnar, P. & Chen, W. P. 1984 *J. geophys. Res.* **89**, 6911–6917.

Ni, J. & Barazangi, M. 1983 *Geophys. Jl R. astr. Soc.* **72**, 665–689.

Powell, C. M. & Conaghan, P. J. 1973 *Earth planet. Sci. Lett.* **20**, 1–12.

Romanowicz, B. 1982 *J. geophys. Res.* **87**, 6865–6883.

Sapin, M., Wang, X. J., Hirn, A. & Xu, Z. X. 1985 *Ann. Geophys.* **3**, 637–646.

Tang, B. X., Liu, Y. L., Zhang, L. M., Zhou, W. H. & Wang, Q. S. 1981 In *Geological and ecological studies of Qinghai-Xizang Plateau*, pp. 683–689. Beijing: Science Press.

Van de Meulebrouck, J., Tarits, P., Le Mouel, J. L. & Men, L. S. 1983 *Terra cogn.* **3**, 272.

Williams, G. D. & Fischer, M. W. 1984 *Tectonics* **3**, 773–780.

*Phil. Trans. R. Soc. Lond.* A **326**, 33–88 (1988)     [ 33 ]

*Printed in Great Britain*

# A review of geophysical constraints on the deep structure of the Tibetan Plateau, the Himalaya and the Karakoram, and their tectonic implications

By P. Molnar†

*Laboratoire de Géophysique Interne et Tectonophysique (CNRS U.A. 733), Institut de Recherches Interdisciplinaires de Géologie et Mécanique, B.P. 68, Université Scientifique, Technique, et Médicale de Grenoble, 38402 Saint Martin d'Hères Cèdex, France*

The Tibetan Plateau, the Himalaya and the Karakoram are the most spectacular consequences of the collision of the Indian subcontinent with the rest of Eurasia in Cainozoic time. Accordingly, the deep structures beneath them provide constraints on both the tectonic history of the region and on the dynamic processes that have created these structures.

The dispersion of seismic surface waves requires that the crust beneath Tibet be thick: nowhere less than 50 km, at least 65 km, in most areas, but less than 80 km in all areas that have been studied. Wide-angle reflections of P-waves from explosive sources in southern Tibet corroborate the existence of a thick crust but also imply the existence of marked lateral variations in that thickness, or in the velocity structure of the crust. Thus isostatic compensation occurs largely by an Airy-type mechanism, unlike that, for instance, of the Basin and Range Province of western North America where a hot upper mantle buoys up a thin crust.

The P-wave and S-wave velocities in the uppermost mantle of most of Tibet are relatively high and typical of those of Precambrian shields and stable platforms: $V_p = 8.1$ km s$^{-1}$ or higher, and $V_s \approx 4.7$ km s$^{-1}$. Travel times and waveforms of S-waves passing through the uppermost mantle of much of Tibet, however, require a much lower average velocity in the uppermost mantle than that of the Indian, or other, shields. They indicate a thick low-velocity zone in the upper mantle beneath Tibet, reminiscent of tectonically active regions. These data rule out a shield structure beneath northern Tibet and suggest that if such a structure does underlie part of the plateau, it does so only beneath the southern part.

Lateral variations in the upper-mantle structure of Tibet are apparent from differences in travel times of S-waves from earthquakes in different parts of Tibet, in the attenuation of short-period phases, Pn and Sn, that propagate through the uppermost mantle of Tibet, and in surface-wave dispersion for different paths. The notably lower velocities and the greater attenuation in the mantle of north–central Tibet than elsewhere imply higher temperatures there and are consistent with the occurrence of active and young volcanism in roughly the same area. Surface-wave dispersion across north–central Tibet also requires a thinner crust in that area than in most of the plateau. Consequently the relatively uniform height of the plateau implies that isostatic compensation in the north–central part of Tibet occurs partly because the density of the relatively hot material in the upper mantle is lower than that elsewhere beneath Tibet, the mechanism envisioned by Pratt.

Several seismological studies provide evidence consistent with a continuity of the Indian Shield, and its cold thick lithosphere, beneath the Himalaya. Fault-plane solutions and focal depths of the majority of moderate earthquakes in the Himalaya

† Permanent address: Department of Earth, Atmospheric, and Planetary Sciences, Massachusetts Institute of Technology, Cambridge, Massachusetts 02139, U.S.A.

are consistent with their occurring on the top surface of the gently flexed, intact
Indian plate that has underthrust the Lesser Himalaya roughly 80–100 km or more.
P-waves from explosions in southern Tibet and recorded in Nepal can be interpreted
as wide-angle reflections from this fault zone. P-wave delays across the Tarbela
network in Pakistan from distant earthquakes indicate a gentle dip of the Moho
beneath the array without pronounced later variations in upper-mantle structure.
High Pn and Sn velocities beneath the Himalaya and normal to early S-wave arrival
times from Himalayan earthquakes recorded at teleseismic distances are consistent
with Himalaya being underlain by the same structure that underlies India.

Results from explosion seismology indicate an increase in crustal thickness from the
Indo–Gangetic Plain across the Himalaya to southern Tibet, but Hirn, Lépine, Sapin
and their co-workers inferred that the depth of the Moho does not increase smoothly
northward, as it would if the Indian Shield had been underthrust coherently beneath
the Himalaya. They interpreted wide-angle reflections as evidence for steps in the
Moho displaced from one another on southward-dipping faults. Although I cannot
disprove this interpretation, I think that one can recognize a sequence of signals on
their wide-angle reflection profiles that could be wide-angle reflections from a
northward-dipping Moho.

Gravity anomalies across the Himalaya show that both the Indo–Gangetic Plain
and the Himalaya are not in local isostatic equilibrium. A mass deficit beneath the
plain is apparently caused by the flexure of the Indian Shield and by the low density
of the sedimentary rock in the basin formed by the flexure. The mass excess in the
Himalaya seems to be partly supported by the strength of the Indian plate, for which
the flexural rigidity is particularly large.

An increase in the Bouguer gravity gradient from about 1 mGal km$^{-1}$ (1 mGal =
10$^{-3}$ cm s$^{-2}$) over the Indo–Gangetic Plain to 2 mGal km$^{-1}$ over the Himalaya
implies a marked steepening of the Moho, and therefore a greater flexure of the
Indian plate, beneath the Himalaya. This implies a northward decrease in the
flexural rigidity of the part of the Indian plate underlying the range. Nevertheless,
calculations of deflections of elastic plates with different flexural rigidities and flexed
by the weight of the Himalaya show larger deflections and yield more negative
gravity anomalies than are observed. Thus, some other force, besides the flexural
strength of the plate, must contribute to the support of the range. A bending moment
applied to the end of the Indian plate could flex the plate up beneath the range and
provide the needed support. The source of this moment might be gravity acting on
the mantle portion of the subducting Indian continental lithosphere with much or all
of the crust detached from it.

Seismological studies of the Karakoram are consistent with its being underlain by
particularly cold material in the upper mantle. Intermediate-depth earthquakes
occur between depths of 70 and 100 km but apparently do not define a zone of
subducted oceanic lithosphere. Rayleigh-wave phase velocities are particularly high
for paths across this area and imply high shear wave velocities in the upper mantle.
Isostatic gravity anomalies indicate a marked low of 70 mGal over the Karakoram,
which could result from a slightly thickened crust pulled down by the sinking of cold
material beneath it.

Geophysical constraints on the structure of Tibet, the Himalaya and the
Karakoram are consistent with a dynamic uppermost mantle that includes first, the
plunging of cold material into the asthenosphere beneath southern Tibet and the
Karakoram, as the Indian plate slides beneath the Himalaya, and second, an
upwelling of hot material beneath north–central Tibet. The structure is too poorly
resolved to require such dynamic flow, but the existence for both a hot uppermost
mantle beneath north–central Tibet and a relatively cold uppermost mantle beneath
southern Tibet and the Karakoram seem to be required.

## Introduction

The deep structure of the Himalaya and Tibet is poorly known, compared with that of much of the world, but by the same comparison, its geophysical study has a tradition of importance exceeded by few areas. The concept of isostasy (Airy 1855), which perhaps represents the birth of geophysics as a subdiscipline of and a contributor to the geological sciences, grew from geodetic work in India and from the realization that the deep structure of the Himalaya and Tibet had contributed to large discrepancies in surveyed positions (Pratt 1855). With the invention of gravity meters, the Himalaya became an important testing ground for theories of isostasy, and deviations from local isostatic equilibrium as large as those of the Himalaya are found in only a few other regions of the Earth. Now in the 1980s, with a generation of Earth scientists educated on the premises that geophysics can be used to solve geological problems, that plate tectonics is a fact, and that dynamic processes controlling the tectonic evolution of the Earth are driven largely by gravity acting on density differences within the Earth, the Himalaya and Tibet once again are testing grounds for theories of orogenesis. Accordingly, considerable progress in defining the deep structure of the Himalaya and Tibet has been made on the last 10 years, and a review at this time seems appropriate given both the likelihood that little new information will be gathered in the next few years and the result that most existing data have been analysed with modern techniques.

The physiographic scale of the Himalaya and Tibet make them unusually prominent features on the Earth (figures 1 and 2), and as such an understanding of the processes that created them is likely to carry with it the understanding of how many smaller mountain belts

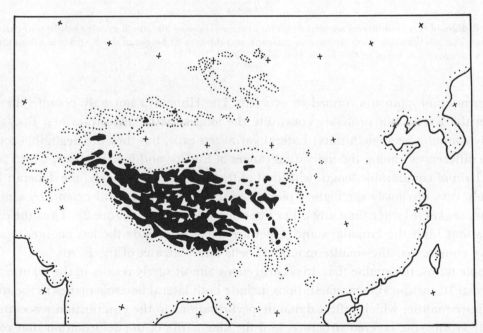

FIGURE 1. Simple contour map of part of Asia, with areas higher than 5000 m shaded and with the 2500 m contour dotted (drawn from Cauët *et al.* 1957, pp. 34–35). The Tibetan Plateau comprises the area that is mostly higher than 5000 m. A similar map for any other continent would be nearly entirely white, with only specks of black here and there. No part of Europe or Australia lies above 5000 m. The Himalaya follows the southern margin of Tibet, and the Karakoram constitute the large area higher than 5000 m in the northwest corner of Tibet (see figure 4).

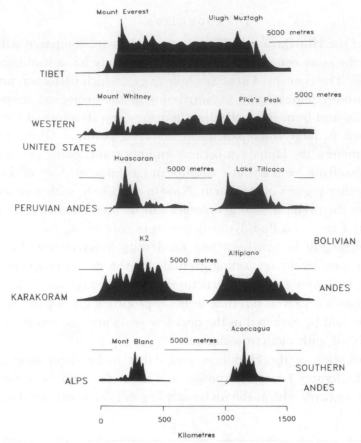

FIGURE 2. Profiles of elevation across selected mountain ranges, showing the much greater height and extent of the Tibetan Plateau than other intermountain plateaux and the greater height of the Karakoram than most other ranges (drawn from maps in the *Times Atlas of the World*).

and intermontane plateaux formed or evolved. The Himalaya not only contain the highest peaks on the Earth, but probably constitute the longest active mountain belt that shares a relatively common geologic history. Lateral variations exist, but they are negligible compared with the differences among the individual ranges of Europe and North Africa, which together form a chain of comparable length with that of the Himalaya. Similarly, the Tibetan Plateau is not only very obviously the highest plateau on Earth, but its lateral extent, by almost any definition, makes it larger than any other intermontane plateau (figure 2). Thus the dynamic processes that built the Himalaya and Tibet are likely to include the less energetic processes that have constructed the smaller mountain belts and plateaux of the Earth.

In simple terms, the engine that drives orogenesis almost surely resides in the mantle and not in the crust. Its measurable manifestations include both lateral heterogeneities in the structure of the upper mantle, which reflect dynamic perturbations to the equilibrium associated with a radially symmetric, layered structure, and the kinematics of the deformation that results in the structures visible at the surface of the Earth. Although geophysical methods have contributed to an increased knowledge of the kinematics of deformation, they are virtually the only methods capable of resolving the deep structure of orogenic belts.

This review focuses on the structure of the upper mantle and on its boundary with the overlying crust, for it is in this depth range that the dynamic processes affecting orogenesis are most likely to reveal themselves in the internal structure of the Earth. Among the many subdisciplines of geophysics, the two that contribute most of the constraints on lateral variations in the deep structure are gravity and seismology. Variations in the Earth's gravity field are caused by heterogeneities in the distribution of mass, and therefore density, in the Earth. The analysis of variations in the gravity field, or of gravity anomalies, can be a very expedient method for proving that a particular hypothetical structure (or model) is wrong. Because of their inherent non-uniqueness, however, only when constrained by simplifying assumptions can gravity anomalies demonstrate that a particular structure is required. Variations in seismic wave velocities in the Earth, in general, can be resolved with much greater precision than those of density, and accordingly most of what we know of the deep interior and its lateral variations derives from studies of seismic waves. The present review concentrates on these two sources of information.

Because both the geologic evolution and the deep structure of the Himalaya and of the Karakoram differ from those of Tibet, I consider these three areas separately. This is facilitated by the relative importances of the two types of data in these areas.

This review begins with Tibet, briefly noting the constraints of the one published gravity datum from its interior and allusions to unpublished gravity data, followed by a summary of results from explosion seismology, from seismic surface waves, and from regionally and teleseismically recorded body waves. Seismic studies with explosive sources yield weak constraints on the deep structure of the interior of Tibet, but tight constraints for the area south of the Indus–Tsangpo Suture Zone. It is important to realize that most geologists not concerned with Quaternary tectonics place the boundary of the Himalaya at this suture, but the Quaternary and active tectonics of the area between the suture and the Greater Himalaya is similar to that of Tibet and very different from that of the Himalaya farther south (Molnar & Tapponnier 1978). Similarly the deep structure of this area suggests a continuity with Tibet. Accordingly, I include the area north of the Greater Himalaya within the Tibetan Plateau.

Seismic experiments in the Himalaya yield data that either are ambiguous or yield surprising results, depending upon one's point of view. Although Hirn et al. (1984a–c; Hirn & Sapin 1984; Lépine et al. 1984; Sapin et al. 1985) have stressed the differences in what these data suggest from what has been inferred from gravity anomalies in the Himalaya, I focus this review more on the gravity anomalies than on the seismic data. Finally, I consider tentative implications of gravity anomalies and aspects of seismology for the structure of Tibet's western and northwestern margins, across the Karakoram and western Kunlun belts, which may offer a key to understanding the dynamics of the continuing collision and the penetration of India into the rest of Asia.

## DEEP STRUCTURE OF THE TIBETAN PLATEAU

### Introduction

Beginning with Argand (1924), and perhaps before his work, the supposition that the high elevation of the Tibetan Plateau is compensated by a thick crust has underlain virtually all ideas concerning the geologic evolution of the plateau. There now seems little doubt that the crust of the Tibetan Plateau is thick: the thickness is at least 50 km everywhere and closer to 65–70 km in most of the plateau. At the same time the evidence that demonstrates the existence of thick crust, while convincing, is not simple and has required approaches that are somewhat different from those used elsewhere. In my opinion, the proof that the crust is thicker than 55 km throughout most of the plateau was obtained only in 1979 by Wang-ping Chen.

Argand (1924) envisioned the crustal thickening to have occurred by the underthrusting of the Indian subcontinent beneath the entire plateau (figure 3). This idea has received support

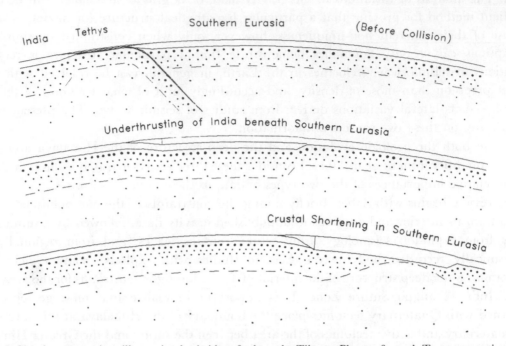

FIGURE 3. Simple cross sections illustrating basic ideas for how the Tibetan Plateau formed. Top cross section shows oceanic lithosphere, carrying India, being subducted beneath southern Eurasia before the collision of India with Eurasia. No attempt has been made to illustrate possible heterogeneities in the Asian lithosphere before the collision. Middle cross section illustrates Argand's (1924) concept of the Indian subcontinent underthrusting southern Eurasia, with the corollary that the entire Indian lithosphere has been underthrust (Barazangi & Ni 1982; Ni & Barazangi 1983, 1984; Powell & Conaghan 1973, 1975; Seeber et al. 1981). No attempt has been made to distinguish variations on this idea such as that of Zhao & Morgan (1985). The bottom cross section indicates an evolution of Tibet in which the crust of southern Asia has shortened in its north–south dimension (Dewey & Burke 1983; England & Houseman 1986) as India has penetrated into Eurasia. The evolution of the lithosphere beneath southern Eurasia is drawn as if it has undergone convective response to the thickening of it (see figure 22 for more discussion).

from extrapolations of geologic cross sections in the Himalaya (Klootwijk et al. 1985; Powell & Conaghan 1973, 1975), from extrapolations of seismological constraints on the orientations of active faults at the Himalaya (Ni & Barazangi 1984; Seeber & Armbruster 1981; Seeber et al. 1981), from the areal extent of the plateau compared with the areal extent of material presumed to have lain northeast of India and west of Australia before the dispersal of Gondwana (Curray & Moore 1974; Johnson et al. 1976), and finally from some seismological studies of Tibet itself (Barazangi & Ni 1982; Ni & Barazangi 1983). Given the convictions of so many other Earth scientists that India has not underthrust the whole of Tibet, however, it is clear that none of the evidence presented in support of this concept has been overwhelmingly convincing.

Seismic studies of the lower crust and upper mantle of Tibet inherently lack the information needed to decide how the structure has evolved, but such studies can be used to ask whether or not the structure that presently underlies Tibet is similar to that presently underlying India. Because much of the Indian subcontinent consists of a Precambrian shield, for which the upper-mantle structure is, in general, very distinctive, seismological studies can resolve whether Tibet is at present underlain by a shield, and the answer to this question is clearly no (Lyon-Caen 1986). This does not, however, prove that Tibet was not underthrust by a shield at some earlier date and that subsequently the structure changed.

Finally, seismological studies are now capable of resolving lateral variations in crustal and upper-mantle structure of Tibet, and despite the remarkably uniform height of the plateau, surprisingly large variations in upper-mantle shear-wave velocity have been found (see, for example, Brandon & Romanowicz 1986; Molnar & Chen 1984), and they, in turn, imply large variations in temperature beneath the plateau.

This review addresses the evidence concerning these three aspects of Tibet's structure: the thickness of the crust beneath the plateau, the average velocity structure in its upper mantle, and lateral variations in the structure of the crust and upper mantle.

### The crustal thickness of the Tibetan Plateau

*Gravity anomalies*

To my knowledge there is only one published gravity measurement from the interior of the Tibetan Plateau, made by Amboldt (1948) in the 1920s with a pendulum gravity meter. The free-air anomaly of 18 mGal† implies a state of nearly isostatic equilibrium, but clearly alone it places no constraint on the mechanism of isostatic compensation.

Profiles of gravity anomalies along the main road in eastern Tibet, from the Himalaya through Lhasa to Golmud and north of the plateau (figure 4) are apparently characterized by the large negative Bouguer anomalies ($-500$ mGal) that must exist for isostatic equilibrium to exist, and if an Airy-type mechanism prevails (Tang *et al.* 1981; Teng 1981; Teng *et al.* 1980; Wang *et al.* 1982; Zhou *et al.* 1981). Variations of gravity over distances of 50–200 km along the profiles are shown to be small, at most only 30 mGal. The largest such variation lies near the Indus–Tsangpo Suture (figure 4) and corresponds to a 30 mGal low (Tang *et al.* 1981; Zhou *et al.* 1981). Given the lack of information describing how these plots were obtained, however, one cannot know if such a low is defined by more than one measurement or if it is a consequence of how the data were projected onto the profile, for instance from an area farther west where altitudes are higher. In my opinion, the absence of raw data and a published discussion of them makes it risky to conclude more than that deviations from isostatic equilibrium are not enormous.

*Explosion seismology in Tibet*

*Seismic refraction.* Profiles have been shot in Tibet both by Chinese scientists from the Academy of Sciences (Institute of Geophysics, Academy of Sciences 1981; Teng *et al.* 1980, 1981, 1983) and jointly by French and Chinese scientists, in this case mostly from the Ministry of Geology (Hirn *et al.* 1984 *a–c*; Sapin *et al.* 1985; Teng *et al.* 1985).

In 1976 and 1977, Chinese scientists carried out a seismic refraction program along a profile from the Great Himalaya into southern Tibet, essentially along the road from the border of Sikkim, India to the lake Nam Tso (figure 4). The total length is about 450 km. Teng *et al.* (1980, 1981, 1983; Institute of Geophysics, Academy of Sciences 1981) inferred a thick crust

$$\dagger \ 1 \text{ mGal} = 10^{-3} \text{ cm s}^{-2}.$$

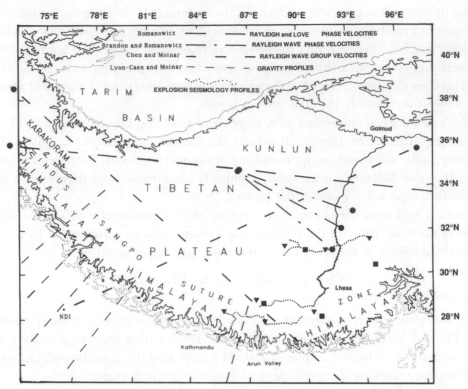

FIGURE 4. Map of the Tibetan Plateau, the Himalaya and the Karakoram, showing locations of seismic refraction and wide-angle reflection profiles and shotpoints (inverted triangles), relevant paths for measured surface-wave dispersion with black circles showing epicentres of earthquakes used, temporary long-period seismograph stations (squares) (Jobert *et al.* 1985), gravity profiles, and the road from Lhasa to Golmud. Contours for 1500 m (dotted) and 3000 m (solid line) outline the plateau.

of 70–73 km beneath Tibet and decreasing beneath the Himalaya. They reported a structure with five layers, including a pronounced low-velocity zone, overlying a mantle with a P-wave velocity of 7.95 km s$^{-1}$ in one area, of 8.4 km s$^{-1}$ in another, and with an average of 8.15±0.03 km s$^{-1}$. Beneath the Himalaya, the interfaces of these layers are shown to dip toward the north, with slightly different velocities from those further north (Teng *et al.* 1981, 1983).

Although Teng *et al.* (1981, 1983; Institute of Geophysics, Academy of Sciences 1981) show some particularly clear seismograms and plots of observed and calculated travel time curves, I found the discussion of the data too brief for me to be able to evaluate the conclusions. In particular, I am unable to recognize clear signals that might be Pn phases on the plot showing seismograms recorded at distances greater than 250 km (figure 3 of Institute of Geophysics, Academy of Sciences 1981; reproduced as figure 4 of Teng *et al.* 1981). The other well-illustrated profiles (figure 2 of Institute of Geophysics, Academy of Sciences, 1981; or figure 3 of Teng *et al.* 1981, 1983) show clear signals but only from distances smaller than 250 km, for which refractions from the Moho would not be the first arrivals.

The joint Chinese and French studies represent well-planned experiments designed to elucidate the deep crustal structure and, in particular, the shape of the Moho, but in my opinion too little space was devoted either to showing or to discussing the seismograms. Moreover, following long traditions in the presentation of seismic refraction results, some traces are obscured by dark black lines that show interpretations, and in the shorter of the papers discussing the results, only one interpretation that fits the data is shown, making an evaluation of the non-uniqueness or the uncertainties difficult. As a result, my review of their results is clouded by an inability to decide which aspects of their inferences are required, as well as by prejudices that makes me dubious of some of them.

These studies fall into two types: reversed refraction profiles with stations spaced at different distances between the two shot points at the ends of the profile, and profiles recorded at roughly constant distances from sources (fan-profiles), designed to record strong wide-angle reflections from interfaces with abrupt velocity contrasts, such as the Moho. Of the two refraction profiles, one was 300 km north of the Indus–Tsangpo Suture Zone, and the other about 80 km south of it (figure 4).

The profile shot parallel to and south of the Indus–Tsangpo Suture Zone yields the better constrained deep structure (Hirn et al. 1984a, c). Seismographs were distributed over a distance of 500 km, 60–100 km south of the suture zone, approximately halfway between the high peaks of the Greater Himalaya and the suture zone (see also Teng et al. 1985). Explosives were detonated at both ends and approximately in the middle of the line, and correlatable signals were recorded at distances greater than 400 km. In addition, one of the shots was recorded along the suture zone at distances of 170–270 km from the eastern shot point.

Teng et al. (1985) used travel times from these explosions to infer flat-layered velocity structures along four segments of the profile. Their inferred structures include from five to eight layers overlying a halfspace (mantle) with a velocity of 8.1 or 8.2 km s$^{-1}$. A low-velocity zone (6.0 or 6.1 km s$^{-1}$) is reported at the base of the crust just above the Moho. On two segments, a second low-velocity zone is shown in the upper crust. The existence of marked velocity contrasts is implied by the strong, apparently reflected, signals that define travel time curves with decreasing apparent velocity with increasing distance. The clearest is that from the Moho and was analysed in detail by Hirn et al. (1984c). Refracted signals, however, are less clear, and the reported velocities of 8.1 and 8.2 km s$^{-1}$ for the mantle are very different from the velocity inferred from the same data by Hirn et al. (1984c), discussed below. Thus it is difficult for me to know which aspects of these multilayered structures are required.

Hirn et al. (1984c) abstained from deducing, or at least reporting, a layered velocity structure, and instead they focused attention on the strong signals at distances of 200–300 km attributed to reflections from the Moho. The arrival times at these distances imply a thick crust (thickness about 75 km), but the strength of the signals at such long distances together with the arrival times cannot be matched by an abrupt increase in velocity at the Moho, as would be required by the low-velocity zone inferred by Teng et al. (1985). With a series of calculations, they showed how an increase in the thickness of a transition zone between crust and mantle affects both the distance at which strong wide-angle reflections are recorded and the times when these reflected phases arrive (figure 4 of Hirn et al. 1984c).

Seismograms recorded along the suture zone from the eastern shot also show clear sharp signals at the six stations between 230 and 260 km from that shot (figure 6 of Hirn et al. 1984c). The travel times of these signals imply that the depth to the reflector is similar to that observed along the more continuous profile farther south, but the relative differences in arrival times at neighbouring receivers are sufficiently large that the lateral variations in crustal structure must exist. In any case, the depth of the Moho of about 75 km beneath the southern edge of the Tibetan Plateau, between the Greater Himalaya and the Indus–Tsangpo Suture Zone, seems to be well established.

Before reviewing Hirn's and his colleagues' constraints on the variations in the depth of the

Moho, two other results derived from their profiles just south of the suture are worth noting: one concerning the lower crust and the other the upper mantle.

First, to fit both the amplitudes of the reflections from the Moho and another set of signals, which they attribute to reflections within the crust, Hirn & Sapin (1984) suggested that attenuation in the lower crust is very high. The relative amplitudes of the signals could be matched if there were a very large difference in velocity across the intracrustal reflector, but the smaller amplitudes and the lower predominant frequencies of the more deeply reflected phases concurs with the alternative suggestion that $Q$ in the lower crust is very low (about 100), and therefore attenuation is very high. A unique explanation for such a low $Q$ cannot be given, but one obvious explanation is that temperature in the lower crust is high. This possibility is consistent with, but also not required by, the high heat flow measured in lakes at the eastern end of the profile (Francheteau et al. 1984; Jaupart et al. 1985). It is not obviously in accord with magnetotelluric studies, which indicate shallow but not deep layers of low electrical conductivity, and which therefore suggest high temperature at shallow depth (Pham et al. 1986).

Second, Hirn et al. (1984c) recorded weak, but discernable, signals from 300 km to more than 400 km from the eastern shot point, and they pointed out that these signals seem to define refractions from the Moho. They stated that signals from the western shot point are less clear, but that 'clear onsets at 300 km, 340 km, and 400 km correspond within 0.1 s' with the travel-time curve for the eastern shot point. The velocity defined by this travel-time curve is unusually high, 8.7 km s$^{-1}$. They noted that only a very unusually low (unreasonably low) temperature could explain such a velocity, but that 8.7 km s$^{-1}$ is not unreasonable if anisotropy exists.

Sapin et al. (1985) described a profile 500 km in length within the interior of the Tibetan Plateau (figure 4). Explosives were detonated at the two ends and at a site 200 km from the western shotpoint. The major part of the discussion of Sapin et al. (1985) concerns the crustal structure, shallower than about 30 km, for which recordings at distances less than 200 km place strong constraints. The absence of clear arrivals refracted at the Moho do not allow a constraint to be placed on the velocity in the mantle. Clear wide-angle reflections, presumably from the Moho, were recorded from the eastern shot point, at distances greater than 200 km, but those from the western shotpoint are less clear. Such signals were recorded clearly at some recording sites, but not at neighbouring sites 10–20 km away. Sapin et al. (1985) noted the following cautious conclusion. 'The relevant observation is that reflections from 70 km depth from a crust–mantle boundary exist in places but the continuity and the geometry of the Moho along the...line cannot be assessed.' Thus, what is the best-documented seismic refraction study of Tibet north of the Indus–Tsangpo Suture places no constraint on the P-wave velocity in the mantle and yields only a weak evidence for the crustal thickness of about 70 km.

*Wide-angle reflection profiles.* Among the data presented by Hirn and his collaborators, the fan profiles, designed to record wide-angle reflections from the Moho, provide the most tantalizing information and therefore, perhaps the most difficult to interpret. Their studies of wide-angle reflections can be conveniently grouped into three sets: one beneath the Himalaya (that I defer to the discussion of the Himalaya below): a second, beneath the area just south of the Indus–Tsangpo Suture Zone: and a third, north of the suture.

Wide-angle reflections from the Moho are particularly strong from an area near the suture, where the Moho is marked by a thick transition zone, thicker than 6 km and of the order of 12 km (Hirn et al. 1984c). Whereas the wide-angle reflections imply a crustal thickness of about

75 km along the main reversed refraction profile, signals also recorded at distances of about 200 km from the eastern shotpoint but at locations south of this profile show earlier arrival times. Hirn *et al.* (1984*c*) inferred that the crustal thickness decreases to about 70 km, approximately 15 km south of the main refraction profile, suggesting the possibility of a relatively steep northward dip of the Moho of about 25° (figure 5). Reflections recorded north

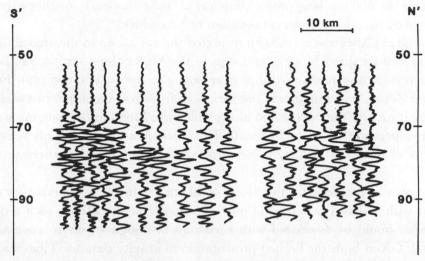

FIGURE 5. Seismograms recorded on a north–south profile roughly equidistant and 200 km west of the eastern shotpoint of the east–west profile between the Indus–Tsangpo Suture and the Greater Himalaya (figure 4). The timescale has been converted into a depth scale after correcting for different distances. The prominent phases mark strong wide-angle reflections from the Moho and indicate depths of it of 70–75 km. (From Hirn & Sapin 1984.)

of this profile show a scatter in onset times consistent with a crustal thickness between 70 and 75 km, assuming the same velocity structure of the crust as that deduced from the long refraction profile through this area (Hirn *et al.* 1984*c*). A similar fan-profile made from the shot point in the middle of this profile and recorded on a north–south line at the west end of the profile reveals sharp onsets on seven seismograms at the times corresponding to a Moho at 70 km (figure 15, discussed in more detail below). These reflection points lie between 30 and 50 km north of the crest of the High Himalaya Range but south of the main refraction line. Thus wide-angle reflections at various localities between the suture and the High Himalaya indicate a mean depth of the Moho transition zone of about 70–75 km, either with definite, if small (5 km), variations in depth, or with heterogeneity in the crustal velocity structure that translates into apparent variations in depth.

Using two shot points roughly 250 km west of the main north–south road in eastern Tibet, Hirn *et al.* (1984*b*; Allègre *et al.* 1984) made fan profiles designed to record wide-angle reflections from the Moho beneath southern Tibet. Reflections from the southern shot recorded near the suture imply that the Moho beneath the suture zone also lies about 70 km below the surface (Hirn *et al.* 1984*b*).

The recordings of wide-angle reflections farther north of the suture provide evidence for why neither the Moho nor the uppermost mantle could be studied well on the long refraction profile of Sapin *et al.* (1985) across central Tibet. The great variety in recordings requires marked variations in the structure west of the main road (figure 2 of Hirn *et al.* 1984*b*). Along some parts of the profile, strong signals arrive at the times expected for reflections from a Moho at a depth of 70 km, but at others no clear signal can be seen at these times. In some portions, clear signals are recorded at times suggesting that they represent phases reflected at depths of only 50 km, and sharp signals recorded at progressively different times on neighbouring profiles can be most simply interpreted as reflections from dipping interfaces. Thus given the complexity of the crustal structure revealed by these profiles, the failure to record continuous reflections along the 500 km long profile (Sapin *et al.* 1985) is clearly neither exceptional nor likely to be caused by poor experimental design or procedure.

Hirn *et al.* (1984*b*; Allègre *et al.* 1984) interpreted the variations in the arrival times of these clear signals as being caused by steps and dips of the Moho along the profile. Because they recognized such steps and northward dips near both the Indus–Tsangpo and the Bangong–Nujiang Suture Zones, they interpreted the variations in arrival times in terms of slip on thrust faults that penetrate into the mantle and along which the Moho in the hanging wall has been elevated with respect to that of the foot wall. This interpretation is certainly reasonable, and I do not know of data that refute it. Nevertheless, I also think that there are reasons to doubt it.

First, differences in the depth of the Moho of 20 km are inferred to persist for distances of 100 km. Even with a density contrast of only 400 kg m$^{-3}$ at the Moho, such a difference in crustal thickness should be associated with variations in Bouguer gravity anomalies of more than 150 mGal. Given both the limited presentation of gravity data for Tibet and that those measurements were made roughly 125 km east of where the steps in the Moho have been inferred to lie, the published plots of gravity anomalies (Tang *et al.* 1981; Teng *et al.* 1980; Wang *et al.* 1982; Zhou *et al.* 1981) certainly do not disprove the inferred steps in the Moho of Hirn *et al.* (1984*b*), but they give no hint of such steps elsewhere in Tibet.

Second, because the recordings clearly show lateral heterogeneity within the crust, it is reasonable to expect there to be lateral heterogeneity in the crust–mantle transition. Because the nature of that transition affects the strength of wide-angle reflections, the absence of a signal at a distance of 250 km could be caused by variations in the structure near the Moho, which nevertheless lies at a relatively uniform depth.

The dependence of the strength of the reflection on the velocity gradient at the Moho is shown most clearly by synthetic seismograms (figure 4 of Hirn *et al.* 1984*c*) for a set of plausible transitions at the crust–mantle interface. For the profile south of the Indus–Tsangpo Suture, it was necessary to postulate a very thick transition zone (12 km) between the crust and mantle beneath this area. For an abrupt step in velocity at the same depth, centred at 76 km, the strongest wide-angle reflections would be recorded at distances less than 200 km. For a shallower Moho, these signals would be strong at yet shorter distances. Most, but not all, seismograms from the fan profiles were recorded at distances between 200 and 250 km, but they are corrected for distance and plotted (figure 5; figure 2 of Hirn *et al.* 1984*c*) so that reflections from the same depths are aligned. Hirn *et al.* (1984*b*) noted that such corrections are based on the assumption that the velocity structure of the crust is the same everywhere and that 'the largest uncertainty in the conversion of time into depth results from the lack of control on possible variations in the velocity contrast across the Moho along the section'. For me, this uncertainty, coupled with the likely possibilities that the velocity structure of the Moho beneath Tibet is variable and that the unusually thick transition south of the suture does not prevail throughout Tibet, casts doubts on the inference of 20 km steps in the Moho.

My comments here are not meant to be a criticism of the work done by Hirn *et al.* (1984*b*), of the experimental design, of the field investigation, or of their analysis. Nor do I assert that their inferences of steps in the Moho or of major thrust faults offsetting it are incorrect. Nevertheless, I think that their presentation of their data is inadequate to allow the reader to decide whether the inferred steps and steep dips of the Moho are required. Thus, I conclude that data from explosion seismology show a complex crustal structure, with dipping interfaces that could well be the manifestations of major thrust faults; but for the area north of the Indus–Tsangpo Suture Zone, I do not think that these results prove that the Moho lies at a depth less than about 65 km.

*Surface-wave propagation across Tibet*

*Introduction.* The velocities of seismic surface waves can provide important constraints on the seismic wave velocity structure of large areas (at least several hundred kilometres in dimension) such as Tibet, but for the non-seismologist there must seem to be numerous obstacles to appreciating these constraints. An intuitive understanding of the analysis and interpretation of surface waves is much more difficult to grasp than that of travel times of body waves. First, the waves are dispersed with longer periods being more sensitive to deeper structure and, in general, travelling faster than shorter periods. Second, the velocities can be described in two ways. Phase velocities describe the velocity of a wave train with a specific period. They are more difficult to measure but contain more information than group velocities, which describe the velocity of a packet of waves spanning a narrow range of periods. Third, there are two types of surface waves, Rayleigh and Love waves, both of which depend primarily on the shear-wave velocity in the Earth, but in different ways. Fourth, for a region the size of Tibet, where until recently all nearby permanent long-period seismographs have lain outside the plateau, isolating the portion of the path within Tibet has been a serious problem. Finally, even when reliable surface-wave velocities exist, the non-uniqueness of the constraints that they impose on the Earth structure is often obscured in the presentation of the results; all too often only one model that fits the data is shown, leaving the reader at a loss as to how to decide what other structures also are allowed by the data.

Most early studies of Tibet considered only group velocities of Rayleigh waves, and such studies, by themselves, did not place tight constraints on the crustal thickness or the shear-wave velocity in the upper mantle. This is illustrated clearly by two experiments: one inadvertent and the other specifically addressing Tibet.

In the 1950s, Maurice Ewing and Frank Press measured low Rayleigh-wave phase velocities for paths across the Basin and Range Province of the western United States, where a broad area of high elevation is associated with large negative Bouguer gravity anomalies that indicate an overall state of isostatic equilibrium. They concluded that the crust beneath the Basin and Range Province is thick and noted the following. 'The correlation of phase-velocity variations with crustal-thickness changes is justified, and permits specification of the mechanism of (isostatic) compensation as the regional Airy system' (Ewing & Press 1959). Whereas this conclusion is correct for most of the Earth, the Basin and Range Province looms as the Earth's most glaring example of where Airy's model of isostasy fails, for the crust is thinner, not thicker, than normal in that region. Recognizing both that none of the scientists who have studied surface-wave dispersion across Tibet have attained the stature of either Ewing or Press, and that until the 1980s surface-wave phase velocities for Tibet could not be measured as reliably as those for western United States in 1959, readers should be very cautious about believing many of the inferences drawn solely from early studies of surface-wave dispersion across Tibet.

In presenting surface-wave group velocities for Tibet, Bird (1976) and Chen (1979; Chen & Molnar 1981) compared them with calculations for different crustal thicknesses and for different shear-wave velocities in the uppermost mantle (Sn velocities). The measured dispersion curves could be matched satisfactorily by structures with crustal thicknesses between 55 and 85 km, for different, but reasonable, Sn velocities. Thus for group velocities to provide a tight constraint on the crustal thickness, not only must the dispersion curve be measured accurately, but also the Sn velocity should be obtained independently.

To some extent this non-uniqueness associated with group velocities is removed by accurate measurements of phase velocities at sufficiently long periods, but the study of phase velocities across Tibet required the development of techniques to circumvent the absence of long period seismographs on both sides of the plateau (Patton 1980a; Romanowicz 1982) or the (temporary) installation of long period seismographs within Tibet (Jobert et al. 1985). Although studies by Brandon & Romanowicz (1986), Jobert et al. (1985) and Romanowicz (1982) provide relatively tight constraints on the crustal and upper-mantle structure, these studies not only have considered relatively small portions of Tibet. Moreover, they demonstrate clear lateral heterogeneity in the structure. Consequently, I also review briefly the less-definitive work based on group velocities, which covers parts of Tibet not studied with phase velocities.

*Group velocities.* Among the many studies of surface-wave dispersion across Tibet, a relatively small area to be studied with surface waves, two approaches have been taken in gathering data.

Most studies used long paths for which the portion beneath Tibet itself was less than 70 % of the overall path (Chun & Yoshii 1977; Feng 1982; Feng & Teng 1983; Gupta & Narain 1967; Patton 1980b; Pines et al. 1980; Tung & Teng 1974). The first such study (Gupta & Narain 1967) used paths for which Tibet constitutes only 10 % of the path. Clearly, to isolate the part of the path appropriate for Tibet, it was necessary to know, or to assume, the dispersion appropriate for paths outside of Tibet. Some later studies used numerous multiple paths and solved simultaneously for dispersion curves in different areas (Chun & McEvilly 1986; Feng 1982; Feng & Teng 1983; Patton 1980b), but for most such studies, it is difficult to know both how well resolved the dispersion curves for Tibet are and, given the existence of lateral heterogeneity, what are the precise regions represented by those curves.

All of these studies show that surface waves are delayed significantly when paths cross the Tibetan Plateau, and all reported crustal thicknesses of about 70 km and low shear-wave velocities (4.4–4.5 km s$^{-1}$ or less) for the mantle immediately below the Moho. Because these upper-mantle velocities are distinctly lower than those determined by using travel times of Sn (discussed below), these inferences of a thick crust must also be viewed with scepticism.

The other approach to determining dispersion curves for Tibet has been to consider, in so far as possible, paths confined to the plateau. Initially this approach was realized by the analysis of surface waves from earthquakes in or on the margins of the plateau recorded at stations close to the edge of the plateau (Bird 1976; Chen 1979; Chen & Molnar 1981). The principal disadvantage of this approach has been that there have been few earthquakes sufficiently large, without being too large, and in appropriate locations to define the entire dispersion curve for periods between 20 and 90 s. Chen relied primarily on Rayleigh-wave group velocities measured for two paths, one from northeast Tibet to New Delhi and the other from the Pamir across southern Tibet to Shillong (figure 4). Chen's data, coupled with Sn velocities determined independently (and discussed below), were sufficient to show that the average crustal thickness for Tibet is about $70(\pm5\text{–}10)$ km.

Recently, Chun & McEvilly (1986) presented group velocities for both Rayleigh and Love waves crossing Tibet, determined by comparing measured values for paths with different fractions across Tibet and by extrapolating to paths purely across Tibet. Implicit in this exercise was the assumption that for each period and for each path, the velocity is the same throughout the area traversed by these paths outside of Tibet. They obtained very low group velocities in the period range from 25 to 45 s, and they inferred a zone of very low velocities within the crust. They also inferred a crustal thickness of $74\pm10$ km and an uppermost mantle shear velocity of $4.50\text{–}4.55$ km s$^{-1}$, but they did not show whether their data also allow a thinner crust and a higher upper-mantle velocity.

*Phase velocities.* In my opinion, the most definitive studies of surface-wave dispersion across Tibet are those of Brandon & Romanowicz (1986), Jobert *et al.* (1985), Patton (1980*b*) and Romanowicz (1982), who determined phase velocities for surface waves across parts of Tibet. Except for relatively short periods, the first phase velocities measured for the Tibetan Plateau were determined by Patton (1980*b*), using earthquakes northwest of Tibet and recorded at stations in southeast Asia and the Philippines. Some arbitrariness was required both to assign boundaries to regions with different structures and then to decide how much of each path should be assigned to which province. Given the existence of lateral heterogeneity within Tibet and the likelihood that paths crossing the margins of Tibet will be refracted and will not follow great circles, I have greater doubts about dispersion curves determined by such regionalization than those determined by using strictly pure paths. Nevertheless, the measurements obtained later by Romanowicz (1982) lie within the uncertainties quoted by Patton (1980*b*) and clearly give credibility to his analysis.

Romanowicz (1982) used pairs of earthquakes that lie along great circles passing very close to stations in opposite directions from Tibet. She first measured dispersion curves for each earthquake at each station. Then for each station, in essence, she took the differences in propagation times for phases from the two earthquakes and obtained average phase velocities for the portion of the paths between the earthquakes and within Tibet. Thus although the structure between the earthquakes and for which the dispersion curve was obtained need not be laterally homogeneous, she could avoid the risky assumptions that are necessary in regionalizing paths that do not follow the same great circles.

Romanowicz (1982) determined phase velocities for both Rayleigh and Love waves, and thus obtained tighter constraints on the structures than had been possible. Her data could not be fitted with most previously published structures, but they corroborated the necessity of both a thick crust of about 65–70 km and a high Sn velocity of about $4.65\text{–}4.7$ km s$^{-1}$ (figure 6). The agreement between the structures inferred by Chen and by Romanowicz is, in fact, somewhat surprising because the paths that Chen studied were principally in the southern and western parts of the high plateau, but hers cross the northern and the eastern parts of Tibet (figure 4). The lower elevations of eastern Tibet are likely to be compensated by thinner crust than that beneath the high parts of the plateau, where Chen concentrated his study.

Jobert *et al.* (1985) installed a temporary array of four seismographs in southern Tibet, from

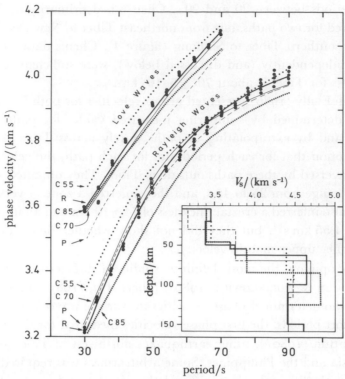

FIGURE 6. Comparisons of Romanowicz's (1982) observed and calculated Rayleigh- and Love-wave phase-velocity
dispersion curves showing both poor and satisfactory agreement, and corresponding shear-wave velocity
structures. Points show measurements, with the higher velocities defining the curve for Love waves and with
the lower velocities for Rayleigh waves. The path used to measure the phase velocities crosses northern and
eastern Tibet (figure 4). The dispersion curves calculated for Romanowicz's (1982) structure are labelled R.
Note that Chen & Molnar (1981) could not rule out crustal thickness of 55 km (curve C55) or 85 km (curve
C85) with group velocities alone, but phase velocities do not permit such average crustal thicknesses. Patton's
(1980 b) proposed structure, P, based solely on Rayleigh waves, can be rejected only because of a poor fit to
Love waves. (Redrawn from Romanowicz 1982.)

300 to 580 km apart, surrounding Lhasa, and spanning the Indus–Tsangpo Suture Zone
(figure 4). They measured phase velocities for four separate paths, and for two cases they
reported measurements for the relatively long periods of 150 s. The velocities for periods less
than 100 s require a thick crust, about 70 km but not as much as 80 km. On the basis of the
dispersion of the longest periods, 100–150 s, they inferred a thin (30 km) high-velocity layer
overlying a low-velocity zone in the mantle beneath their array. I suspect, however, that
because their array would span only one wavelength at these long periods, uncertainties both
in the phase velocities and in the assumption of lateral homogeneity do not permit the resolution
of the shear velocity below a depth of about 100 km.

In an extension of Romanowicz's (1982) study but using more paths, Brandon &
Romanowicz (1986) compared Rayleigh-wave phase velocities across north–central and
northernmost Tibet. They corroborated Romanowicz's (1982) measurements for the western
70 % of the path that she had studied across the northernmost part of Tibet. Despite a large
scatter, measured phase velocities for shorter paths across north–central Tibet and for periods
between 30 and 60 s (figure 7) are consistently larger than those for the paths across
northernmost Tibet. These different phase velocities require a thinner crust for north–central
Tibet (50–60 km) than for the rest of Tibet (65–70 km) but with a lower upper-mantle shear

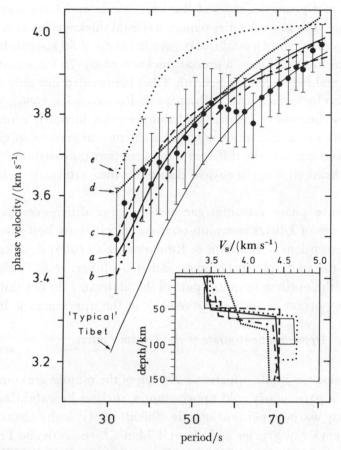

FIGURE 7. Comparisons of Brandon & Romanowicz's (1986) observed phase velocities for short paths across north–central Tibet (figure 4) and calculated curves for layered structures yielding both satisfactory and unsatisfactory fits. 'Typical Tibet' corresponds to curve R in figure 6. Note the good fits for structures $a$, $b$, and $c$ with thinner crusts (50–60 km) than normal for Tibet and with lower-mantle S-wave velocities (4.5 km s$^{-1}$ or less), but poor fits for structures with a crust thick as 65 km ($d$ and 'Typical Tibet') or with uppermost-mantle velocities higher than 4.5 km s$^{-1}$ ('Typical Tibet' and $e$). (Redrawn from Brandon & Romanowicz 1986.)

velocity (4.4 km s$^{-1}$ or less) than elsewhere beneath Tibet (about 4.65 km s$^{-1}$). If the crust were nearly as thick as elsewhere (65 km), then the shear velocity in the uppermost mantle must be low ($ca.$ 4.4 km s$^{-1}$) or the average shear-wave velocity in the crust must be distinctly higher than elsewhere in Tibet (figure 7). They stated that a high uppermost mantle shear velocity (4.6 km s$^{-1}$ or more) can be ruled out. Thus the structure of north–central Tibet (or of Chang Thang) must be very different from that to the north and east, from that determined by Jobert $et$ $al.$ (1985) for southernmost Tibet, or from the average for southern and western Tibet, determined from group velocities and Sn travel times (Chen 1979; Chen & Molnar 1981).

In my opinion, the differences in the structures cannot be isolated and resolved well enough to prove both that the crust is much thinner (15 km) instead of only slightly thinner (5 km) than elsewhere in Tibet, and that the S-wave velocity in the mantle is much less (only 4.4 km s$^{-1}$) than that elsewhere (4.7 km s$^{-1}$), instead of only slightly less (4.55 km s$^{-1}$). Brandon & Romanowicz's (1986) results, however, do not seem to allow only slight differences in both crustal thickness and upper-mantle shear-wave velocity. As discussed below, my suspicion is that the crust is slightly thinner (by 5 km or at the most 10 km) than beneath most of the plateau, but the shear velocity is distinctly lower than elsewhere.

4

*Summary*. Both group and phase velocities of Rayleigh waves and Love waves are delayed along paths crossing Tibet. The low velocities require a crustal thickness in excess of 50 km, and for most regions in excess of 60 km. Crustal thicknesses in excess of 80 km can be ruled out for all paths studied, and for most of Tibet, a crustal thickness of 65–70 km seems required.

Clear evidence for lateral heterogeneity beneath Tibet is provided not only by body waves (discussed below) but also by surface waves (Brandon & Romanowicz 1986), which show an area of lower uppermost shear-wave velocity and thinner crust in north–central Tibet than elsewhere in the plateau. These variations might explain the differences in group velocities measured by different workers, and the different structures that they deduced, but if so, they also render the regionalization of surface-wave dispersion into arbitrary tectonic provinces risky.

Although Rayleigh-wave phase velocities can resolve large differences in upper-mantle velocities for regions the size of Tibet, constraints on these velocities are best derived from body waves. Thus, with the exceptions of Brandon & Romanowicz's (1986) detailed investigation of north–central Tibet, the study of southernmost Tibet by Jobert *et al.* (1985) and that of Romanowicz (1982) for the northeasternmost part of the plateau, I do not think that surface waves have placed an important bound on the velocity in the upper mantle beneath Tibet.

## Upper-mantle structure of the Tibetan Plateau

### Introduction

P-wave and S-wave velocities in the uppermost 200 km of the mantle are commonly used to distinguish regions with a particularly cold upper mantle, such as Precambrian shields, from regions with a particularly warm uppermost mantle. Shields are typically characterized by Pn and Sn velocities of 8.1 km s$^{-1}$ or greater and about 4.7 km s$^{-1}$, respectively. Pn velocities less than 8.0 km s$^{-1}$ are typical of regions with high heat flow such as the Basin and Range Province of the western United States (see, for example, Black & Braile 1982) and shear-wave velocities of 4.5 km s$^{-1}$ and lower are typical of the low-velocity zone in the upper mantle, where partial melting is likely.

The Indian Shield is underlain by a thick high-velocity zone. Pn and Sn phases propagate with velocities of 8.4 and 4.7 km s$^{-1}$, respectively (Huestis *et al.* 1973; Ni & Barazangi 1983), and waveforms of SH phases show that the high-velocity layer is thick (*ca.* 200 km) (Lyon-Caen 1986). Thus, if Argand's (1924) hypothesis that India underthrusted the whole of Tibet were correct, then one might expect to find Tibet underlain by a structure similar to that beneath India.

Showing the presence or absence of shield structure beneath Tibet has been difficult, because although the Pn and Sn velocities are high, it has been difficult to measure the thickness of the region characterized by such velocities. For regions the size of entire continents, surface-wave phase velocities at long periods (100 s or more) can resolve thick high-velocity lids on minor low-velocity zones or pronounced low-velocity zones without lids, but, with the possible exception of the study of Jobert *et al.* (1985), phase velocities, and especially group velocities, of surface waves across Tibet have not been measured at sufficiently long periods to resolve such deep structure.

In this section, I review three types of studies using body waves to constrain the upper-mantle structure of Tibet: (1) measurements of Pn and Sn arrival times for earthquakes in different parts of Tibet, which yield estimated P-wave and S-wave velocities in the uppermost mantle; (2) differences in S-wave and P-wave travel times from earthquakes in Tibet and the Himalaya recorded at teleseismic distances, which yield constraints on the average shear wave velocity in the upper several hundred kilometres of the mantle beneath Tibet; and (3) travel times and waveforms of SH phases for paths beneath Tibet and recorded at regional distances, which place constraints on the depth distribution of material with different velocities.

*Propagation of Pn and Sn across the Tibetan Plateau*

Three studies have reported Pn velocities for Tibet, and two of them Sn velocities. For each, the arrival times of these phases from earthquakes in different parts of Tibet and recorded at different distances from recording stations within or near Tibet were used to construct travel time curves (figure 8). Chen & Molnar (1981) used reported arrivals at the station in Lhasa;

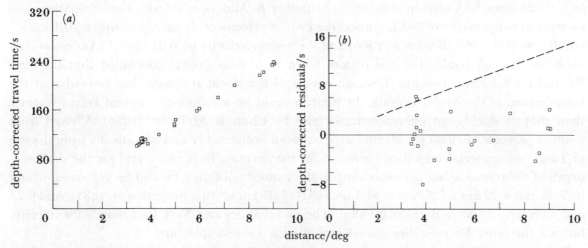

FIGURE 8. Plots of Sn (3–10°) travel times (*a*) and residuals (*b*) about the best fitting linear travel-time curve against distance from earthquakes within the Tibetan Plateau to the station at Lhasa (figure 4). Velocity = $4.77 \pm 0.8$ km s$^{-1}$. The broken line in (*b*) shows slope of travel-time curve if the velocity were 4.5 km s$^{-1}$. *N* is the number of arrival times used; in (*b*) = 23. All origin times were normalized to a common focal depth of 33 km, reducing travel-time differences (scatter) arising from erroneous focal depths and correspondingly erroneous origin times. (From Chen & Molnar 1981.)

Jia Su-juan *et al.* (1981) measured arrival times of Pn at Lhasa and at stations east of the plateau in China; and Ni & Barazangi (1983) measured arrival times at stations of the World-Wide Standardized Seismograph Network (WWSSN) in India and Pakistan. Arrival times measured by Jia Su-juan *et al.* (1981) and by Ni & Barazangi (1983) are probably more reliable than those taken by Chen & Molnar (1981) from bulletins, but there is overall agreement in Pn and Sn velocities: $8.12 \pm 0.06$ km s$^{-1}$ and $4.77 \pm 0.08$ km s$^{-1}$ (Chen & Molnar 1981); $8.11 \pm 0.04$ km s$^{-1}$ for Pn (Jia Su-juan *et al.* 1981); and $8.42 \pm 0.10$ km s$^{-1}$ and $4.73 \pm 0.06$ km s$^{-1}$ (Ni & Barazangi 1983). Paths confined to Tibet may yield less scatter than those for which different path lengths lie outside of Tibet, and the relatively high Pn velocity of Ni & Barazangi (1983) might be a consequence of the smaller fractions of their paths beneath Tibet. The agreement of the measured Sn velocities, however, does not support this inference. Finally, Chen & Molnar (1981), but not Ni & Barazangi (1983), found that adjustments to origin times for different reported focal depths reduced the scatter in travel-time curves.

The apparent velocities of both Pn and Sn are similar to those of shields and markedly different from areas characterized by high heat flow. Barazangi & Ni (1982; Ni & Barazangi 1983) made the step of concluding that Tibet is underlain by a shield structure, but caution should be exercised before accepting this inference.

First, because of curvature of the Earth, waves refracted at deep levels will yield higher apparent velocities than those refracted at shallower depths in material of the same velocity. For a Moho 35 km deeper than normal the apparent velocities should be 0.55% higher than normal (Chen & Molnar 1981).

Second, an increase in pressure increases the material velocity. A difference in crustal thickness of 35 km corresponds to a higher pressure of about 1 GPa at the same distance below the Moho. For Olivine (forsterite) this would cause an increase in the P-wave velocity of 0.1 km s$^{-1}$ (Anderson *et al.* 1968; Schreiber & Anderson 1967*a*) but in the S-wave velocity of only 0.025 km s$^{-1}$ (Anderson *et al.* 1968; Schreiber & Anderson 1967*b*). For comparison, an increase in temperature of 250 K reduces the P-wave velocity of olivine (forsterite) by 0.1 km s$^{-1}$ (Anderson *et al.* 1968; Soga *et al.* 1966) and the S-wave velocity by 0.07 km s$^{-1}$ (Anderson *et al.* 1968; Soga *et al.* 1966). For this reason, Chen & Molnar (1981) concluded that the high Pn and Sn velocities beneath Tibet do not require a shield structure, but instead that the temperature at the Moho beneath the plateau could be a couple of hundred kelvin warmer than that of shields, an inference drawn also by Chun & McEvilly (1986). Viewed from another perspective, if 35 km of crust were removed isothermally and isostatically from the top of Tibet, the material velocities, corrected for the decrease in pressure and for the shallower depth of the refractor but corresponding to the various published Pn and Sn velocities, would be 7.98 and 4.72 km s$^{-1}$ (Chen & Molnar 1981), 7.97 km s$^{-1}$ (Jia Su-juan *et al.* 1981), and 8.27 and 4.68 km s$^{-1}$ (Ni & Barazangi 1983). The Sn velocities and Ni & Barazangi's Pn velocity, but not the other Pn velocities, would accord with a shield structure.

The greatest risk in using the Pn and Sn velocities to infer that Tibet is underlain by a shield-like upper mantle arises from the lack of constraints that measurements of Pn and Sn phases place on the thickness of the layer with such a velocity. The existence of relatively high Pn and Sn velocities does not rule out the possibility that the mean velocities in the upper 300 km of the mantle are low and that a thick, pronounced low-velocity zone exists.

Thus, measurements of Pn and Sn velocities beneath Tibet are relatively high; some of them are consistent with a shield-like structure beneath Tibet, but they do not require such a structure.

*S–P travel-time differences*

To constrain the average shear-wave velocity structure in the upper mantle beneath Tibet, Chen & Molnar (1981; Molnar & Chen 1984) measured S-wave and P-wave arrival times to distant stations (30° or less) from earthquakes in Tibet and the Himalaya (see also Pettersen & Doornbos 1987). With accurate focal depths, determined from the synthesis of P phases (Baranowski *et al.* 1984; Molnar & Chen 1983), corrections for erroneous origin times associated with erroneous focal depths given by the International Seismological Center, allowed the determination of travel times of P and S phases. To reduce the scatter in S-wave travel times, differences in S-wave and P-wave travel times were considered, because at stations where one phase is delayed or advanced, usually the other is as well. Finally, Chen & Molnar considered earthquakes in both Tibet and the Himalaya so that a comparison between the structures beneath the two areas could be made, and so that delays or advances introduced by the different structures near the receivers could be further minimized.

Travel-time differences between S-waves and P-waves from earthquakes in the Himalaya deviated by −1.95 s to +0.65 s from Jeffreys & Bullen's (1940) travel times, after correcting for the crust being somewhat thicker than normal beneath the Himalayan earthquakes (figure 9). If Herrin's (1968) more recent P-wave travel times were used, then S–P travel-time differences would be 2–4 s smaller than normal, which concurs with the Himalaya being underlain by a shield-like structure (see discussion of the Himalaya, below).

S–P travel-time differences from Tibetan earthquakes are typically 2–5 s larger than those for the Himalaya, again after correcting for the thick crust of Tibet (figure 9). The largest

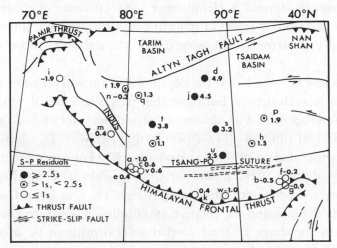

FIGURE 9. Map of earthquakes and mean S–P travel-time residuals to stations 30–80° from them. Times are in seconds and include corrections for excess crustal thickness beneath the Himalaya and beneath Tibet. Residuals were obtained from Jeffreys & Bullen's (1940) P-wave and S-wave travel times. If Herrin's (1968) P-wave travel times were used, the values would all be about 2 s less. Note the large residuals, corresponding to late S-waves, from earthquakes in northern and central Tibet. (From Molnar & Chen 1984.)

differences are from earthquakes in north–central Tibet, a result corroborated by Petterson & Doornbos (1987) for more recent earthquakes. S–P intervals from three events in central Tibet are 2.5–4.5 s greater than those predicted by Jeffreys & Bullen's (1940) tables, or 0.5–2.5 s greater than those calculated from Herrin's (1968) P-wave times, also corroborated by Pettersen & Doornbos (1987). If such differences were caused by differences in shear velocities in the upper 250 km of the mantle beneath Tibet and the Himalaya, then the mean shear-wave velocity beneath central Tibet would be 4–8% lower than that of the Himalaya. Smaller differences in S–P intervals from earthquakes in western, southern, and eastern Tibet than from Himalayan earthquakes imply average velocities 2–3% less than those of the Himalaya (Molnar & Chen 1984). A 2% difference would be only 0.1 km s⁻¹ which is not very significant, but 4% and 8% differences correspond to 0.2 and 0.4 km s⁻¹, which would require a thin high-velocity lid over a thick low-velocity zone beneath Tibet. Thus, in so far as these results apply to the uppermost mantle, they imply that a shield-like structure could underlie only the margins of Tibet.

The weakness of S–P travel-time intervals as a constraint on the upper-mantle velocity beneath Tibet lies in the assumption that the S-waves from earthquakes in Tibet are delayed in the uppermost mantle and not at greater depth beneath Tibet.

*Travel times and waveforms of SH phases*

To investigate variations in seismic-wave velocities with depth in the upper several hundred kilometres of the mantle requires an analysis of phases that penetrate to varying depths in this range. Because of the low-velocity zone and because of rapid increases in velocity near depths of 400 and 650 km, travel times for rays that bottom at 100, 400 and 700 km are comparable to one another. Thus phases travelling along these different paths combine to make relatively complicated signals at distances of 15–25° from the sources, and, in general, the different arrivals can be distinguished and identified only by synthesizing seismograms and comparing them with observed signals (Grand & Helmberger 1984). Because of conversions of vertically polarized S-waves to and from P-waves at reflecting and transmitting interfaces, it is easier to work with horizontally polarized shear waves, SH, than the yet more complex vertically polarized SV.

Clearly a high or a low velocity below the depth of 400 km will advance or delay only those signals that penetrate to such depths, but those that pass above this depth will be unaffected by the velocity structure below it. The dimensions of and the velocity in a shallow low-velocity zone will affect the arrival times of all rays penetrating below it. In addition, because waves refracted both just below depths of 400 km and near 650 km are strongly focused at some distances, the velocity of the material in an overlying low-velocity zone affects the distances at which these signals are recorded with strong amplitudes. Thus, as Lyon-Caen (1986) showed, both the relative arrival times and the relative amplitudes of separate phases, both of which manifest themselves in the shape of the recorded waveforms, can be used to constrain the upper-mantle velocity structure of an area the size of Tibet.

Lyon-Caen's (1986) procedure was first to examine paths across India, to establish that it is underlain by a shield structure, and then to consider Tibet. For epicentral distances of 10–20°, the first signals to arrive pass through the uppermost mantle, above the low-velocity zone. Their arrival times corroborate the existence of high-velocity layers beneath both India and Tibet. A plot of travel times for paths crossing Tibet, however, reveals delays of 5–12 s for rays reflected at the velocity increase at a depth of about 400 km, of which only 4 s can be attributed to the thicker crust of Tibet than India (figure 10). The greater delays imply the existence of lower velocities in the mantle beneath Tibet at depths shallower than 400 km.

Such delays are perhaps best illustrated by a comparison of SS phases that are reflected beneath Tibet and beneath India recorded at epicentral distances of 35–45° (figure 11). At these distances, the direct S-wave penetrates the lower mantle, but the first large signal comprising SS is reflected at 400 km depth. A comparison of the differences in arrival times of SS and S thus provides a strong constraint on the average velocity above 400 km. The SS phase bouncing off the Earth's surface beneath Tibet is delayed 25 s from that bouncing beneath India, an extremely large delay. Roughly half of this delay can be attributed to Tibet's thick crust, but at least 12 s of it in the mantle implies that the average S-wave velocity shallower than 400 km is at least 3 % less than that beneath India.

SH phases from earthquakes in the Tien Shan and passing beneath Tibet to stations in India are not delayed (squares in figure 10). These rays do not sample the upper 250 km of Tibet, but they do pass through the mantle at depths between 250 and 660 km beneath Tibet. In

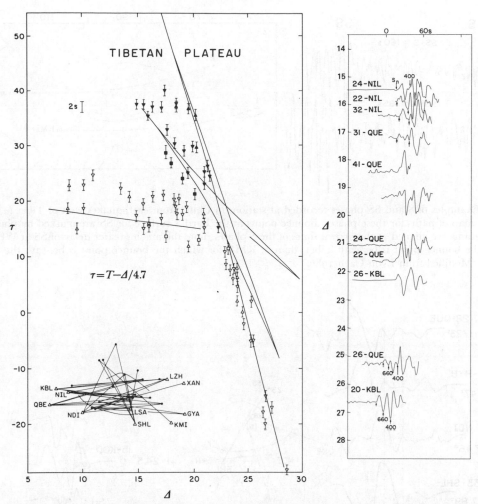

FIGURE 10. Travel times of SH phases against distance for paths across the Tibetan Plateau (left) and examples of seismograms recorded at various distances (right). Different symbols illustrate different paths: upright triangles for recordings at LZH, XAN and LSA, inverted triangles for recordings at KBL, QUE, NDI, KMI, and GYA for roughly east-west paths, and squares for north-south paths (see map in lower left). Open symbols show first arrivals, and closed symbols show later arrivals. Lines show the travel time curves derived by Lyon-Caen (1986) for the Indian Shield but delayed 8 s for the Sn branch and 4 s for other branches to account for the thick crust of Tibet. Note that most arrivals reflected from or refracted just below the 400 km discontinuity (recorded at distances between 15 and 20°) are late, but those from earthquakes in the Tien Shan to stations in India (dark squares) are early; these latter paths pass below the plateau at depths of 250 km or more. (From Lyon-Caen 1986.)

particular, signals reflected at 400 km beneath Tibet arrive at the same time as similar phases travelling only beneath the Indian Shield (figure 12). Thus they too imply that the mantle deeper than about 250 km beneath Tibet is not unusual.

Most of the seismograms shown in figures 10, 11 and 12 sampled the northern part of Tibet, and one might suspect that Lyon-Caen (1986) has shown that a shield structure is absent only from the northern part of Tibet. Notice, however, that the SH phase from an earthquake in central Tibet travelling south to HYB (15-HYB in figure 12) through the uppermost mantle of southern Tibet and bottoming at 400 km is markedly delayed, by almost as much as the signal from an event in northern Tibet to QUE (28-QUE in figure 12). Moreover, notice that the waveform at KOD from this event in central Tibet, which consists of phases refracted

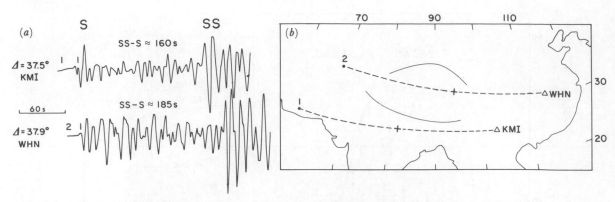

FIGURE 11. Examples of S and SS phases recorded at stations in China from earthquakes west of Tibet (*a*), (top) and (*b*) map of paths for these phases. Bounce points at the Earth's surface for SS are marked by plus signs. Seismograms are aligned at the arrival times of the S phases; note the much greater delay of SS at WHN, for which the bounce point is beneath Tibet, than at KMI, for which the bounce point is beneath the Indian Shield. (Modified from Lyon-Caen 1986.)

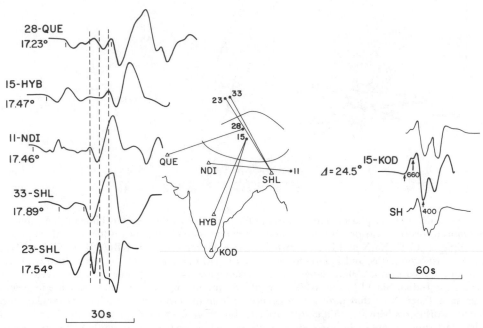

FIGURE 12. Comparison of SH phases recorded at a narrow range of distances for different paths shown on the map. On the left, seismograms are aligned by their reduced times: the measured time minus both the origin time of the earthquake and the epicentral distance divided by 4.7 km s⁻¹. The broken vertical lines are 4 s apart. The first tick marks the Sn phase, and the second tick marks the arrival from the 400 km discontinuity. The first broken line is aligned at the time of this latter arrival at NDI for the path (11-NDI) entirely across the Indian subcontinent. Note the prominent delays at QUE and HYB from earthquakes within Tibet, but the absence of delays at SHL for paths from earthquakes in the Tien Shan and for which rays pass deeper than 250 km beneath Tibet. This implies that the low-velocity material beneath Tibet is shallower than 250 km. On the right, the recorded SH phase from an earthquake within Tibet to KOD is compared with calculated seismograms for a shield structure (SH) and for one with a low-velocity zone shallower than 250 km (top). Seismograms are aligned at the times of refracted phases penetrating below a depth of 660 km and reflected phases from that depth. For a shield structure, the reflection from 400 km is calculated to arrive too early to be distinguished. With a low-velocity zone in the upper mantle, with the velocity equal to the mean of those deduced for Tibet and for a shield, this phase is clearly separate in the synthetic seismogram, and its arrival time correlates with a pulse on the observed seismogram. This suggests that at least part of southern Tibet is underlain by a well-developed low-velocity zone, and not by a shield structure. (From Lyon-Caen 1986.)

below 660 km, reflected at 660 km, and reflected at 400 km, cannot be matched with a shield structure (figure 12). A low-velocity zone in the upper mantle near the source is necessary to refract these rays differently and to cause a separation in the arrival times.

Thus despite the existence of a relatively high shear velocity in the uppermost mantle $(4.7 \text{ km s}^{-1})$, at least the northern half, and possibly all, of the Tibetan Plateau is not underlain by a shield structure. Instead, the structure of the uppermost mantle beneath much of Tibet is more nearly that of the Basin and Range Province of the western United States than that of a stable continental region. A major uncertainty is the location of the transition from the thick low-velocity zone in north–central Tibet to the thick high-velocity zone that underlies the Indian Shield.

### Lateral heterogeneity beneath the Tibetan Plateau

*Seismic-wave velocities and attenuation*

With increasing data of all kinds, the existence of lateral heterogeneity in the crust and upper mantle beneath Tibet has become increasingly clear. Seismic evidence can be interpreted as evidence for a zone of low velocity, and hence of high temperature, in the uppermost mantle beneath north–central Tibet. Less convincing evidence suggests that the coldest temperatures are beneath southern Tibet, excluding the Karakoram discussed below. Rayleigh phase velocities indicate both a thinner crust and a lower upper-mantle shear-wave velocity beneath north–central Tibet (Brandon & Romanowicz 1986) than beneath other parts of Tibet. The largest delays in S-waves, as reflected in S–P residuals (figure 9), are from earthquakes in north–central Tibet (Molnar & Chen 1984; Pettersen & Doornbos 1987), and Lyon-Caen's (1986) largest S-wave delays are also for paths across this part of Tibet. Finally, whereas Ni & Barazangi (1983) found high velocities of Pn and Sn for most of Tibet, Sn was not observed for paths crossing much of north–central Tibet (figure 13). Crudely, efficient and inefficient propagation are correlated with relatively cold and relatively hot uppermost mantle, respectively (see, for example, Molnar & Oliver 1969). Although the precise areas studied with these different techniques are not the same, and the areas of inferred low velocities and high attenuation also are not precisely the same, the overall coherency of the results is clear.

The existence of low velocities in the upper mantle beneath north–central Tibet is important because in this part of Tibet there is abundant evidence for late-Cainozoic volcanism (Burke *et al.* 1974; Deng 1978; Kidd 1975; Molnar *et al.* 1987*a*; Sengör & Kidd 1979). This volcanism includes both basaltic (Deng 1978) and very acidic material (Molnar *et al.* 1987*a*). The basaltic component suggests that an important mantle component is involved. The distribution of young volcanism is not yet well mapped, but the young cones visible on the *Landsat* imagery are confined to north–central Tibet.

There is a suggestion, from S-wave travel times (figure 9), that the average upper-mantle shear-wave velocities increase toward southern, western and eastern Tibet. Although there are too few earthquakes to allow one to contour mean S–P residuals from earthquakes in Tibet, the measured values increase smoothly away from north–central Tibet, where the events with the largest residuals occurred.

*Lateral variation in temperature in the upper mantle*

Two well-located earthquakes have occurred at depths of about 90 km beneath southern Tibet (Chen *et al.* 1981; Molnar & Chen 1983). Intermediate-depth earthquakes are very rare

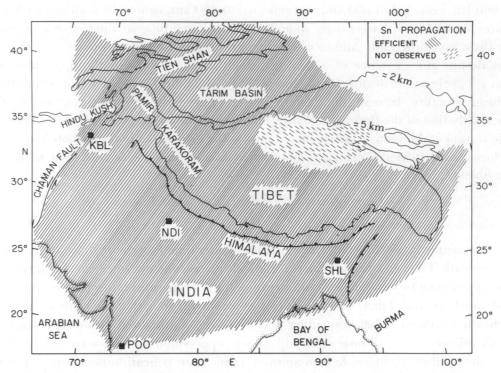

FIGURE 13. Map of the region in which the phase Sn propagates efficiently and inefficiently. Inefficient propagation implies the existence of a zone of low $Q$, or high attenuation. The boundary between areas of high and low $Q$ is not sharply defined. Blank areas indicate areas with few or no observations. (From Ni & Barazangi 1983.)

except at active subduction zones, and Chen & Molnar (1983) infer that their existence implies temperatures of no more than 600–800 °C in the mantle, in this case at a depth of 90 km beneath Tibet. To my knowledge, except for the Karakoram, these are the only reliably located earthquakes at such depths beneath Tibet.

Clearly this evidence is insufficient to prove the existence of a smooth change in upper-mantle structure (and hence in temperature) from low velocities (and high temperatures) in north–central Tibet to higher velocities (and lower temperatures) in southern Tibet, but it seems likely that variations in velocity in the mantle would be smooth instead of abrupt, especially if they were caused by variations in temperature.

The lateral variations in upper-mantle shear velocity coupled with the remarkably uniform elevation of the plateau suggests that if isostatic equilibrium prevails, there must also be variations in crustal thickness. Therefore, in retrospect, the existence of relatively thin crust in north–central Tibet deduced by Brandon & Romanowicz (1986) ought not to be a surprise. It seems to me, however, that the differences in crustal thickness between north–central and southern Tibet cannot be more than 10 km. If the Moho in north–central Tibet were 500 K warmer than elsewhere beneath Tibet, and if lateral variations in the upper mantle were confined to the uppermost 200 km of the mantle, then the mean temperature difference would be 250 K. For a coefficient of thermal expansion of $3 \times 10^{-5} \ \mathrm{K^{-1}}$, the mean density in the uppermost mantle of north–central Tibet would be 25 kg m$^{-3}$ less dense than that in southern Tibet, and a deficit of $5 \times 10^6$ kg m$^{-2}$ would exist in a column of mass in the mantle 200 km thick. For a density contrast of 500 kg m$^{-3}$ at the Moho, this deficit could be compensated if the crust were 10 km thinner.

Note that although a temperature 500 K higher than that beneath southern Tibet is a large difference, it would correspond to a temperature of at least 1000 °C at the Moho beneath north–central Tibet, which is consistent with the partial melting of the mantle to form the basalt erupted at the surface.

## Summary

The seismic data are broadly consistent with partial melting of the uppermost mantle of north–central Tibet, where recent volcanism has been observed. Correspondingly, there is no suggestion of such low velocities, and such high temperatures, in the mantle elsewhere beneath Tibet, for which late-Cainozoic volcanism has not been reported. The results are also consistent with a slightly thinner crust in north–central Tibet than farther south, suggesting that both Airy and Pratt isostasy share compensation for north–central Tibet's great height. Finally, the average shear-wave velocity in the upper mantle of southern Tibet seems to be higher than that in northern Tibet, but neither is the degree of difference well determined, nor is the location of the transition from one to the other well mapped.

## DEEP STRUCTURE OF THE HIMALAYA

### Introduction

When viewed on the scale of continent-sized objects, the Himalaya can appear as little more than the edge of the Tibetan Plateau. Despite the comparatively narrow width of the range, however, both the geologic history and the nature of the transition from the Precambrian shield of India to the very different structure of the plateau require that the Himalaya be treated separately from Tibet. The rocks comprising Tibet north of the Indus–Tsangpo Suture were already part of Asia before the collision with India, and those that form the Himalaya had been part of the separate continent of India. Thus unlike the other margins of Tibet where truly intracontinental deformation has occurred, the Himalaya lie as close, both in proximity and in meaning, to what might be called a plate boundary within a continent as can be found in Asia. In some oversimplified views, the Himalaya and Tibet might be said to lie on different plates. In any case, the imbrication of slices of India's ancient northern margin, as the leading edge of India was subducted beneath the southern margin of Asia, constitute a very different style and history of deformation from what has been found in Tibet. Thus, although the 40–50 Ma of intense tectonic activity have greatly altered the structure of both areas and the boundary between them, it is perhaps wise to begin by assuming their deep structures to be different.

The many tens to perhaps many hundreds of kilometres of underthrusting of India, first beneath southern Tibet and subsequently beneath slivers of India's own ancient northern margin, lead to the logical expectation that the crust thickens smoothly northward. Hirn *et al.* (1984*a*; Hirn & Sapin 1984), however, contend that the Moho may not be a smooth surface and suggest that it has been sliced by southward dipping thrust faults. In this context, the precise nature of the transition from a Moho at a normal depth beneath India to that at 70–75 km beneath southern Tibet is crucial for understanding how the range is supported. Let us first consider various studies of seismic waves from earthquakes, which are consistent with a smooth transition but do not prove it, then continue with the discussion of results from explosion seismology, which led Hirn, Lépine, Sapin and their colleagues to conclude that the transition is not smooth, and finally turn to gravity anomalies, which may place constraints on the forces supporting the range, if the Moho is, in fact, a smooth surface.

### Constraints from earthquake seismology on the deep structure of the Himalaya

The structure of Himalaya has not been studied in detail by using seismic waves from earthquakes, but what has been done is consistent with the structure of the Indian Shield extending beneath the Himalaya.

### Fault-plane solutions and focal depths of earthquakes

The evidence most suggestive of this is provided by studies of moderate sized ($5.5 \leqslant$ magnitude $\leqslant 6.5$) earthquakes in the Himalaya (Baranowski *et al.* 1984; Ni & Barazangi 1984). Fault-plane solutions indicate thrust faulting with one nodal plane dipping gently north or northeast beneath the Himalaya and with the other nodal plane dipping steeply south or southwest away from the range. Clearly it is more sensible to assume that the gently dipping planes are the fault planes. For about half of the earthquakes studied, the plunge of the slip vector is less than $15°$, suggesting underthrusting on very gently dipping planes. For the other half, it is between 20 and $35°$, implying steeper dips of the thrust faults.

Focal depths for most of these earthquakes, below the Earth's surface, are about $15 \pm 3$ km (Baranowski *et al.* 1984; Ni & Barazangi 1984). Thus these earthquakes occur about 13 km below sea level or about $8 \pm 3$ km deeper than the depth of the basement of the Indian plate at the foot of the range, where about 5 km of late-Cainozoic sediment have been deposited. The dip of the basement of the Ganga Basin, at the front of the Himalaya is about $2-4°$. If these earthquakes, which occur about 80–100 km north of the front of the range, occurred on the top surface of the Indian plate, the average dip of that surface beneath the Lesser Himalaya would be

$$\delta = \arctan\,(8 \pm 3 \text{ km}/90 \pm 10 \text{ km}) = 5 \pm 2°.$$

Thus both the focal depths and the fault-plane solutions of roughly half the major earthquakes are consistent with those earthquakes occurring on the top surface of the intact Indian Shield as it plunges beneath the Himalaya (Baranowski *et al.* 1984; Molnar & Chen 1982; Ni & Barazangi 1984).

### Body-wave propagation beneath the Himalaya

Menke (1977) showed that the northward dip of the plate beneath the Himalaya is corroborated by increasingly delayed P-wave arrival times recorded at more northeasterly stations of the Tarbela array in Pakistan. Part of the delay is caused by higher elevations of the more northeasterly stations, but the remaining gradient in residuals is most simply explained by northeastward dip of the Moho of $4°$. This direction is perpendicular to the trend of the range, and the amount is slightly larger than the dip of the basement beneath the neighbouring basin. Moreover, an inversion of relative arrival times across the Tarbela array for lateral heterogeneity in the crust and upper mantle reveals no evidence of other systematic variations in velocity in the mantle (Menke 1977).

Pn and Sn arrival times from earthquakes in the Himalaya and paths along the Himalaya yield high apparent velocities. For paths along the range to the Tarbela network in Pakistan, Menke & Jacob (1976) obtained Pn and Sn velocities of $8.50 \pm 0.35$ km s$^{-1}$ and $5.00 \pm 0.19$ km s$^{-1}$ respectively, but with a large scatter of more than 10 s in travel times. With more data and better located events, Ni & Barazangi (1983) obtained $8.48 \pm 0.05$ km s$^{-1}$ and $4.75 \pm 0.05$ km s$^{-1}$, which are not significantly different from corresponding velocities for the Indian Shield (or Tibet).

As discussed above, S-wave travel times from earthquakes in the Himalaya do not show significant delays (Chen & Molnar 1981; Molnar & Chen 1984). Depending upon whether one uses the P-wave travel times of Jeffreys & Bullen (1940) or Herrin (1968) to determine origin times of earthquakes and P-wave residuals, S-waves from Himalayan earthquakes arrive between 1 s late and 2 s early (figure 9), or between 1 s and 4 s early, relative to Jeffreys & Bullen's (1940) S-wave tables (Molnar & Chen 1984). These residuals are comparable with those at stations on some shields (Doyle & Hales 1967; Sipkin & Jordan 1975, 1980), and thus they are consistent with such a structure beneath the Himalaya. Similarly, Lyon-Caen (1986) found that both arrival times and waveforms of S-waves recorded in India from earthquakes in the Himalaya are consistent with a shield-like structure.

Among the results described in this section, none has the resolution to prove either that the entire Himalaya is underlain by a shield structure or that upper-mantle structure beneath the Himalaya is significantly different from that of India, but none give a hint of an important difference.

### Constraints from explosion seismology on the deep structure of the Himalaya

Although numerous long seismic refraction profiles have been made in different parts of India (see, for example, Kaila 1982), no seismic refraction profile has been shot along the Himalayan chain. What information does exist on the velocity structure of the Himalaya is derived from one long unreversed profile that obliquely crosses the Himalaya (Beloussov et al. 1980; Kaila et al. 1978), from recordings of quarry blasts in the Kathmandu area (Pandey 1981), and from wide-angle reflections generated either by explosions in southern Tibet and recorded in Nepal (Hirn & Sapin 1984; Hirn et al. 1984 a) or by local earthquakes recorded in Nepal (Pandey 1986). As I discuss below, some of these results have been interpreted as evidence that the Indian Shield does not extend far beneath the Himalaya, and these data deserve particular scrutiny.

#### The profile of Kaila et al. (1978) across the Kashmir Himalaya

Apparently, this profile, one of a series of profiles from the Tien Shan in the U.S.S.R. to Kashmir (Beloussov et al. 1980), was the first to use explosive sources and sufficient distances to record clear reflections from the Moho. The explosion used by Kaila et al. (1978) was detonated near Nanga Parbat at the western syntaxis of the Himalaya, and the unreversed profile crossed the Himalaya obliquely.

Kaila et al. (1978) inferred a steep, 15–20°, component of the dip of the Moho in the north–northwest direction of the profile, with the Moho reaching a depth of 70 km beneath the Greater Himalaya. They inferred an abrupt step of 5 km in the depth of the Moho just to the north of this locality, so that a maximum depth of 75 km was reached. They recorded no clear signals refracted by the underlying mantle. Because the profile is not reversed, the velocity structure of the crust cannot be constrained well, and the inferred depths of the Moho must surely be uncertain by at least several kilometres. Similarly the inferred dip of the Moho could be less if the velocity in the lower crust were greater than Kaila et al. (1978) assumed. Thus although a component of northward dip is probably inescapable, its value probably is not well constrained.

*Wide-angle reflections beneath the Himalaya in Nepal*

As discussed briefly above, Hirn and his colleagues carried out an extensive study of the structure of the area between the Indus–Tsangpo Suture Zone and the Greater Himalaya. I discussed these results above because an appreciation of their principal features was necessary for interpreting the seismograms that they obtained farther north in Tibet. While engaged in the recording of these profiles, they also installed stations in Nepal to record wide-angle reflections from the Moho beneath the Himalaya. These observations can be treated as three separate experiments. Four recordings made in the Arun Valley of Eastern Nepal at distances between about 190 and 250 km from the western shotpoint and three recordings near the Kathmandu Valley at distances of about 230, 260 and 305 km from the eastern shotpoint (figure 4) show strong signals that apparently are reflections from the Moho beneath the Greater Himalaya just west of Mount Everest (Lépine *et al.* 1984). Second, a series of recordings made in Nepal at somewhat different distances, of 100–150 km, and at different azimuths south of the western shotpoint reveal strong reflected phases with an apparent velocity of 7 km s$^{-1}$ (figure 3 of Lépine *et al.* 1984). Finally a fan profile, extending southwards from one of those discussed above (figure 5), was constructed by using the middle shotpoint of the profile in southern Tibet and recorded by stations in southern Tibet and northern Nepal, across the Greater Himalaya (Hirn & Sapin 1984; Hirn *et al.* 1984*a*).

Some seismograms reveal more than one strong phase, suggesting a more complicated structure than simply an abrupt increase in velocity at the Moho. Among the four seismograms recorded in the Arun Valley from the western shotpoint, three recorded near one another at distances of about 230, 240 and 250 km show two coherent signals (figure 14). At each site, a clear signal arrives about 1.5 s earlier than would be predicted if the structure were the same as that determined for the region between the suture zone and the Greater Himalaya, and a second, sharper signal was recorded another 2 s earlier. The seismogram recorded in the Kathmandu Valley about 230 km from the middle shotpoint of the profile in southernmost Tibet also shows two phases (figure 14), with the later phase apparently arriving at the approximate time expected from a Moho transition centered at about 75 km depth, and with the second phase arriving 2 s earlier. Two more distant stations recorded only the earlier phase (figure 14), which Lépine *et al.* (1984) inferred to be a Moho reflection at a depth 15 km shallower than that beneath southern Tibet, or at a depth of 55 km (Hirn *et al.* 1984*a*). Lépine *et al.* (1984, p. 120) reported: 'Thus the Tibetan Moho is seen to extend at its 70–82 km depth as far south as the region of Mt. Everest.' With regard to the earlier reflected phase, attributed to a shallower Moho, they concluded: 'The depth of the Everest Moho is about 15 km less than the Tibetan Moho and it extends as far north as Mt. Everest.' Thus they envisage a step in the Moho somewhat north of Mount Everest.

Although the data probably are consistent with these conclusions, I am not convinced that these six seismograms with halfway points near the Greater Himalaya, can provide unique constraints on the dips and depths of the reflector. The existence of the later phase represents a complication, but inconsistencies in the presentation of the data make it very difficult to offer a concrete alternative to the interpretation given by Lépine *et al.* (1984). The broken lines that define arrival times of PmP (Moho) reflections north of the Himalaya and with which comparisons were made, are drawn differently on the two record sections (figure 14). The two seismograms recorded at 230 km from the two shots show travel times of the two later phases

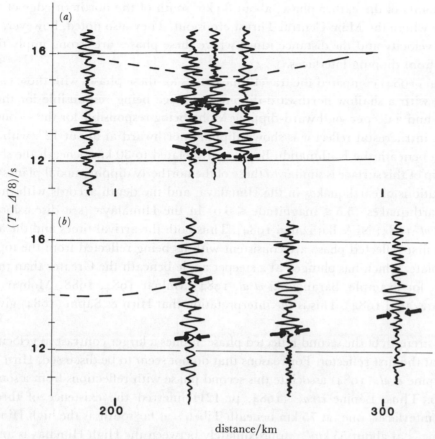

FIGURE 14. Seismograms showing wide-angle reflections, recorded in Nepal from explosions in southernmost Tibet (see figure 4). (*a*) Signals recorded in the Arun Valley from the Western shotpoint. Broken line shows observed arrival times for the east–west profile in southern Tibet, between the Indus–Tsangpo Suture Zone and the Greater Himalaya, where the crustal thickness is 70–75 km. The second strong phase arrives only 1 s earlier, suggesting that if the reflectors were flat, the one responsible for this phase would be nearly 70 km deep. The earlier arrivals presumably mark reflections from a shallower interface, at a depth of about 55 km (Hirn *et al.* 1984*a*). (*b*) Signals recorded near Kathmandu from the middle shotpoint on the profile in southern Tibet. Broken line is reportedly the same as in top figure. Note again the earlier arrivals that suggest a shallower reflector than that farther north. (From Lépine *et al.* 1984.)

that differ by about 0.5 s, but with one shown arriving at the predicted arrival time (figure 14*b*) and the other 1.5 s early (figure 14*a*). Thus, I cannot offer a simple explanation for the later phases that Lépine *et al.* (1984, p. 120) attributed to Moho reflections at '70–82 km depth as far south as the region of Mt. Everest', but I remain unpersuaded that they represent a continuation of an essentially flat Moho at a depth of 70 km as far south as the crest of the High Himalaya. Nevertheless, the relatively early signals almost surely represent reflections from depths many kilometres shallower than 70 km.

Lépine *et al.* (1984, figure 3) showed numerous seismograms recorded at distances of 120–150 km roughly south of the western shotpoint that show pairs of signals arriving with apparent velocities of about 7 km s$^{-1}$. On most seismograms, the earlier signal is the weaker and is followed about 1.3 s later by a stronger signal. Lépine *et al.* (1984) noted that both the strength and arrival times of the signals suggest that they are reflected phases from a major velocity contrast, such as the Moho. Assuming horizontal layers, they obtained a depth of about 35 km, for reflection points of the earlier phase, about 30 km south of the northern edge of the high peaks and near where the Main Central Thrust crops out. They also noted, however, that the high apparent velocity and the distance range where these phases are strong imply that they were reflected from dipping interfaces.

Hirn & Sapin (1984) compared the travel-time curves for these phases with those calculated for a structure with a shallow northward-dipping interface being responsible for the earlier reflected phase and a deeper southward-dipping Moho being responsible for the second phase. The shallower, intracrustal reflector is shown dipping northward at about 12°, with a depth of about 10 km beneath the Kathmandu Basin, extrapolated to 30 km beneath the shotpoint. The inferred dip of this surface is similar to those of the northerly dipping nodal planes of many fault-plane solutions of earthquakes in the Himalaya, and the depth accords with the depths of moderate earthquakes ($5.5 \leqslant$ magnitude $\leqslant 6.5$) in the Himalaya (see discussion above) (Baranowski et al. 1984; Ni & Barazangi 1984). Thus both the arrival times and the apparent velocity of this first reflected phase are consistent with its being reflected from the top surface of the Indian plate, which has plunged at a steeper angle beneath the Greater than the Lesser Himalaya (see, for example, Baranowski et al. 1984; Molnar 1984, 1988; Molnar & Chen 1982; Ni & Barazangi 1984). This is the interpretation that Hirn & Sapin (1984) give to this phase.

The greater strength of the second reflected phase implies a larger contrast in velocity at this reflector than at the first reflector. For reasons that do not seem to be discussed, Hirn & Sapin (1984) and Lépine et al. (1984) associate this second phase with reflections from a southward-dipping Moho. Thus, Lépine et al. (1984, p. 121) inferred the existence of three deep, essentially flat interfaces: one 'at 75 km beneath Tibet...as far south as the high Himalayas', 'a shallower one...at about 55 km', 'approximately between the High Himalayas and 50 km further south', and a third 'at a still shallower level, around 35 km...possibly downdipping towards south...on top of the Everest Moho'. Whereas the presentations given by Hirn & Sapin (1984) and Lépine et al. (1984), as well as a brief discussion by Pandey (1986) of reflected phases from an earthquake in Nepal and recorded there, show their data to be consistent with this structure, it seems to me that they have not eliminated the possibility that the Moho dips smoothly north.

Because of the similar apparent velocity and essentially constant delay of only 1.3 s between the earlier and later reflected phases with apparent velocities of 7 km s$^{-1}$, I find it easy to believe that the later phase is reflected from a surface parallel to, but deeper than, that causing the earlier phase. Recall that the Wind River Thrust Fault, in the western United States, was traced at depth by two strong signals about 1 s apart on a COCORP reflection profile (Smithson et al. 1979). I suspect that both phases observed by Hirn & Sapin (1984) and by Lépine et al. (1984) could be reflected from velocity contrasts within a thick, major fault zone between the Indian plate and the overlying Himalayan Thrust Sheet. If sedimentary rock from the Ganga Basin had been underthrust along this zone, as has been suggested (Acharya & Ray 1982; Lyon-Caen & Molnar 1983), then the fault zone could be associated with a thick zone of low-velocity material. It seems likely to me that the interfaces responsible for both phases with apparent velocities of 7 km s$^{-1}$ dip north, and that the reflected phases responsible for at least one of the inferred steps in the Moho (Lépine et al. 1984) can be explained without a step.

The evidence most suggestive of southward-dipping interfaces beneath the Himalaya is from the fan profile constructed by using seismographs at distances of 170–305 km from the eastern shotpoint and recorded in southernmost Tibet and in Nepal (Hirn & Sapin 1984; Hirn *et al.* 1984*a*). These instruments were placed so as to record clear wide-angle reflections from depths of several tens of kilometres, and in particular from the Moho. In presenting the data, Hirn & Sapin (1984; Hirn *et al.* 1984*a*) corrected the arrival times for the different distances (between 170 and 305 km) at which the recordings were made so that reflections from horizontal interfaces at the same depths of 50–70 km would be aligned. They then plotted the seismograms so that the time axis is translated into depth in kilometres (figure 15). Although this has the effect of vividly portraying the seismograms in a physically interpretable manner, the absence of specific information about arrival times at different stations, the locations of the stations, and the timescale in seconds render any further quantitative analysis by the reader difficult.

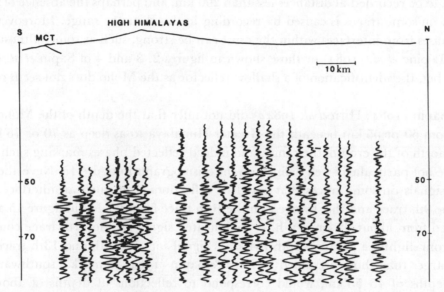

FIGURE 15. Wide-angle reflections recorded on a north–south profile across the Greater Himalaya, from Nepal to Tibet, as distances of 170–305 km from the middle shotpoint along the southern refraction profile (see figure 4). Arrival times have been adjusted for different recording distances and plotted to illustrate probable depths of reflectors. Coherent reflections from about 70 km on the right (north) and from about 55 km on the left (south) are clear. Note the possible northward continuity of the shallower reflector to shallower depths, marked by strong signals corresponding to depths of 55–45 km on the middle traces. Note also strong signals that follow these signals, corresponding to depths of 62 km on the 9th and 10th traces from the left, of 65 km, 69 km, and 72 km on the 11th, 12th, and 13th traces, and of 75 km on the 14th and 15th traces from the left. I suspect that these signals might be reflections from a northward-dipping Moho. (Modified from Hirn & Sapin 1984.)

The coherency of signals recorded north and south of the Greater Himalaya are consistent with their being reflections from a strong continuous reflector, such as the Moho. As noted above, the first strong signal from the eastern shotpoint and recorded at stations tens of kilometres north of the Greater Himalaya corresponds to a reflection from about 70 km (seven traces on the right in figure 15). The arrival times of the first strong signals recorded from stations south of the Greater Himalaya are appropriate for reflections at depths of 50–55 km (eight traces on the left of figure 15). As plotted, the arrival times, adjusted for different distances, suggest a slight southward dip of this interface, but this effect probably could be caused by differences in corrections for distance and by lateral variations in the local velocity structure.

Between these groups of recording sites, the arrival times of the first strong phase at the more northerly sites suggest a shallower depth of this reflector; the apparent depth of about 50 km beneath the crest of the Greater Himalaya decreases to 45 km some 30 km farther north (figure 15). These phases can be seen clearly; they are the strongest signals on three of seven traces (the 9th, 11th and 15th traces from the left in figure 15). Stronger signals are recorded later on the 10th, 12th, 13th and 14th traces from the left in figure 15, but on the 10th, 13th and 14th traces, signals can be observed at the times corresponding to depths of about 55, 45 and 45 km, respectively.

Hirn & Sapin (1984; Hirn *et al.* 1984*a*) did not give sufficient information for me to be convinced that these arrivals represent an interface dipping south or to show that such an interface is the Moho. Note that on the basis of both their observed and their calculated seismograms (figure 2 of Hirn *et al.* 1984*a*), reflections from a Moho at a depth of 70 km might be too weak to be recorded at distances less than 200 km, and perhaps the absence of reflections from 70 km on some traces is caused by recording in this distance range. Moreover, because reflected phases from interfaces within the crust can be strong, such as some of those shown in figure 3 of Lépine *et al.* (1984) or those shown in figures 2, 3 and 4 of Sapin *et al.* (1985) for southern Tibet, the identification of a shallow reflector as the Moho does not seem required to me.

Hirn & Sapin (1984; Hirn *et al.* 1984*a*) did not infer that the depth of the Moho increases smoothly from 50 or 55 km beneath the Greater Himalaya to as deep as 70 or 75 km tens of kilometres north of the crest of the range, and indeed reflected phases marking such a dipping interface are not particularly well defined on the seismograms in figure 15. Nevertheless, there are strong signals on some traces that could be reflections from such an interface. A strong signal on the 9th trace and the strongest on the 10th trace from the left in figure 15 are plotted as if reflected from about 62 km depth. A complicated signal on the 11th trace could mark a reflection from slightly a greater depth (64 km). Signals on the 12th and 13th traces, both of which are larger than those presumed by Hirn & Sapin (1984) to mark a southward dipping Moho at depths of 45–55 km, would correspond to reflections at depths of about 69 and 72 km, respectively. Finally, signals can be seen on the 14th and 15th traces at times corresponding to depths of about 75 km. These signals are not as coherent on neighbouring traces as those arriving earlier and interpreted by Hirn & Sapin (1984; Hirn *et al.* 1984*a*) as reflected from a shallow south-dipping interface. Nevertheless, the existence of these phases suggests to me that these authors have not eliminated the possibility that the Moho does dip smoothly beneath the Himalaya, but that it is masked somewhat by complexity in the crust at shallower depths, and by the effects of recording these signals at different distances.

In reviewing the seismic refraction studies of the Himalaya, I have taken a prejudiced view. With no preconceived notions of what Hirn and his colleagues should have found, much of the structure that they inferred is as reasonable as any. Accordingly, I certainly cannot assert that they are wrong in what they deduced. At the same time I feel strongly that the brevity of the data presentation allows one the freedom to doubt their interpretations. Specifically, I do not think that they have eliminated the possibility that the Moho does dip smoothly to the north beneath the range. First, I think that the pair of phases with apparent velocities of 7 km s$^{-1}$ (figure 3 of Lépine *et al.* 1984), which comprise some of the evidence for one of their inferred steps in the Moho, can be interpreted as reflections from the northward dipping, top surface of the Indian plate plunging beneath the Himalaya. Second, I can see a hint of a northward-dipping reflector between depths of 50 and 70 km on traces 9–15 in figure 15, which could mark a smoothly northward-dipping Moho. Thus I proceed with the assumption that the Moho beneath the Himalaya is not sliced by southward-dipping faults, but is a continuous, smooth, northward-dipping surface.

### Gravity anomalies across the Himalaya

Gravity anomalies lack the resolution to define details of deep crustal structure, and because of their inherent non-uniqueness, they are best used to answer more specific questions than 'What is the deep structure?' They can be much more effectively used to disprove, than to prove, particular hypothesized structures. At the same time, gravity anomalies, unlike other observable quantities, are directly sensitive to mass distributions, and because the force of gravity acting on lateral variations in density is the ultimate force that drives most tectonic processes, gravity anomalies can provide direct constraints on such processes. My review will focus more on this aspect of gravity anomalies than on their use solely as a constraint on hypothetical density structures.

Gravity measurements are most easily compared with calculations by reducing them to the values that would be measured at a common altitude, usually sea level, either to free air or to Bouguer anomalies. Terrane corrections are commonly added to Bouguer anomalies, but rarely, if ever, to free-air anomalies. One difficulty in using gravity anomalies from the Himalaya is that terrain corrections can be enormous (up to 100 mGal) and cannot be ignored. Therefore, one must consider Bouguer anomalies.

Because of the great height of the Himalaya and because of approximate, but incomplete, isostatic equilibrium, the crust beneath the Greater Himalaya is thicker than that of the Lesser Himalaya. Consequently, Bouguer anomalies decrease northward with increasing elevation, and their negative values become large over the Greater Himalaya (Choudhury 1975; Das *et al.* 1979; Duroy *et al.* 1987; Kono 1974; Marussi 1964; Mishra 1982; Qureshy 1969; Qureshy & Warsi 1980; Qureshy *et al.* 1974; Teng *et al.* 1980; Warsi & Molnar 1977). This northward decrease in Bouguer anomalies is characteristic of the Himalaya, and in early studies of gravity it was used as an argument in support of the concept of isostasy. Now, with more and better data, the subtleties in the northward gravity gradient, rather than its mere existence, are the focus of most modern analyses. The easiest way to appreciate the significance of a profile of gravity anomalies is compare it with a reference; in this case the best reference, or the simplest model, is that of local Airy-type isostatic equilibrium.

### Deviations from local isostatic equilibrium

For local Airy isostatic equilibrium, the thickness of the crustal root should be proportional to the elevation. To examine whether the Himalaya are in local isostatic equilibrium, Lyon-Caen & Molnar (1985) compared measured Bouguer anomalies with those calculated from the corresponding average topographic profiles, assuming local isostatic equilibrium (figure 16). Systematic deviations from isostatic equilibrium can be seen for each of the five profiles considered. Observed gravity anomalies over the Indo–Gangetic Plain are as much as 100 mGal more negative than those calculated. Thus a large mass deficit must be present beneath the flat plain. Over the transition from the Lesser to the Greater Himalaya, observed anomalies are as much as 100 mGal less negative than those calculated (see also Tang *et al.* 1981). Thus, a mass excess (with respect to local isostatic equilibrium) must underlie this segment of the

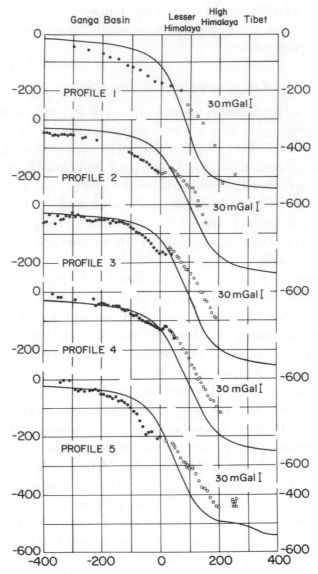

FIGURE 16. Bouguer gravity anomalies for five profiles shown on the map in figure 4. The edge of the Ganga Basin next to the Lesser Himalaya defines the origin, $x = 0$. Solid curves show calculated Bouguer gravity anomalies for structures in isostatic equilibrium with the topographic profiles. Note the poor fit in all cases, with calculated anomalies too small over the Ganga Basin and too large over the Lesser Himalaya. Thus, a mass deficit underlies the basin, and a mass excess underlies the mountains. (From Lyon-Caen & Molnar 1985.)

range. The Himalaya and the neighbouring Indo–Gangetic Plain clearly are markedly out of local isostatic equilibrium.

The reason for examining whether or not the Himalaya is in local isostatic equilibrium is not so much to test that idea but more to gain insight into how deviations from isostatic equilibrium might occur. The well-known existence of a basin beneath the Indo–Gangetic Plains, the Ganga Basin, which is filled with several kilometres of late-Cainozoic sedimentary rock (Karunakaran & Ranga Rao 1979; Lillie & Yousuf 1986; Raiverman *et al.* 1983; Sastri *et al.* 1971), offers an explanation for where at least some of the mass deficit lies. Correspondingly, some authors have postulated the existence of dense rock within the crust of the Lesser

Himalaya to explain the more positive observed anomaly than would be predicted from local isostatic equilibrium (see, for example, Qureshy 1969; Qureshy *et al.* 1974). Gravity anomalies alone, however, can neither prove nor disprove such hypothetical density distributions, and thus to gain insight into the significance of the deviations from local isostatic equilibrium, one must look beyond density as a physical parameter to be adjusted in calculations.

Because deviations from isostasy imply deficits and excesses of mass in the crust and/or mantle, and because such deficits and excesses must be supported by stress differences within the crust and mantle, deviations from isostatic equilibrium can place bounds on the strength of the crust and mantle (see, for example, McKenzie 1967). In the limiting case of a very strong lithosphere, such as on the Moon or Mars, large deficits and excesses of mass, and hence large variations in density, can be supported and maintained for long periods of time. In the other limit of a very weak lithosphere, the buoyancy associated with mass deficits would cause them to rise, and mass excesses would sink, until isostatic equilibrium prevailed. When viewed in this context, the simplest questions to be asked address the 'strength' of the lithosphere, or more precisely, its flexural rigidity. A plate with a large flexural rigidity can support heavy localized masses with only a small deflection, but one with a small flexural rigidity will bend substantially, so as to allow the mass anomalies to be nearly in isostatic equilibrium (Turcotte & Schubert 1982, pp. 104–133). Thus most modern analyses of gravity anomalies over regions with dimensions of hundreds of kilometres pay little attention to the distribution of density within the crust; instead they consider the implications of the gravity anomalies for the physical mechanisms that support various possible distributions of density that, in turn, are allowed by the gravity measurements.

In the context of the Himalaya and the Indo–Gangetic Plains, the lithosphere seems to be flexed down beneath the plains, which are filled with sediment, but the strength of the plate supports a heavy load over the Lesser Himalaya. The abundant evidence for thrust faulting within the Himalaya implies that slices of crust have been thrust southwards onto intact Indian crust. Thus the load consists of those slices. In addition, the sequence of ages and sedimentary facies of the Siwalik sequence, which fills the Ganga Basin, imply both a tilting and a subsidence of the Ganga Basin consistent with its being underthrust beneath the Himalaya (Lyon-Caen & Molnar 1985). Thus the mass deficit beneath the Indo–Gangetic Plain, inferred from the deviation from isostatic equilibrium, probably exists because the Indian plate is flexed down, and the excess mass over the Lesser Himalaya would be part of the mass of the load that flexes the plate down. The strength of the plate contributes to the support of that excess mass.

*Flexure of a simple elastic plate*

The simplest mechanical model commonly used to quantify the flexure of the lithosphere is an elastic plate overlying an inviscid fluid and loaded by masses above, within, and below the plate. Such a model is an obvious oversimplification both because the lithosphere probably does not deform as an elastic plate and because it is underlain by material with finite viscosity. Moreover, although this simple model is quite effective for some situations, it clearly fails in others, so that some workers have replaced it with viscoelastic plates, elastic–perfectly plastic plates with finite yield stresses, plates with creep–strength profiles based on laboratory measurements of flow laws of the constituent minerals, and others. All such models also are approximations, especially given that stresses created by the flow of material at the base of the lithosphere and within the asthenosphere are likely both to warp the overlying plate and to support lateral variations in density. Thus in my view, the value of all of these models is revealed less by their ability to offer simple physical explanations for some observations than by the understanding of how the flexure of such plates fails to account for all of these data.

As of 1987, analyses of gravity anomalies over the Himalaya have considered only elastic plates. The first step in such analyses was to consider a semi-infinite elastic plate with a constant flexural rigidity beneath the Central Indian Highlands, the Indo–Gangetic Plain, and the Himalaya (Duroy *et al.* 1988; Karner & Watts 1983; Lyon-Caen & Molnar 1983, 1985). With such a simple model (figure 17), six parameters are left free to be adjusted: the flexural rigidity

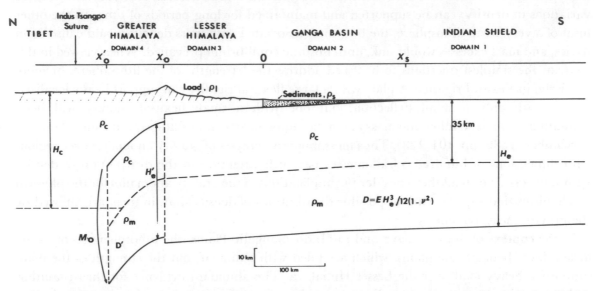

FIGURE 17. Simple plate model used to examine the gravity anomalies. An elastic plate underlies the Indian Shield, the Ganga Basin, and the Himalaya and is loaded by the weight of the Himalaya. The initial analysis by Lyon-Caen & Molnar (1983) examined a plate of constant flexural rigidity, $D$, and extending to the coordinate $X_0$. Later analyses considered a segment with a lower flexural rigidity, $D'$, attached at $X_0'$. Final analyses included the effects of a bending moment and a vertical force (per unit length) acting at the end of the plate. (From Lyon-Caen & Molnar 1983.)

of the plate, the coordinate of the end of the plate $(X_0)$, the force per unit length $(F_0)$ and the bending moment per unit length $(M_0)$ applied to the end of the plate, and the two independent density differences among the three relevant densities: mantle, crust and sedimentary fill in the Ganga Basin. (Note that Karner & Watts ignored the density difference between crust and sedimentary fill and assumed the same average value for both the sediment and the overthrust mass comprising the Himalaya, and that Duroy *et al.* specified numerous different densities for different units in the foreland basin of Pakistan.) Initially, calculations were made ignoring the force and the bending moment at the end of the plate. For the cases where the end of the plate was taken to lie south of the Indus–Tsangpo Suture Zone, Lyon-Caen & Molnar (1983) assumed that the depth of the Moho increased linearly from its calculated depth at its northern end to a depth of 65 km beneath southern Tibet, where local isostatic equilibrium was assumed to exist. Lyon-Caen & Molnar (1983) then performed numerical experiments to examine the sensitivity of calculated gravity anomalies to those four remaining parameters: flexural rigidity, position of the end of the plate, and the two density differences.

Variations among reasonable density differences were found to have little effect on calculated gravity anomalies. For example, if the assumed density of the sediment in the Ganga Basin were increased, the calculated deflection of the plate also increased so that the increased gravitational attraction of the more dense sediment was compensated by its increased thickness and the decreasing amount of underlying mantle.

The effect of varying the flexural rigidity of the plate was most clearly seen in the calculated gravity anomalies over the Indo–Gangetic Plain and the Lesser Himalaya. Tight constraints could not be placed on its value, and the uncertainty in the flexural rigidity is at least a factor of three for each profile. Moreover, flexural rigidities deduced for different profiles across the Himalaya differ by at least three times, and possibly by 10 times or more (Duroy et al. 1987; Lyon-Caen & Molnar 1985). The large uncertainties in all estimates makes it impossible to map variations in flexural rigidity precisely, but clearly a variation of three times inferred for profiles only 200 km apart calls for caution in interpreting the inferred values, as well as in trusting too much the simple model of an elastic plate. In any case, the inferred flexural rigidities are much larger (10–1000 times) than those deduced for the Alps (Karner & Watts 1983) or the Apennines (Royden & Karner 1984), or the transverse ranges of California (Sheffels & McNutt 1986).

The parameter that affects the calculated gravity anomalies most and that also is most difficult to interpret is the position of the end of the plate. Ignoring any force applied to the end of the plate, the load that flexes the plate down was taken to be the weight of the portion of the Himalaya that overlies the plate. Thus, in Lyon-Caen & Molnar's (1983) calculations, only the portion of the Himalaya that directly overlies the plate is supported by it. The support of the Himalaya between the end of the plate and southern Tibet was, initially, ignored. The farther north that the plate is assumed to extend beneath the Himalaya, the greater is the assumed load on it, and the greater is the calculated deflection of the plate (figure 18). If only the southernmost hundred kilometres of the Lesser Himalaya comprised the load, this load would be insufficient to flex the plate down enough to create a basin as deep as the Ganga Basin; calculated gravity anomalies are too small. In the case where the elastic plate extends beneath 250 km of the Himalaya to the Indus–Tsangpo Suture Zone, the weight is so large that the calculated depth of the Ganga Basin is too large, and the calculated gravity anomalies are much too negative. Thus, Lyon-Caen & Molnar (1983, 1985) concluded that the loading and flexing of an elastic plate of constant flexural rigidity cannot account for the gravity anomalies across the entire Himalaya or for how the range is supported. Accounting for both the gravity anomalies and the support of the range, therefore, requires a more complicated mechanical model.

*Elastic plate with a variable flexural rigidity*

The gradient in the Bouguer gravity anomalies steepens markedly over the Himalaya, with gradients of less than 1 mGal km$^{-1}$ across the Indo–Gangetic Plain and as much as 2 mGal km$^{-1}$ across the Greater Himalaya (figure 16). This steepening of the gradient is most simply explained by a steepening of the Moho, from a dip of only 2–4° beneath the Indo–Gangetic Plains to a dip of 10–20° beneath the Greater Himalaya (Choudhury 1975; Das et al. 1979; Kono 1974; Lyon-Caen & Molnar 1983, 1985; Mishra 1982; Teng et al. 1980; Warsi & Molnar 1977).

Clearly, other distributions of density could explain the gravity gradient, but an increase in

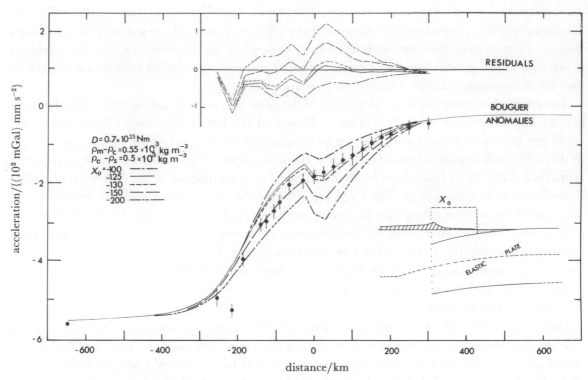

FIGURE 18. Comparisons of measured Bouguer anomalies along the easternmost profile in figure 4 with calculated anomalies for plates extending different distances beneath the Himalaya. Only the material between the Ganga Basin and $X_0$ comprises this load. In these calculations, the Moho north of $X_0$ was drawn as a northward-dipping plane to $x = 250$ km, where isostatic equilibrium is presumed to prevail and where the depth of the Moho below sea level was taken to be 65 km. Note the excessively large calculated gravity anomalies for $X_0 = 150$ km, because the load is too heavy and the plate is flexed too much, and the small values for $X_0 = 100$ km, for which the load of the Himalaya is too small. (From Lyon-Caen & Molnar 1983.)

the average depth of the Moho is also required by Hirn et al.'s (1984a, Hirn & Sapin 1984) demonstration that the Moho is at a depth of 70–75 km north of the Greater Himalaya and at shallower depths farther south. For a depth of 43 km at the foot of the Himalaya (38 km of typical crust of the Indian Shield (Kaila 1982) plus 5 km of Siwalik sedimentary rock (Karunakaran & Ranga Rao 1979; Raiverman et al. 1983; Sastri et al. 1971)) and for 70 km below the area 150 km further north, the average dip of the Moho beneath the Lesser and Greater Himalaya is 10°. Obviously, if the average dip of the Moho is gentler beneath the Lesser Himalaya, it must be steeper beneath the Greater Himalaya. Thus, in so far as the Moho is a relatively smooth surface beneath the Himalaya, a steepening of it beneath the Greater Himalaya seems to be required.

The following discussion proceeds from the supposition that the Moho is a smooth surface and is not cut by southward dipping faults as inferred by Hirn et al. (1984a; Hirn & Sapin 1984). Gravity anomalies, however, cannot prove that the Moho is smooth; in fact, Hirn & Sapin (1984) demonstrated that their inferred structure with steps in the Moho is consistent with observed gravity measurements in the Himalaya. If their inferred structure is shown to be correct, most of the following discussion of gravity anomalies will be rendered worthless.

The steepening of the Moho from ca. 3° to ca. 15° in a distance of approximately 100 km implies that in cross section the average radius of curvature of the Moho is about 150–200 km, and therefore that the Indian plate is much more severely bent beneath the Himalaya than it is beneath the Indo–Gangetic Plain. For the plate to bend sharply beneath the range, its flexural rigidity there must be much less than that beneath the plains.

Recognizing this, Lyon-Caen & Molnar (1983) calculated gravity anomalies across the Himalaya by using a simple model of an elastic plate with a northward decrease in the flexural rigidity. This was implemented by treating the Indian plate as a semi-infinite plate with a constant flexural ridigity beneath the Indo–Gangetic Plains and the Lesser Himalaya and with a short segment of plate with much lower flexural rigidity attached to it and underlying the Greater Himalaya. This model contains eight free parameters: the same two density differences as before, two flexural rigidities, the position where the two plates are joined, the position of the end of the plate, and the force and moment at the end of the plate. Initially the force and the moment were ignored. Calculations show that the addition of a segment of plate with a small flexural rigidity yields steep dips of the Moho and steep gravity gradients beneath the Greater Himalaya (figure 19). As with a single plate of constant flexural rigidity, however, the weight of the Greater Himalaya is so large that in all calculations, it causes a very large downward bending of the plate, and calculated gravity anomalies over the Lesser Himalaya and the Ganga Basin are significantly more negative than observed. Thus some additional force must contribute to the support of the Himalaya; in the absence of that force, the entire Himalayan Range would be lower than it is by a couple of kilometres, the Ganga Basin would be deeper by one or two kilometres, and the Moho beneath both would likewise be deeper.

*Additional forces acting on the Indian plate*

It might be tempting to infer that thermal expansion of material in a hot upper mantle beneath the Himalaya has caused the surface of the range to stand higher than it would if the underlying mantle were cold. Although one cannot rule this possibility out on a local scale, the mantle beneath the Indo–Gangetic Plains and Lesser Himalaya is not likely to be hot. The seismic observations discussed above seem to rule that out. Moreover, were it hot, although the low density of the hot upper mantle would buoy up the crust and make the Moho shallower than normal, it would also contribute compensating negative gravity anomalies.

A sufficiently large bending moment applied to the end of the plate can flex the plate up beneath both the Lesser Himalaya and the Indo–Gangetic Plains and down beneath the Greater Himalaya. Thus its inclusion in the calculations can contribute to the support of the range as well as cause a steepening of the Moho, and hence of the Bouguer anomaly gradient, beneath the Greater Himalaya. A large range of possible values for the bending moment can be found for an allowable range of possible flexural rigidities of the short segment of plate beneath the Greater Himalaya (Lyon-Caen & Molnar 1983). Consequently, useful constraints cannot be placed on the value of such a bending moment (per unit length) except that it must be at least $5 \times 10^{17}$ N (figure 20).

Two possibilities for the origin of a bending moment suggest themselves: a horizontal force (per unit length) acting on a deflected plate and a vertical force (per unit length) acting on a cantilever attached to the end of the plate. The horizontal force is not a likely source. For a plate deflected $y_0 = 30$ km, the horizontal force per unit length, $F_0 = 2 M/y_0$, necessary to create a moment per unit length of $M = 5 \times 10^{17}$ N, must be at least $3.3 \times 10^{13}$ N m$^{-1}$. For a plate 100 km thick, this would correspond to an excess horizontal compressive stress of at least 330 MPa (or 3.3 kbar). Such a large regional horizontal compressive stress seems unlikely.

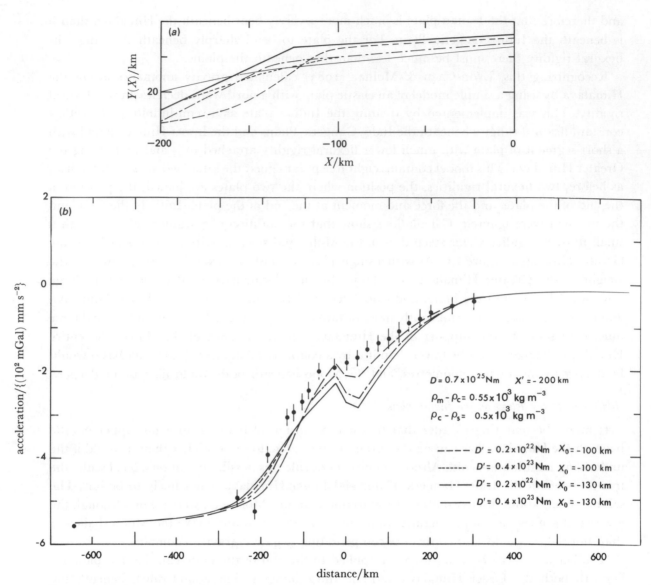

FIGURE 19. Deflections of the Indian plate (a) and Bouguer gravity anomalies (b) assuming the two plate model in figure 21. (a) Dark solid line shows the cross-sectional shape of the Moho that yields a good fit to gravity anomalies, and other lines show calculated cross-sectional shapes for different flexural rigidities (D') of the short segment of plate and different positions ($X_0'$) where it is joined to the semi-infinite plate. (b) Calculated gravity anomalies are shown for the same parameters (listed on the figure) used in calculating the deflections. Note poor fits in all cases. The weight of the Himalaya is so large that calculated deflections of the plate are too large to yield a match to the observed gravity anomalies. Thus the weight of the Himalaya is too large to be supported by any combination of elastic plates without some additional force acting on the plate. (From Lyon-Caen & Molnar 1983.)

If the crust (or part of it) were scraped off the underthrusted continental lithosphere, then gravity acting on the mantle lithosphere, particularly if the lower crust attached to it underwent a phase change to eclogite facies (Richardson & England 1979), could contribute a substantial vertical force per unit length. The force (per unit length) acting on a slab of lithosphere more dense than the asthenosphere by $\delta\rho = 50$ kg m$^{-3}$ (ca. 500 K colder than surrounding asthenosphere) and with a thickness of 50 km and a length of 200 km would be

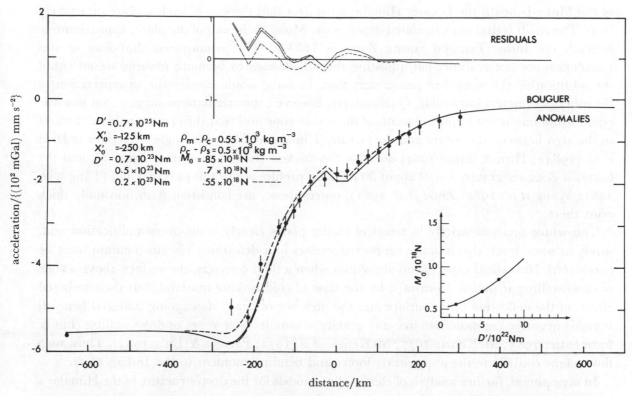

FIGURE 20. Comparison of calculated and observed Bouguer anomalies for models (figure 17) with two segments of plate with different flexural rigidities and with bending movements applied to the end of the plate. Plot in lower right shows the trade-off between values of $D'$ and the bending moment. Note the wide range of values for both parameters and the clear impossibility of resolving their values with additional, or better, measurements of gravity. (From Lyon-Caen & Molnar 1983.)

$5 \times 10^{12}$ N m$^{-1}$. If centred 100 km from the end of the elastic plate, the bending moment would be $5 \times 10^{17}$ N. Thus, gravity acting on a slab of mantle lithosphere with crust either detached from it or having undergone a phase change to a more dense mineralogy could cause the necessary bending moment.

Proving such a simple explanation is virtually impossible, at least at present. Lyon-Caen & Molnar (1983, 1985) were content with finding one set of reasonable parameters that allow a fit of observed and calculated gravity anomalies and depths to the basement of the Ganga Basin. The trade-offs among flexural rigidities, positions of the ends of the plates, and values of the forces and bending moments make deducing a unique, or even well constrained, set of parameters impossible. Moreover, because the postulated cold mass that would cause the bending moment should lie 100 km or more north of the suture zone, direct evidence, either with gravity anomalies or with seismic wave propagation, for its existence or absence does not yet exist. The evidence most suggestive of cold material north of the Himalaya is from the Karakoram region, discussed below, but it too is inconclusive.

The analysis of a flexed plate loaded by a force and a bending moment on its end led to one prediction that, if taken literally, appears absurd but might nevertheless be qualitatively observable in the seismic and gravity data. In all Lyon-Caen & Molnar's calculations, the flexed plate is taken to include a section of crust with constant thickness, and the steep gradient of the Moho beneath the Greater Himalaya requires that the end of such a plate be severely bent. The result is that our calculated depth of the Moho at the end of the plate, approximately beneath the Indus–Tsangpo Suture Zone, is 115 km! The assumptions that lead to this calculation are not realistic, but adjusting the calculations to be more realistic would entail the addition of yet more free parameters that, in turn, would render the interpretation of the other parameters impossible. Qualitatively, however, the calculations suggest that isostatic equilibrium ought not to prevail south of the suture zone and that the crust should be thickest in the area between the suture and the Greater Himalaya. The wide-angle reflections of Hirn *et al.* (1984a; Hirn & Sapin 1984) do show the thickest crust in that area (figure 5), and the hint of a Bouguer gravity low of about 30 mGal in profiles across this part of Tibet (Tang *et al.* 1981; Wang *et al.* 1982; Zhou *et al.* 1981), noted above, are consistent with unusually thick crust there.

The whole analysis strictly in terms of elastic plates clearly is an oversimplification and, surely at some level, dynamically supported stresses in a deforming viscous medium must be considered. Numerical experiments show that when a fluid convects, the surface above a zone of downwelling is pulled downward by the flow of cold sinking material, but the combined effects of the deflection of the surface and the presence of heavy downgoing material beneath it yields negative (isostatic and free-air) gravity anomalies over zones of downwelling (Lin & Parmentier 1985; McKenzie 1977; McKenzie *et al.* 1974; Parsons & Daly 1983). Thus, such flow might contribute the appropriate forces and bending moment to the Indian plate.

In my opinion, further analysis of elastic-plate models for the deep structure of the Himalaya is not likely to yield important constraints on the structure or on the dynamics of mountain building. Although the model of an elastic plate beneath the Indo–Gangetic Plains and the Lesser Himalaya accounts for the deep structure of these regions, surely other models, with more realistic rheologies, will also do so. To explain the gravity anomalies beneath the Greater Himalaya in terms of a flexed plate, the flexural rigidity must be considerably smaller in that area than beneath the Lesser Himalaya and farther south. Moreover, an elastic-plate model is unable to account for the support of the high range of mountains without an additional force being applied to the plate. Thus, in my opinion, to match the gravity anomalies with a structure satisfying mechanical equilibrium of a simple flexed elastic plate requires a knowledge of too many independent parameters; the trade-offs among them, especially given the mechanical oversimplification of a purely elastic plate, make the introduction of additional complexity, describable by additional parameters, unlikely to allow additional understanding without independent constraints on some of the parameters.

The value of the simple model of an elastic plate, like most simple physical models, lies as much in the aspects in which it fails as in those in which it succeeds. In this context, the recognition of the need for an additional force acting on the Indian plate and contributing to the support of the Himalaya is the most important implication.

## Deep structure of the Karakoram

The Karakoram Range lies at the narrow western end of the Tibetan Plateau, midway between the Indian Craton on the south and Tarim Basin on the north (figures 1 and 4). The Indian subcontinent is actively underthrusting the Himalaya, and the Tarim Basin has been (and probably continues to be) underthrust southward beneath the western Kunlun, the range on the northern edge of the plateau. Thus in the area of the Karakoram, crustal shortening occurs both north and south of the range and one might expect the convergence of the subcrustal portions of lithosphere to manifest itself as measurable geophysical anomalies. Because there is some evidence suggesting that the upper mantle is unusual beneath the Karakoram, I isolate this area from the discussions of Tibet and the Himalaya, given above, and describe these features.

### Gravity anomalies across the Karakoram

As part of the Sino–Swedish expeditions in the 1920s and 1930s, Amboldt (1948) measured gravity at a number of sites in the Karakoram, Kunlun and Tarim Basin. At present, his measurements are the only published ones north of the Karakoram. Later Marussi (1964) reanalysed the measurements, added terrain corrections, and also computed isostatic anomalies, for various assumed reference crustal thicknesses. He found the pattern described above for the Himalaya to be mirrored on the northern edge of Tibet: a negative isostatic anomaly marked the southern edge of the Tarim Basin and a positive isostatic anomaly characterized the northern edge of the Kunlun.

This pattern of mass deficit over a foreland basin and mass excess over the neighbouring range is reminiscent of the Himalaya (discussed above) and other ranges. Thus, aware that Norin (1946) had mapped folded Tertiary rock, overthrust by Cretaceous rock on the northern margin of the Kunlun, Lyon-Caen & Molnar (1984) pursued an analysis of the gravity anomalies in terms of the simple model of a semi-infinite elastic plate over an inviscid fluid, loaded solely by the weight of the range above the plate. To match the gravity anomalies with such a model, the plate must extend 80 km southward beneath the Kunlun Range, and probably 100 km or more. No upper limit could be placed on the southward extent of such a plate. Taken literally, this suggests a comparable amount of underthrusting, of at least 80 km, of the Tarim Basin beneath the Kunlun. Both because Amboldt's (1948) measurements of gravity and elevation are relatively inaccurate, and because the elastic plate model is an oversimplification, however, it probably is wiser merely to conclude that underthrusting of the Tarim Basin of some tens of kilometres (or more) has occurred.

Not only do the gravity anomalies on and adjacent to the ranges north and south of the Karakoram indicate marked deviations from local isostatic equilibrium, but so do those from the Karakoram itself. As noted above, the negative isostatic anomalies shown on Marussi's (1964) cross section over Indo–Gangetic Plains and the southern margin of the Tarim Basin probably can be explained by flexure of the Indian and Tarim plates to form foreland basins. The positive isostatic anomalies would be caused by the strength of the plates supporting the mountains and redistributing the buoyant response of their loads. Notice, however, that a negative isostatic anomaly characterizes the Karakoram Range (figure 21), implying that a mass deficit underlies the Karakoram. Marussi (1964) attributed this anomaly to a massive low-density granitic intrusion in the Karakoram, but I suspect that its source might be deeper.

We can estimate the mass deficit per unit length ($\delta M$) from the integral of the isostatic gravity anomaly ($\delta g$) over the horizontal distance

$$2\pi G \delta M = \int_{-\infty}^{\infty} \delta g(x) \, \mathrm{d}x,$$

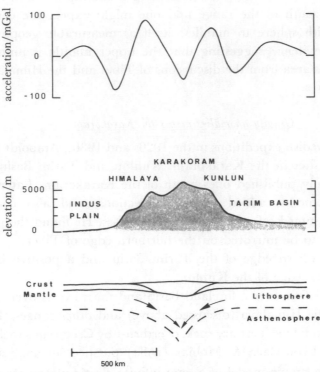

FIGURE 21. Sketches of isostatic anomalies, of topography and of a simple cross section across the Himalaya, the Karakoram and the Kunlun. Note the negative isostatic anomalies over the foreland basins of the Himalaya and the Kunlun, positive values over the Himalaya and the Kunlun, and negative anomalies over the Karakoram. I suspect that downwelling in the mantle holds the Karakoram out of isostatic equilibrium. (Gravity profile redrawn from Marussi 1964.)

where $2\pi G = 4.18 \times 10^{-5}$ mGal m$^2$ kg$^{-1}$. Ignoring other contributions to the gravity field, the minimum value of $-60$ mGal and the width of about 120 km for this anomaly would imply that the value of the integral is about $4 \times 10^6$ mGal m, corresponding to a mass deficit per unit length parallel to the chain of about $10^{11}$ kg m$^{-1}$. This mass deficit would contribute force per unit length of $10^{12}$ N m$^{-1}$, which is comparable in magnitude to the force applied to the end of the Indian Plate in the calculations of Lyon-Caen & Molnar (1983, 1985). Thus, neither the force nor the mass deficit is small.

The mass deficit per unit length, if caused by granite with a density 50 kg m$^{-3}$ less than the surrounding rock would imply a large cross-sectional area of 2000 km$^2$. Were it caused by an unusually deep Moho, with a density contrast of 500 kg m$^{-3}$, it would be consistent with an average excess thickness of crust of 4 km over a width of 50 km.

The narrow width of isostatic anomaly might at first suggest that its source could not be as deep as 70–80 km, near the Moho; but the data defining it are neither abundant nor as accurate as modern data. Moreover, if the crust were unusually deep beneath the Karakoram, there must be a force holding it down, and stresses induced by the downwelling of cold material would be a likely force (figure 21). The mass excess associated with this cold material would contribute a positive anomaly. Thus the measured isostatic anomaly would be the difference between two larger anomalies: a negative one due to the overthickening (with respect to isostatic equilibrium) of the crust, and a positive one due to the cold, thickened downwelling root of mantle lithosphere. Thus, both the dimensions and the amplitude of this negative isostatic anomaly might give underestimates of the possible deflection of the Moho.

Clearly the explanation of Marussi's (1964) negative isostatic anomaly over the Karakoram in terms of a thickened crust drawn down by a cold mantle root is not unique. Moreover, the anomaly is not well enough defined to make quantitative analysis useful. Nevertheless, I think it worth considering.

### Seismic studies of the Karakoram

The Karakoram is unusual by being underlain by a zone of intermediate depth earthquakes (Chen & Roecker 1981), and there is a hint that the shear wave velocity in the upper mantle is high (Brandon & Romanowicz 1986). Both suggest that the uppermost mantle is colder than normal.

#### Seismicity

Intermediate depth earthquakes (70 km $\leqslant$ depth $\leqslant$ 300 km) are very rare except where oceanic lithosphere has clearly been subducted since 10 Ma or so. Consequently, those few narrow and well-defined belts of intense seismic activity at intermediate depths in continental regions (Hindu Kush, Pamir, Romania and Burma) are usually ascribed to late-Cainozoic subduction of oceanic lithosphere, even where convincing geologic evidence for the suturing of continental fragments in late-Cainozoic time is absent. In addition, there are few areas of the earth where intermediate depth earthquakes have unequivocally occurred in the last 20 years, but where subduction of oceanic lithosphere since 20 Ma almost surely has not occurred (Chen & Molnar 1983; Chen et al. 1981; Hatzfeld & Frogneux 1981). Above, we noted two such earthquakes in southern Tibet. The Karakoram appears to be the intracontinental region with the most active diffuse intermediate depth seismicity, possibly with the largest of such earthquakes, and with many of the deepest (up to 100 km) (Chen & Molnar 1983; Chen & Roecker 1981). This zone does not appear to be a continuation of the seismic zone that dips southward beneath the Pamir and that may represent late-Tertiary subduction of oceanic lithosphere (see, for example, Billington et al. 1977; Roecker et al. 1980).

Estimates of temperatures of the material in which earthquakes in the mantle occur are virtually all less than 800 °C, and most are less than 600 °C (see, for example, Chen & Molnar 1983; Molnar et al. 1979). In so far as those estimates are reasonable, the occurrence of earthquakes at depths of 100 km beneath the Karakoram would imply very low average geothermal gradients of less than 8 K km$^{-1}$, and probably less than 6 K km$^{-1}$. Thus, the occurrence of these earthquakes implies a relatively cold upper mantle beneath the Karakoram.

#### Seismic-wave velocities

In addition to studying Rayleigh-wave phase velocities across Tibet (discussed above), Brandon & Romanowicz (1986) analysed paths across the Karakoram and Pamir. Although the structure along these paths is probably complex, and their main focus was not on this area, they could resolve unusually high phase velocities for paths across this area. In fitting measured and calculated phase velocities, they concluded that the shear wave velocity is approximately 5 km s$^{-1}$ in the uppermost mantle. This is an extremely high velocity, probably too high to be explained solely by a low temperature of the material. This I suspect that the velocity is not tightly constrained. Nevertheless, the relatively high phase velocity seems to be required, and therefore, so is a relatively high shear-wave velocity in the mantle.

In addition, among the earthquakes in the Himalaya and Tibet for which Molnar & Chen (1984) measured S-wave residuals, the earthquake with the largest negative residual occurred in the western Himalaya just southwest of the Karakoram (figure 9). These relatively early arrivals are also consistent material with relatively high velocities underlying the northwestern Himalaya (and the Karakoram); one earthquake, however, constitutes only a suggestion and certainly not proof.

### Summary of constraints on the deep structure of the Karakoram

Taken together, the existence of intermediate depth earthquakes, the high phase velocity of Rayleigh waves (Brandon & Romanowicz 1986), and one earthquake with relatively early S-waves suggest that the mantle beneath the Karakoram is unusually cold, and the gravity anomalies along a profile crossing the Karakoram can be interpreted in terms of cold material sinking beneath the range and drawing the Moho down (figure 21). This interpretation is certainly not unique, and the seismic data do not place a quantitative constraint on the degree to which the temperature in the mantle beneath the Karakoram is lower than that beneath a shield. What can be said is that the deep structure of the Karakoram is different both from that of most of Tibet (if not all of Tibet) and from that of the Himalaya, and that it is consistent with what one might expect of the uppermost mantle beneath a region of intense crustal shortening. As usual, there are enough data to suggest a plausible interpretation, but not to prove or to refine it.

### Summary and synthesis

Like that of all regions, the deep structure of Tibet, the Himalaya and the Karakoram is a consequence of geologic processes operating for millions to tens of millions of years. Because we can know the deep structure at only a brief instant in geologic time, the knowledge of it for any region can rarely be used to decide uniquely which among a number of proposed tectonic evolutions of the region is correct. Nevertheless, given the rather limited knowledge of the geologic history of the Himalaya and especially that of Tibet and the Karakoram, the deep structure of these regions provides an important constraint on their possible tectonic evolutions. Moreover, because gravity acting differently on density variations within Earth is almost surely the principal driving force of large-scale tectonic processes, a knowledge of lateral variations in structure is fundamental for understanding dynamic processes in the earth. This review has focused on these two aspects of the deep structure: (1) as a constraint on suggested and plausible tectonic evolutions of the world's highest plateau and mountain chain and (2) as a clue to understanding the dynamic processes that have built these extraordinary features.

Virtually all proposed scenarios for the evolution and for the support of the Himalaya and Tibet include the presumption that the crustal thickness increases from a relatively normal value beneath India to a value of about 70 km beneath the plateau. Geophysical evidence of a variety of kinds supports this presumption, in a general form. Seismic refraction profiles across the stable parts of India reveal an average crustal thickness of about 38 km (Kaila 1982). Bouguer-gravity anomalies become increasingly negative from the Central India Highlands, across the Indo–Gangetic Plains and the Himalaya, to southern Tibet (Das et al. 1979; Kono

1974; Qureshy *et al.* 1974). The few published measurements from Tibet and its margins indicate large negative values and are consistent with a very thick crust (Amboldt 1948; Marussi 1964). Wide-angle reflections from the Moho indicate an increasing depth of the Moho from about 55 km beneath the Greater Himalaya, and apparently shallower farther south, to 70–75 km beneath the southern edge of Tibet (Hirn & Sapin 1984; Hirn *et al.* 1984*a*; Lépine *et al.* 1984). Clear wide-angle reflections from studies farther north, in the southern half of Tibet, indicate a depth of the Moho of about 70 km in some, but not all, areas (Hirn *et al.* 1984*b*, *c*). Hirn, Lépine, Sapin and their colleagues have inferred the existence of steps, instead of smooth variations, in the Moho beneath the Himalaya and beneath Tibet, but as noted above, I remain unconvinced of these steps. Finally, surface-wave dispersion across most of Tibet corroborates the existence of a thick crust over most of Tibet (Chen 1979; Chen & Molnar 1981; Jobert *et al.* 1985; Romanowicz 1982). Thus, to a good first approximation, it is reasonable to assume that Tibet is underlain by a thick crust.

The average crustal thickness of Tibet appears to be 65–70 km. It clearly is not as great as 80 km or as little as 55 km. Locally the thickness seems to reach 75 km, just north of the Greater Himalaya and south of the Indus–Tsangpo Suture Zone (Hirn & Sapin 1984). In north–central Tibet, in an area a few hundred kilometres in dimension, the thickness appears to be only 50–60 km (Brandon & Romanowicz 1986). Hirn *et al.* (1984*b*) recorded strong variations in the wide-angle reflections that they associate with lateral variations in crustal thickness of as much as 15–30 km over distances of tens of kilometres. The crustal thickness beneath the Greater Himalaya, where the highest elevations are found but also where the mean elevation is similar to that beneath southern Tibet (Bird 1978), is about 55 km (Hirn & Sapin 1984; Lépine *et al.* 1984), much smaller than that farther north. Thus, although to a good first approximation the crustal thickness of Tibet is double that of most areas and this large thickness compensates for the remarkably uniform, high elevation of the plateau, both marked lateral heterogeneity and large deviations from isostatic equilibrium must exist beneath the Himalaya and Tibet.

The recognition of marked lateral heterogeneity in the mantle beneath Tibet is one of the important discoveries of the structure of Tibet made in the 1980s and suggests that simple Airy-type isostatic compensation does not prevail over all of Tibet. The variations in surface-wave dispersion that indicate a relatively thin crust beneath north–central Tibet also imply that the shear wave velocity in the underlying uppermost mantle is much lower there than elsewhere beneath Tibet and the Himalaya (Brandon & Romanowicz 1986). Large delays of S-waves from earthquakes in north–central Tibet and recorded at teleseismic distances indicate that the mean shear-wave velocity in the uppermost mantle is 4–8% lower than that beneath the Himalaya (Molnar & Chen 1984; Pettersen & Doornbos 1987). Waveforms and travel times of shear waves recorded at regional distances (less than 30°) corroborate the inference that the material with the low shear-wave velocity is in the upper 250 km of the Earth and not deeper (Lyon-Caen 1986). Finally, the phase Sn, which propagates in the upper 100 km or so of the mantle, is attenuated on paths crossing north–central Tibet, suggesting very low Q in the uppermost mantle of this area (Ni & Barazangi 1983). Taken together, the low shear-wave velocities and the high attenuation in the mantle suggest that the mantle beneath north–central Tibet is unusually hot, an inference that accords with the existence of recent basic volcanism in this part of Tibet (Burke *et al.* 1974; Deng 1978; Kidd 1975; Molnar *et al.* 1987*a*).

Except for the volcanism, the topography of northern Tibet gives no hint of a relatively thin crust or a relatively hot upper mantle. Thus I presume that the high, uniform elevation is isostatically compensated. (There are no published gravity measurements to test this.) Clearly the thinner crust suggests that if Airy isostasy prevails elsewhere in Tibet, it does not do so in north–central Tibet. Correspondingly the inference of high temperature in the uppermost mantle suggests that relatively light material in the mantle contributes to this compensation. Thus, in my view, the relatively uniform height of the Tibetan Plateau masks an important feature: the existence of relatively thin crust overlying relatively hot uppermost mantle beneath north–central Tibet.

The variation in crustal thickness implies that whatever the process is that built the Tibetan Plateau, it did not do so with the uniformity implied by the uniform elevation. If the Indian crust slid beneath Tibet, it probably has not slid beneath northern Tibet. If crustal shortening, within what was southern Eurasia before the collision, built the plateau, then the shortening has not been uniform across the plateau. In general, the existence of lateral variations in temperature in the mantle is both a prerequisite for and a direct manifestation of thermal convection. Thus I suspect that north–central Tibet is the locus of convective upwelling of hot material in the upper mantle (figure 22). The corresponding downwelling would lie astride the upwelling, probably beneath southern Tibet and perhaps also beneath northernmost Tibet.

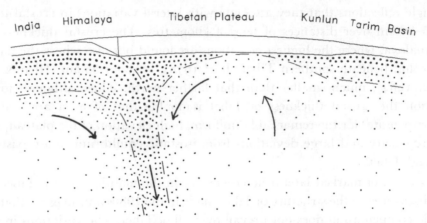

FIGURE 22. Idealized cross section across the Himalaya and Tibet illustrating differences in crustal thickness and origin, as well as hypothetical subduction of Indian mantle lithosphere and induced convective circulation in the asthenosphere.

The existence of cold material in the upper mantle is implied by the shield structures beneath India and beneath the Tarim Basin. Moreover, the evidence for active crustal shortening in the Himalaya and on the northern margin of Tibet (Molnar *et al.* 1987*a*, *b*) imply an active underthrusting of the shields beneath these margins of Tibet. Thus we might look to the transitions from these shield structures at the margins of Tibet to search for evidence of cold downwelling material in the mantle.

A study of shear waves crossing the Indian Shield confirms that it is underlain by a thick high-velocity layer with only a minor low-velocity zone beneath it (Lyon-Caen 1986). Several observations imply that this thick lithosphere has been thrust beneath the Himalaya, at least 80 km and probably more than 100 km. Fault-plane solutions and focal depths of many of the moderate earthquakes suggest that these events have occurred on the top surface of the intact Indian plate, some 80–100 km north of the front of the range (Baranowski *et al.* 1984; Ni &

Barazangi 1984). Gravity anomalies over the Indo–Gangetic Plain imply that the underlying Ganga Basin has formed by the flexure of the Indian plate, and those from the Lesser Himalaya imply that the plate extends 100 km beneath this part of the range, with the Moho dipping at the same gentle angle (*ca.* 3–5°) as beneath the Ganga Basin (Lyon-Caen & Molnar 1983, 1985). Travel times of P-waves recorded at the Tarbella array, which spans the Lesser Himalaya in part of Pakistan, also imply a gentle dip of about 4° of the Moho (Menke 1977). Finally, arrival times of S-waves from earthquakes in the Himalaya are consistent with that area being underlain by a shield-like upper mantle (Molnar & Chen 1984).

The structure of the mantle beneath the Greater Himalaya and southernmost Tibet is less unequivocally defined. There is little doubt that crustal thickness increases from south to north. Whereas Hirn & Sapin (1984) and Lépine *et al.* (1984) inferred that this increase is manifested by steps in the Moho at southward-dipping thrusts faults, I think that some of their data, at least, can be interpreted in terms of a smooth increase in the depth of a northward dipping Moho. In either case, gravity anomalies imply that the average dip of the Moho is steeper beneath the Greater Himalaya ($\delta \approx 15°$) than beneath the Lesser Himalaya ($\delta \approx 3–5°$) (Kono 1974; Lyon-Caen & Molnar 1983, 1985; Warsi & Molnar 1977). If the Indian plate extends beneath the Greater Himalaya, this increase in dip requires that the plate be bent more sharply than it is beneath the Lesser Himalaya. The bending of the plate seems to be the result of dynamic forces acting on the mantle lithosphere north of the Greater Himalaya (Lyon-Caen & Molnar 1983, 1985).

The upper-mantle structure of southern Tibet near the Indus–Tsangpo Suture Zone is uncertain because of limited data and conflicting interpretations. There is a suggestion of a very high Pn velocity (8.5–8.7 km s$^{-1}$), which might indicate a deep high-velocity layer (Hirn & Sapin 1984; Menke & Jacob 1976). The occurrence of two earthquakes at depths of 90 km beneath southern Tibet suggests the existence of relatively low temperatures there (Chen *et al.* 1981; Molnar & Chen 1983). It is my prejudice that mantle beneath southernmost Tibet is relatively cold (figure 22), but neither of these observations are strong evidence for this prejudice. Moreover, it conflicts with the inference from long-period surface-wave dispersion that the high-velocity layer in the uppermost mantle is thin and extends to a depth of only 100 km or so (Jobert *et al.* 1985).

The southern margin of the low-velocity region of north–central Tibet is also poorly defined. The phases Pn and Sn propagate efficiently and with high velocities through the uppermost mantle of central, southern, eastern and western Tibet (Ni & Barazangi 1983). Delays of S-waves, both to distant stations (more than 30°) (Molnar & Chen 1984; Pettersen & Doornbos 1987) and to regional distances (less than 30°) from the few earthquakes studied in central and southern Tibet (Lyon-Caen 1986), suggest that the average upper-mantle velocity beneath central and southern Tibet is lower than that beneath the Himalaya and higher than that beneath north–central Tibet. Although these observations could be taken as evidence that the thickness of the high-velocity layer in the uppermost mantle increases southward from central to southern Tibet (figure 22), the sparsity of observations does not require this to be so.

The lateral variations in the mantle beneath Tibet are too poorly resolved to allow a quantitative, or even qualitatively definitive, constraint on the processes that built Tibet. The S-wave travel times, in particular, indicate that the structure beneath most of Tibet (more than half and probably at least two thirds) is not that of a shield. Therefore, if the Indian Shield did underthrust most or all of Tibet, the mantle portion of that lithosphere has been modified substantially since that underthrusting occurred. Neither the absence of a shield-like upper-mantle structure, nor any of the arguments that follow prove that India has not underthrust all of Tibet, but obviously they require that proponents of that view explain where their shield has gone.

The present deep structure of Tibet and the Himalaya are consistent with the Indian Shield underthrusting only the southernmost 100–300 km of Tibet and with the thick crust being a consequence of crustal shortening within the ancient southern margin of Eurasia (Dewey & Burke 1973; England & Houseman 1986). When crustal shortening and crustal thickening occur, the underlying mantle lithosphere initially attached to the crust, must also undergo horizontal shortening and thickening. The thickening of the mantle lithosphere advects cold isotherms downwards, and therefore produces horizontal temperature gradients. The advection of this cold material, therefore, is likely to induce thermal convection in the asthenosphere that will draw the lower part of the neighbouring lithosphere into the downgoing limb of the cell (Fleitout & Froidevaux 1982; Houseman et al. 1981). This material will be replaced by hotter material from below. England & Houseman (1988) infer that recent uplift of Tibet is caused by such a replacement of the cold, thickened lithospheric root by such material. My image of the mantle dynamics beneath Tibet includes downwelling beneath southern Tibet, upwelling beneath north–central Tibet, and possibly downwelling also beneath the northern margin of Tibet (figure 22).

The downwelling on the southern margin of Tibet should be enhanced by the continued underthrusting of the Indian Shield beneath the Himalaya. As noted above, both gravity anomalies over the Lesser Himalaya and wide-angle reflection profiles over the Greater Himalaya indicate that the crustal thickness beneath the Himalaya is much smaller than it would be if local Airy-type isostatic equilibrium prevailed. The flexure of a strong Indian plate helps to support the high range of mountains, but numerical experiments with plates of various flexural rigidities show that the strength of the plate cannot by itself support the range; an additional force is necessary (Lyon-Caen & Molnar 1983, 1985). A bending moment applied to underthrust the India plate, or perhaps some other dynamic process, could flex the plate beneath the Lesser Himalaya up a few kilometres to shallower depths than it would be if it were depressed solely by the weight of the Himalaya. The moment could be produced by the downward force of gravity acting on the cold slab of India's mantle lithosphere from which all or much of the crust has been detached at the Himalaya (figure 22). The plunging of this mantle lithosphere, in turn, would help to draw the lower part of southern Tibet's mantle lithosphere into the asthenosphere.

This interpretation of the mantle dynamics beneath the Himalaya and Tibet is by no means proven and is surely an oversimplification. The only evidence even suggestive of subduction of Indian mantle lithosphere beneath southern Tibet is the indirect evidence from gravity anomalies interpreted in terms of a flexed plate acted upon by an additional force to support the Himalaya.

If both of these processes have occurred in Asia (shortening of Tibet's underlying mantle lithosphere and underthrusting of India's mantle lithosphere beneath southern Tibet), the area where manifestations of these processes are likely to be clearest is near the Karakoram. In that area, both northward underthrusting of the India Shield beneath the Himalaya and southward underthrusting of the Tarim Basin shield beneath the Kunlun (Lyon-Caen & Molnar 1984; Norin 1946) seem to continue. This is the area where diffusely distributed intermediate depth seismicity is most prevalent on Earth (Chen & Roecker 1981), suggesting both rapid deformation and low temperatures in the upper mantle. Moreover, shear-wave velocities in the upper 100 km of the mantle appear to be unusually high (Brandon & Romanowicz 1986), also suggesting low temperatures there. Thus, what little evidence exists is consistent with the presence of a root to the bottom of the mantle lithosphere, if not active downwelling.

The hypothesized dynamics of the mantle described above do not constitute original ideas; variations of them have been proposed before both as general processes (Fleitout & Froidevaux 1982; Houseman *et al.* 1981) and specifically for the Alps (Panza & Mueller 1979), the Apennines (Royden & Karner 1984), the Pyrenees (Brunet 1986) and the transverse ranges of California (Humphreys *et al.* 1984; Sheffels & McNutt 1986). In these areas where the geologic histories and geologic structures are relatively well known, the geophysical signatures of the structure of the mantle, however, appear to be more subtle than in the Himalaya, the Karakoram or Tibet, where the geologic history is still only crudely known. Thus, because only cruder methods than can be applied to these other areas, nevertheless, reveal large lateral variations in the deep structure of Asia, it is likely that with newer techniques, the resolution of such inhomogeneities beneath the Himalaya, the Karakoram, and Tibet will enhance our understanding of the dynamics of mountain building more than the refinement of the deep structure of more minor mountain belts. Thus continued scientific pilgrimages to the birthplace of geophysics as a geological or tectonic tool and the furtherance of a tradition established by Airy, Everest and Pratt more than 100 years ago are likely to provide a key to the understanding of the dynamics of mountain building.

I thank M. Barazangi, H. Lyon-Caen, and B. Romanowicz for providing me with figures from their published papers, and Dan McKenzie, F.R.S., for critically reading the manuscript. This research was supported in part by the Centre National de Recherches Scientifiques of France, by the National Science Foundation under grant 8500810-EAR and by the National Aeronautical and Space Administration under grant NAG4-795.

## REFERENCES

Acharya, S. K. & Ray, K. K. 1982 *Bull. Am. Ass. Petrol. Geol.* **66**, 57–70.

Airy, G. B. 1855 *Phil. Trans. R. Soc. Lond.* **145**, 101–104.

Allègre, C. J. *et al.* 1984 *Nature, Lond.* **307**, 17–22.

Amboldt, N. 1948 *Reports from Scientific Expedition to Northwestern Provinces of China under the Leadership of Dr. Sven Hedin, Publ. 30, Geodes.* Stockholm: Tryckeri Aktiebolaget Thule. (112 pages.)

Anderson, O. L., Schreiber, E., Liebermann, R. C. & Soga, N. 1968 *Rev. Geophys.* **6**, 491–524.

Argand, E. 1924 *Int. Geol. Cong. Rep. Sess.* **13** (1), 170–372.

Baranowski, J., Armbruster, J., Seeber, L. & Molnar, P. 1984 *J. geophys. Res.* **89**, 6918–6928.

Barazangi, M. & Ni, J. 1982 *Geology* **10**, 179–185.

Beloussov, V. V., Belyaevsky, N. A., Borisov, A. A., Volvovsky, B. S., Volkovsky, I. S., Resvoy, D. P., Tal-Virsky, B. B., Khamrabaev, I. Kh., Kaila, K. L., Narain, H., Marussi, A. & Finetti, J. 1980 *Tectonophysics* **70**, 193–221.

Billington, S., Isacks, B. L. & Barazangi, M. 1977 *Geology* **5**, 699–704.

Bird, G. P. 1976 Ph.D. thesis, Massachusetts Institute of Technology, U.S.A.

Bird, P. 1978 *J. geophys. Res.* **83**, 4975–4987.

Black, P. R. & Braile, L. W. 1982 *J. geophys. Res.* **87**, 10557–10568.

Brandon, C. & Romanowicz, B. 1986 *J. geophys. Res.* **91**, 6547–6564.

Brunet, M. F. 1986 *Tectonophysics* **129**, 343–354.

Burke, K. C. A., Dewey, J. F. & Kidd, W. S. F. 1974 *Geol. Soc. Am. Abstr. Programs* **6**, 1027–1028.

Cauët, R., Serryn, P. & Vincent, M. 1957 *Petit Atlas Bordas: la France, le Monde.* Paris: SCIP.

Chen, W.-P. 1979 Ph.D. thesis, Massachusetts Institute of Technology, U.S.A.

Chen, W.-P. & Molnar, P. 1981 *J. geophys. Res.* **85**, 5937–5962.

Chen, W.-P. & Molnar, P. 1983 *J. geophys. Res.* **88**, 4183–4214.

Chen, W.-P., Nabelek, J. L., Fitch, T. J. & Molnar, P. 1981 *J. geophys. Res.* **86**, 2863–2876.

Chen, W.-P. & Roecker, S. W. 1981 *Eos, Wash.* **61**, 1031.

Choudhury, S. K. 1975 *Geophys. Jl R. astr. Soc.* **40**, 441–452.

Chun, K.-Y. & McEvilly, T. V. 1986 *J. geophys. Res.* **91**, 10405–10411.

Chun, K.-Y. & Yoshii, T. 1977 *Bull. seism. Soc. Am.* **67**, 737–750.

Curray, J. R., & Moore, D. G. 1974 In *The geology of continental margins* (ed. C. A. Burk & C. L. Drake), pp. 617–628, New York: Springer-Verlag.

Das, D., Mehra, G., Rao, K. G. C., Roy, A. L. & Naraya, M. S. 1979 *Himalayan Geology Seminar, Section III, Oil and Natural Gas Resources, Geol. Surv. India,* misc. publ. **41**, 141–148.

Deng Wanming 1978 *Acta geol. sin.* **2**, 148–162.

Dewey, J. F. & Burke, K. C. A. 1973 *J. Geol.* **81**, 683–692.

Doyle, H. A. & Hales, A. L. 1967 *Bull. seism. Soc. Am.* **57**, 761–771.

Duroy, Y., Farah, A., Malinconico, L. L. & Lillie, R. J. 1988 In *Tectonics and geophysics of the Western Himalaya* (ed. L. L. Malinconico & R. J. Lillie). Geol. Soc. Amer. Spec. Pap. (In the press.)

England, P. C. & Houseman, G. A. 1986 *J. geophys. Res.* **91**, 3664–3676.

England, P. C. & Houseman, G. A. 1988 *J. geophys. Res.* (In the press.)

Ewing, M. & Press, F. 1959 *Bull. geol. Soc. Am.* **70**, 229–244.

Feng, C.-C. 1982 Ph.D. thesis, University of Southern California, Los Angeles.

Feng, C.-C. & Teng, T.-L. 1983 *J. geophys. Res.* **88**, 2261–2272.

Fleitout, L. & Froidevaux, C. 1982 *Tectonics* **1**, 21–56.

Francheteau, J., Jaupart, C., Shen Xianjie, Kang Wenhua, Lee Defu, Bai Jiachi, Wei Hungpin & Deng Hsiayeu 1984 *Nature, Lond.* **307**, 32–36.

Grand, S. & Helmberger, D. V. 1984 *Geophys. Jl R. astr. Soc.* **76**, 399–438.

Gupta, H. K. & Narain, H. 1967 *Bull. seism. Soc. Am.* **57**, 235–248.

Hatzfeld, D. & Frogneux, M. 1981 *Nature, Lond.* **292**, 443–445.

Herrin, E. 1968 *Bull. seism. Soc. Am.* **58**, 1193–1241.

Hirn, A. *et al.* 1984*a Nature, Lond.* **307**, 23–25.

Hirn, A., Jobert, G., Wittlinger, G., Xu Zhongxin & Gao Enyuan 1984*c Ann. Geophsicae* **2**, 113–117.

Hirn, A., Nercessian, A., Sapin, M., Jobert, G., Xu Zhongxin, Gao Enyuan, Lu Deyuan & Teng Jiwen 1984*b Nature, Lond.* **307**, 25–27.

Hirn, A. & Sapin, M. 1984 *Ann. Geophysicae* **2**, 123–130.

Houseman, G. A., McKenzie, D. P. & Molnar, P. 1981 *J. geophys. Res.* **86**, 6115–6132.

Huestis, S., Molnar, P. & Oliver, J. 1973 *Bull. seism. Soc. Am.* **63**, 469–475.

Humphreys, E., Clayton, R. W. & Hager, B. H. 1984 *Geophys. Res. Lett.* **11**, 625–627.

Institute of Geophysics, Academy of Sciences 1981 *Acta geophys. Sin.* **24**, 155–170.

Jaupart, C., Francheteau, J. & Shen Xian-jie 1985 *Geophys. Jl R. astro. Soc.* **81**, 131–155.

Jeffreys, H. & Bullen, K. E. 1940 *Seismological tables.* British Association Gray–Milne Trust.

Jia Su-juan, Cao Xue-feng & Yan Jia-quan 1981 *N West. seism. J.* **3**, 27–34.

Jobert, N., Journet, B., Jobert, G., Hirn, A. & Zhong, S. K. 1985 *Nature, Lond.* **313**, 386–388.

Johnson, B. D., Powell, C. M. & Veevers, J. J. 1976 *Bull. geol. Soc. Am.* **87**, 1560–1566.

Kaila, K. L. 1982 *Bull. geophys. Res.* **20**, 309–328.

Kaila, K. L., Krishna, V. G., Choudhury, K. & Narain, H. 1978 *J. geol. Soc. India* **19**, 1–20.

Karner, G. D. & Watts, A. B. 1983 *J. geophys. Res.* **88**, 10449–10477.

Karunakaran, C. & Ranga Rao, A. 1979 *Contributions to Stratigraphy and Structure, Himalayan Geology Seminar, Section III, Oil and Natural Gas Resources, Geol. Surv. India,* misc. publ. no. 41, pp. 1–66.

Kidd, W. S. F. 1975 *Eos, Wash.* **56**, 453.

Klootwijk, C. T., Conaghan, P. J. & Powell, C. M. 1985 *Earth planet. Sci. Lett.* **75**, 167–183.

Kono, M. 1974 *Geophys. Jl R. astr. Soc.* **39**, 283–300.

Lépine, J.-C., Hirn, A., Pandey, M. R. & Tater, J. M. 1984 *Ann. Geophysicae* **2**, 119–121.

Lillie, R. J. & Yousuf, M. 1986 In *Reflection seismology: the continental crust.* Am. Geophys. Un., Geodynamic Ser., vol. 14 (ed. M. Barazangi & L. Brown), pp. 55–65.

Lin Jian & Parmentier, E. M. 1985 *Geophys. Res. Lett.* **12**, 357–360.

Lyon-Caen, H. 1986 *Geophys. Jl R. astr. Soc.* **86**, 727–749.

Lyon-Caen, H. & Molnar, P. 1983 *J. geophys. Res.* **88**, 8171–8191.

Lyon-Caen, H. & Molnar, P. 1984 *Geophys. Res. Lett.* **11**, 1251–1254.

Lyon-Caen, H. & Molnar, P. 1985 *Tectonics* **4**, 513–538.

Marussi, A. 1964 *Geophysics of the Karakorum, Italian Expeditions to the Karakorum (K²) and Hindu Kush, Sci. Rept. 2, 1.* (242 pages.)

McKenzie, D. P. 1967 *J. geophys. Res.* **72**, 6261–6273.

McKenzie, D. 1977 *Geophys. Jl R. astr. Soc.* **48**, 211–238.

McKenzie, D. P., Roberts, J. M. & Weiss, N. O. 1974 *J. Fluid Mech.* **62**, 465–538.

Menke, W. H. 1977 *Bull. seism. Soc. Am.* **67**, 725–734.

Menke, W. H. & Jacob, K. H. 1976 *Bull. seism. Soc. Am.* **66**, 1695–1711.

Mishra, D. C. 1982 *Earth planet. Sci. Lett.* **57**, 415–420.

Molnar, P. 1984 *A. Rev. Earth planet. Sci.* **12**, 489–518.

Molnar, P. 1988 *Ann. Geophys.* (In the press.)

Molnar, P., Burchfiel, B. C., Liang K'uangyi & Zhao Ziyun 1987*b Geology* **15**, 249–253.

Molnar, P., Burchfiel, B. C., Zhao Ziyun, Liang K'uangyi, Wang Shuji & Huang Minmin 1987*a Science, Wash.* **235**, 299–305.

Molnar, P. & Chen, W.-P. 1982 In *Mountain building processes* (ed. K. Hsü). London: Academic Press.

Molnar, P. & Chen, W.-P 1983 *J. geophys. Res.* **88**, 1180–1196.

Molnar, P. & Chen, W.-P. 1984 *J. geophys. Res.* **89**, 6911–6917.

Molnar, P., Freedman, D. & Shih, J. S. F. 1979 *Geophys. Jl R. astr. Soc.* **56**, 41–54.

Molnar, P. & Oliver, J. 1969 *J. geophys. Res.* **74**, 2648–2682.

Molnar, P. & Tapponnier, P. 1978 *J. geophys. Res.* **83**, 5361–5375.

Ni, J. & Barazangi, M. 1983 *Geophys Jl R. astr. Soc.* **72**, 665–689.

Ni, J. & Barazangi, M. 1984 *J. geopys. Res.* **89**, 1147–1163.

Norin, E. 1946 *Reports from the Scientific Expedition to the Northwestern Provinces of China under the Leadership of Dr. Sven Hedin, Publ. 29, (III), Geology 7.* Stockholm: Tryckeri Aktiebolaget Thule. (214 pages.)

Pandey, M. R. 1981 *J. Nepal geol. Soc.* **1**, 29–35.

Pandey, M. R. 1986 *J. Nepal geol. Soc.* **6**, 1–11.

Panza, G. F. & Mueller, S. 1979 *Mem. Sci. Geol.* **33**, 43–50.

Parsons, B. & Daly, S. 1983 *J. geophys. Res.* **88**, 1129–1144.

Patton, H. 1980*a J. geophys. Res.* **85**, 821–848.

Patton, H. 1980*b Rev. Geophys. Space Phys.* **18**, 605–625.

Pettersen, Ú. & Doornbos, D. J. 1987 *Phys. Earth planet. Int.* **47**, 125–136.

Pham, V. N., Boyer, D., Therme, P., Xue Cheng-yuan, Li Li & Guo Yuan-jin, 1986 *Nature, Lond.* **319**, 310–314.

Pines, I., Teng, T., Rosenthal, R. & Alexander, S. 1980 *J. geophys. Res.* **85**, 3829–3844.

Powell, C. M. & Conaghan, P. J. 1973 *Earth planet. Sci. Lett.* **20**, 1–12.

Powell, C. M. & Conaghan, P. J. 1975 *Geology* **3**, 727–731.

Pratt, J. H. 1855 *Phil. Trans. R. Soc. Lond.* **145**, 53–100.

Qureshy, M. N. 1969 *Tectonophysics* **7**, 137–157.

Qureshy, M. N. & Warsi, W. E. K. 1980 *Geophys. Jl R. astr. Soc.* **61**, 235–242.

Qureshy, M. N., Venkatachalam, S. V. & Subrahmanyam, C. 1974 *Bull. geol. Soc. Am.* **85**, 921–926.

Raiverman, V., Kunte, S. V. & Mukherjee, A. 1983 *Petrol. Asia J.* **6**, 67–92.

Richardson, S. W. & England, P. C. 1979 *Earth planet. Sci. Lett.* **42**, 183–190.

Roecker, S. W., Soboleva, O. V., Nersesov, I. L., Lukk, A. A., Hatzfeld, D., Chatelain, J.-L. & Molnar, P. 1980 *J. geophys. Res.* **85**, 1358–1364.

Romanowicz. B. 1982 *J. geophys. Res.* **87**, 6865–6883.

Royden, L. & Karner, G. D. 1984 *Nature, Lond.* **309**, 142–144.

Sapin, M., Wang Xiangjing, Hirn, A. & Xu Zhongxin 1985 *Ann. Geophysicae* **3**, 637–646.

Sastri, V. V, Bhandari, L. L., Raju, A. T. R. & Datta, A. K. 1971 *J. geol. Soc. India* **12**, 223–233.

Schreiber, E. & Anderson, O. L. 1967*a J. geophys. Res.* **72**, 762–764.

Schreiber, E. & Anderson, O. L. 1967*b J. geophys. Res.* **72**, 3751.

Seeber, L. & Armbruster, J. 1981 In *Earthquake Prediction: An International Review, Maurice Ewing Series 4* (ed. D. W. Simpson & P. G. Richards), pp. 259–277. Washington, D.C.: Am. Geophys. Un.

Seeber, L., Armbruster, J. & Quittmeyer, R. 1981 In *Zagros, Hindu-Kush, Himalaya, Geodynamic Evolution, Geodyn. Ser.*, vol. 3, pp. 215–242. Washington, D.C.: Am. Geophys. Un.

Sengör, A. M. C. & Kidd, W. S. F. 1979 *Tectonphysics* **55**, 361–376.

Sheffels, B. & McNutt, M. 1986 *J. geophys. Res.* **91**, 6419–6431.

Sipkin, S. A. & Jordan, T. H. 1975 *J. geophys. Res.* **80**, 1474–1484.

Sipkin, S. A. & Jordan, T. H. 1980 *J. geophys. Res.* **85**, 853–861.

Smithson, S. B., Brewer, J. A., Kaufman, S., Oliver, J. E. & Hurich, C. A. 1979 *J. geophys. Res.* **84**, 5955–5972.

Soga, N., Schreiber, E. & Anderson, O. L. 1966 *J. geophys. Res.* **71**, 5315–5320.

Tang Bo-xiong, Liu Yuan-long, Zhang Li-min, Zhou Wen-hu & Wang Qian-shen 1981 In *Geological and ecological studies of Qinghai–Xizang Plateau*, pp. 683–689. Beijing: Science Press.

Teng Ji-wen 1981 In *Geological and ecological studies of Qinghai–Xizang Plateau*, pp. 633–649. Beijing: Science Press.

Teng Ji-wen *et al.* 1981 In *Geological and ecological studies of Qinghai–Xizang Plateau*, pp. 691–709. Beijing: Science Press.

Teng Ji-wen, Sun Ke-zhong, Xiong Shao-bai, Yin Zhou-xun, Yao Hung & Chen Li-fang 1983 *Phys. Earth planet. Int.* **31**, 293–306.

Teng J., Wang S., Yao Z., Xu Z., Zhu Z., Yang B. & Zhou W. 1980 *Acta Geophys. Sin.* **23**, 254–268.

Teng Ji-wen, Xiong Shao-bai, Yin Zhou-xun, Xu Zhong-xin, Wang Xiang-jing & Lu De-yuan 1985 *J. Phys. Earth* **33**, 157–171.

Tung, J. P. & Teng, T. L. 1974 *Eos, Wash.* **55**, 359.

Turcotte, D. L. & Schubert, G. 1982 *Geodynamics: applications of continuum mechanics to geological problems*, pp. 104–133. New York: John Wiley and Sons.

Wang, Chi-yuen, Shi, Y. & Zhou, W. 1982 *J. geophys. Res.* **87**, 2949–2957.

Warsi, W. E. K. & Molnar, P. 1977 In *Himalaya: sciences de la terre* (Colloques Internationaux du CNRS, no. 268), pp. 463–478. Paris: Editions du Centre National de la Recherche Scientifique.

Zhao, W.-L. & Morgan, W. J. 1985 *Tectonics* **4**, 359–369.

Zhou Wen-hu, Yang Zhan-shou, Zhu Hong-bin & Wu Li-gao 1981 In *Geological and ecological studies of Qinghai–Xizang Plateau*, pp. 673–682. Beijing: Science Press.

Phil. Trans. R. Soc. Lond. A **326**, 89–116 (1988)     [ 89 ]

Printed in Great Britain

# Folding and imbrication of the Indian crust during Himalayan collision

By M. P. Coward[1], R. W. H. Butler[2], A. F. Chambers[1], R. H. Graham[3], C. N. Izatt[1], M. A. Khan[1], R. J. Knipe[4], D. J. Prior[4], P. J. Treloar[1] and M. P. Williams[1]

[1] *Department of Geology, Imperial College, London SW7 2BP, U.K.*
[2] *Department of Earth Sciences, Open University, Walton Hall, Milton Keynes MK7 6AA, U.K.*
[3] *B.P. Petroleum, Britannic House, Moor Lane, London EC2, U.K.*
[4] *Department of Earth Sciences, University of Leeds, Leeds LS2 5JJ, U.K.*

India collided with a northern Kohistan–Asian Plate at about 50 Ma ago, the time of ocean closure being fairly accurately defined from syntectonic sediments as well as the effect on magnetic stripes on the Indian Ocean floor. Since collision, Asia has over-ridden India, developing a wide range of thrust scrapings at the top of the Indian Plate. Sections through the imbricated sedimentary cover suggest a minimum displacement of over 500 km during Eocene to recent plate convergence. This requires the Kohistan region to the north to be underlain by underthrusted middle to lower Indian crust, deformed by ductile shears and recumbent folds. These structures are well seen in the gneisses immediately south of the suture, where they are uplifted in the Indus and Nanga Parbat syntaxes. Here there are several phases of thrust-related small-scale folding and the development of a large folded thrust stack involving basement rocks, the imbrication of metamorphic zones and the local development of large backfolds. Some of the important local structures: the large late backfolds, the Salt Ranges and the Peshawar Basin, can all be related to the necessary changes in thrust wedge shape as it climbs through the crust and the three dimensional nature of the thrust movements associated with interference between the Kohistan and western Himalayan trends.

## INTRODUCTION

The Kohistan region of northern Pakistan (figure 1) is one of the best regions in which to study Himalayan tectonics, because (i) the area is relatively accessible, in that it is crossed by numerous valleys including those of the major Indus and Swat Rivers and (ii) there is a full section of the mountain belt preserved within one country, from the suture zone between Indian and Asian Plates to frontal thrust ranges. The area has been studied by reconnaissance mapping and detailed surveys over the past eight years, by research teams from the U.S.A. (notably Dartmouth College and Corvallis, Oregon) and from the U.K. (notably Leicester, Leeds and Imperial College), working in close collaboration with the Universities of Peshawar and the Punjab and the Pakistan Geological Survey. This paper summarizes some aspects of structure of the overthrust Indian Plate, mainly with the results of the Anglo–Pakistani research. Summaries of the deformation in the northern plates are given by Coward *et al.* (1986, 1987) and Pudsey *et al.* (1986).

The suture between Indian and Asian Plates, often known as the Main Mantle Thrust or

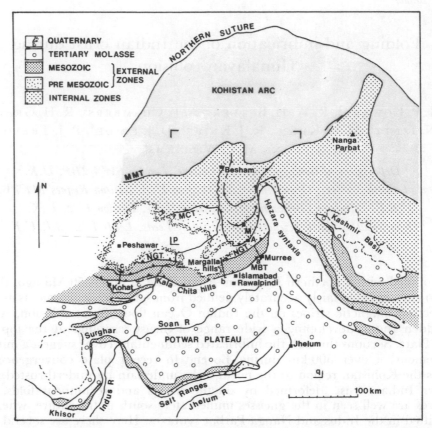

FIGURE 1. Location map for northern Pakistan, showing the distribution of the main thrusts (MMT, Main Mantle Thrust; MCT, Main Central Thrust; MBT, Main Boundary Thrust; NGT, Nathia Gali Thrust), the Kohistan island arc and the internal and external zones; M, Mansehra, A, Abbottabad. Box p shows the location of figure 5, box q shows figure 15. Sections lines a, b and c are shown in figure 11.

MMT (Tahirkheli & Jan 1979; Bard *et al.* 1980) can be mapped as an irregular but generally northward-dipping thrust, folded around later structures developed in the underlying Indian Plate (figure 1). On its hanging-wall are metamorphosed basic and ultrabasic rocks of the Kohistan island arc (Bard *et al.* 1980; Bard 1983), which form a terrane accreted to the Asian Plate during the middle Cretaceous (Coward *et al.* 1986, 1987).

The docking of the Kohistan island arc with the Asian Plate was followed by major folding and shearing of the lower part of the arc and then ocean subduction south of the arc, generating the early Tertiary Kohistan calc-alkaline batholith (Coward *et al.* 1986). Ocean closure occurred at about 50 Ma, the date being constrained by (i) magnetic anomaly patterns of the Indian Ocean, which show a dramatic slowing in the spreading rate at this time (Molnar & Tapponnier 1975; Patriat & Achache 1984) and (ii) the ages of the earliest post-collisional sediments in the suture zone and on the Indian Plate (Bossart & Ottiger 1988; Searle *et al.* 1987). There was subsequent obduction of the Kohistan arc, together with the adjacent part of the Asian Plate, over Indian continental crust.

Throughout most of its length, the MMT forms a discrete ductile fault zone, although in the Swat region there is a ten kilometre wide zone of blueschists, greenschists and ophiolites of the Indian Suture Mélange (Kazmi *et al.* 1984, 1986). This high-pressure assemblage is generally thought to be associated with the collisional event, representing rocks trapped between the Asian and Indian Plates. However, mineral ages of about 75 Ma have been obtained on phengites from blueschists near the Shangla pass, east of Swat (Maluski & Matte 1984), much earlier than the proposed age of collision. Hence, the high-pressure rocks may record an earlier phase of deformation, possibly related to subduction of the Tethyan ocean beneath Kohistan or one of the deformation events within the Kohistan arc (Coward et al. 1987).

In the Nanga Parbat area, the MMT carries rocks of the upper part of the Kohistan arc onto basement gneiss (the Nanga Parbat gneisses) and cover metasediments of the Indian Plate. The lower part of the Kohistan arc must have been transported to the north by a process of subduction shearing or break-back thrusting, so that the upper part of the Kohistan arc lies directly against the Indian continent. Details of this geometry are provided by Butler & Prior (1988a).

The hanging-wall geometry of the MMT is carried intact onto Indian continental crust which subsequently deformed at various $P-T$ conditions. On the basis of 'Himalayan' age metamorphism this deformed Indian crust can be divided into internal (metamorphosed) and external (non-metamorphosed) zones. The internal zone is dominated by a folded and imbricated basement-cover sequence, deformed generally in the amphibolite facies, although locally preserving only greenschist facies assemblages. These are intruded by abundant Himalayan age granites and pegmatites. Some of the pegmatites contain garnet and tourmaline and appear superficially similar to leucogranites to the east in the main Himalayan ranges of India and Nepal, which gives young Tertiary ages (Scharer et al. 1984) with high $^{87}Sr/^{86}Sr$ initial ratios, suggesting that they were derived by partial melting of the subducting Indian Plate. The majority of the granites, however, appear mineralogically and texturally distinct from these Himalayan leucogranites although, like them, they are of young Tertiary age derived by partial melting of the local Indian Plate. In Pakistan, the southern boundary of the internal zone is marked by the Mansehra and Panjal thrusts (Coward & Butler 1985), which form an equivalent structure to, but not necessarily coeval with, the Main Central Thrust (MCT) in the Himalayan ranges to the east (Gansser 1964). The southern thrust sheets near Mansehra, carry a granite dated at about 500 Ma (Le Fort et al. 1980), intruded into already metamorphosed sediments. This dates the basement rocks of the internal zones as Cambrian or older.

The external zone consists of only slightly metamorphosed or non-metamorphosed Precambrian to Recent sediments. Precambrian slates and greywackes form a basement to a Lower Palaeozoic quartzite and clastic sequence of variable thickness, overlain by a thick carbonate sequence of the Abbottabad Formation (Calkins et al. 1975; Latif 1974). To the south, the thrust sheets involve dominantly younger sediments; the southernmost thrust which involves Precambrian basement can be mapped as the Nathia–Gali thrust (Coward & Butler 1985). In the hills north of Islamabad (figure 1), the thrusts involve mainly Mesozoic to Lower Tertiary carbonates with overlying foreland basin sediments of Upper Eocene to Miocene age known as the Murree Formation of the Rawalpindi Group. Along the mountain front, from Islamabad to the Kohat area, the Mesozoic and Cainozoic carbonates are carried on to this Cainozoic molasse on a series of faults that have been correlated with the Main Boundary Thrust (MBT) in India (Windley 1983). South of the MBT in Pakistan there is a thick imbricated sequence of Miocene to Quaternary molasse, termed the Siwalik Group (Pilgrim 1910). This forms part of the foreland basin to the Western Himalayan arc and the Kohistan ranges, but unlike the foreland basin of India and Nepal, it is here disrupted by thrusts that detach above the basement in Cambrian salt to emerge along the Salt Ranges and Surghar and Khisar Ranges (figure 1). Here Mesozoic to Cainozoic rocks have been emplaced on river gravels and fanglomerates as young as 400000 years BP (Johnson *et al.* 1979; Yeats *et al.* 1984).

## DEFORMATION AND THE THRUST WEDGE

Preliminary cross sections through the thrust belt from the MMT to the Salt Ranges, have been published by Coward & Butler (1985) and Coward *et al.* (1987). These were the result of an essentially two-dimensional study and were constructed in such a way that all the thrust geometries were restorable on to an initial stratigraphic template. To gain minimum estimates of orogenic shortening, the original stratigraphic thicknesses were maximized while still being compatible with observed geology. Therefore, a maximum-thickness layer cake stratigraphy was adopted for the restored template. Thinner or more complex stratigraphic geometries would have required necessary increases in orogenic contraction. The section in Coward & Butler (1985) also assumed a consistent SSE direction of thrust transport, a feature in accord with the knowledge of thrust zone kinematics at that time. More recent studies indicate these assumptions to be locally invalid, in that there are thickness and facies changes and some variations in, and interference between, different transport directions. Nevertheless, we consider these sections to give a reasonable indication of the structure so that Coward & Butler's (1985; see also Butler & Coward 1988) estimates of the amount of shortening appear realistic. They calculated that the shortened Indian Plate rocks between the foreland and the MMT, restored to about 730 km, for a present distance of 260 km, indicating a shortening of 470 km. The decoupling surface at the base of the thrust zone must lie within the upper crust and is locally uplifted and eroded as, for example, around the Hazara syntaxis. The thrusts in the Mesozoic to Cainozoic carbonates north of Islamabad appear to curve on the map, to join the fault at the western margin of the Hazara syntaxis (figure 1). This fault is considered to be the decoupling surface to the imbricate sequence and to have been uplifted and folded by later deeper structures developed from the main NW–SE trending Himalayan thrusts of Kashmir (Seeber *et al.* 1981; Coward 1983). The decoupling surface for the Kohistan thrusts gradually cuts deeper in the crustal profile towards the internal zone, presumably to carry lower crustal rocks at depth in the north. However, recent work around the most northerly part of the Indian continent, exposed around the Nanga Parbat syntaxis (Butler & Prior 1988*a*) has found a probable Phanerozoic cover sequence, directly under the MMT. Not only does this imply that the MMT detachments run in upper crustal levels at least this far north, but it also directly requires an additional 150 km displacement greater than that estimated by Coward & Butler (1985) for shortening of the Indian continental crust. We suggest that over 600 km shortening is accommodated beneath the suture (the MMT).

In general, the crustal load resulting from stacked Indian continental crust propagated systematically towards the foreland, to generate flexural subsidence and the southward migration of the Indus–Ganges foredeep with time (Burbank 1983; Lyon-Caen & Molnar 1985). Some thrust movement may be attributable directly to plate movements between India and Asia, the thrust displacement directions being parallel to the NNW–SSE relative plate movement vector (cf. Patriat & Achache 1984). However, many of the structures in both the internal and external zones appear more complex and it is tempting to relate these to the bulk strain necessary to maintain the critical taper of the thrust wedge. Davis *et al.* (1983) have analysed thin-skinned thrust wedges in terms of Coulomb failure criteria, where the wedge shape of a thrust mass largely depends on its internal strength and its basal shear strength. Such theory may be applied qualitatively to larger regions of mountain belts, where Coulomb criteria are not applicable (Platt 1986). Thrust wedges with low basal shear strength, for example wedges moving over sediments with a high fluid pressure, or over salt deposits, will require only a low critical taper, that is, there will be a very gentle slope to the mountain front (Davis & Engelder 1985; Butler *et al.* 1987). However, thrust wedges moving over rocks with a high shear strength, such as crystalline basement gneisses or granites, will require a higher wedge taper and hence a steeper mountain slope.

The frontal ranges of the Himalayas illustrate this principle. Along the main ranges in India and Nepal, the mountain front is steep with a vertical elevation change of 5 km in 100 km horizontally (figure 2). In this region, the basal thrusts lie in foreland basin molassic sediments, passing back into older basement beneath the main Himalayan mountains. This region shows active seismicity, which, from local depth determinations and first motion studies, suggests displacements on thrust faults inclined at about 30° to the north (Seeber *et al.* 1981).

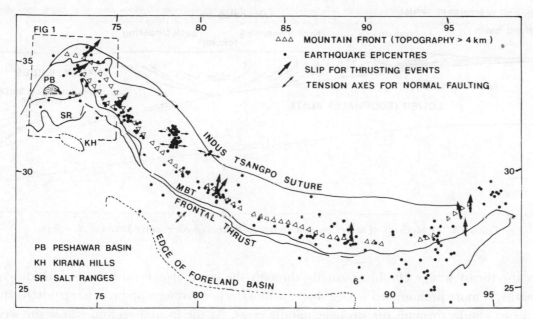

FIGURE 2. Map of the Himalayan mountain front throughout India, Nepal and Pakistan, showing the position of the main topographic slope, the associated sedimentary basins and the zones of seismic activity (partly after Seeber *et al.* (1981)). Earthquake epicentres are shown only for the Indian Plate.

In northern Pakistan, however, there is a much lower critical taper, with a change in elevation from the Kohistan ranges to the foothills of up to 4 km in 200 km horizontally. Indeed, there is an anomalous topographic low, occupied by Plio-Pleistocene sediments of the Peshawar Basin (figure 2). This area of low flat ground is in the same structural position, north of the MBT, as the major mountain front in Nepal. A possible reason for the preservation of this low ground in northern Pakistan during latest thrust movements, is that the thrust wedge probably had low basal shear strength. South of Peshawar, the thrusts detach in Cambrian salt and the very gentle taper to the wedge is probably allowed by displacement on this weak salt layer. We note that deep lithospheric processes, related to crustal loading and subsequent flexure may partly control the origin of the Peshawar depression, as is discussed later. Seeber *et al.* (1981) point out that the frontal regions of the mountain belt in North Pakistan are aseismic; presumably displacement occurs by ductile creep in the weak sediments. The dominant seismic activity in Pakistan occurs in NW–SE trending zones, N and NW of the Hazara syntaxis (figure 2), which from first motion studies, dominantly indicate thrusting to the SW, with a local depth range of between 35 and 70 km (Jackson & McKenzie 1984; Seeber *et al.* 1981). This activity probably represents displacement on the lateral continuation of the main Himalayan thrusts and will be discussed in more detail later.

Exploring the notion further, the critical taper of a thrust wedge can be changed by deformation within the wedge, by accreting material as thrust slices to the front of the wedge, or by the accretion of fault-bounded packages of material from beneath the wedge (figure 3). As erosion removes material from the surface, more internal deformation is needed, or more material must be added to the wedge, to maintain its critical taper (Davis *et al.* 1983).

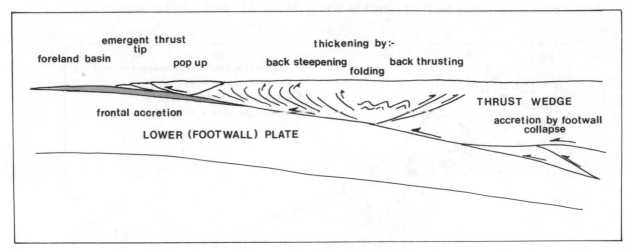

FIGURE 3. (*a*) Methods of accretion of material to a thrust wedge (after Davis *et al.* 1983).

As the thrust wedge climbs gradually through the crust, it overrides rocks with different strengths. A more pronounced wedge shape with a steeper surface slope will be required where the thrust climbs through the stronger middle crust. At the frontal region, where the wedge overthrusts uncompacted sediments, the taper may be too large; that is, the surface slope may be too steep (figure 4). The slope may be reduced by erosion, by the addition of frontal thrust packages or sometimes by collapse in a series of foreland-directed extensional faults, similar to large landslides (cf. Coward 1982).

In the deeper parts of the thrust wedge, the basal shear strength may decrease as the thickened crust heats up. In the Himalayas, the internal zone also shows major extensional structures, where the large-scale normal faults dip down in the same direction as the thrusts and may link with the zone of weakened lower crust at depth. Large-scale extensional faults have been recognized around much of the Himalayan belt, with large normal faults mapped along the northern edge of the Crystalline Zone, north of Everest (Burg & Chen 1984; Royden &

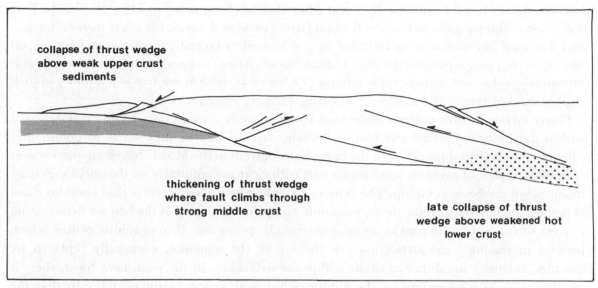

FIGURE 4. Showing how extensional structures may develop in a major thrust wedge (see text for discussion).

Burchfiel 1987) and in Kashmir (Herren 1987). However, these heating processes will weaken the supporting lithosphere, generating steeper gradients that could cause the thrust wedge to lock up, generating breakback structures and refolding earlier thrust sheets (cf. Butler 1987).

Because of the problems in evaluating rigidity decay during lithospheric flexing, in the next section we will continue exploring the simple concept that structural developments, including fault kinematics and geometries, can be related to progressive wedge shape.

## INTERNAL THRUST ZONES

### Stratigraphy

The crystalline rocks of the internal zones can be divided into several different basement sequences, metamorphosed and intruded by granites during the Precambrian or lowermost Palaeozoic, overlain by a cover sequence of uncertain, but probably Phanerozoic, age. Basement rocks include quartzo-feldspathic gneisses, termed the Besham Group (Treloar *et al.* 1988*a*) (figure 5*a*), which consist of metamorphosed granitoid rocks and metasediments, including psammites, pelites, often sulfidic and/or graphitic, and minor marble horizons. They are of unknown age and are confined to the lower thrust sheets that have been uplifted in a broad, dome-like, late fold at Besham.

To the east, the Besham Group is bounded by a major steep ductile shear zone (the Thakot Shear), which shows now a dextral sense of displacement, but is interpreted as a tilted thrust. Above this, to the east, the Tanawal Formation (Treloar *et al.* 1988*a*) is a sequence of dominantly semi-pelitic and psammitic metasediments, possibly correlatable with a series of lowgrade metasediments, the Hazara slates, south of Mansehra (Calkins *et al.* 1975). The stratigraphic relationship between the Tanawal Formation and the metasediments of the Besham Group is unknown, although both form a basement to the Phanerozoic cover. A sheared biotite granite with large orthoclase porphyroclasts (the Swat or Mansehra Granite) intrudes the Tanawal Formation and near Mansehra has been dated at $516 \pm 16$ Ma (Le Fort *et al.* 1980). Mapping shows that the Besham Group rocks and associated cover metasediments, and Tanawal Formation rocks intruded by the Mansehra Granite, form two distinct crustal blocks, or nappes, separated by the Thakot Shear. These nappes, distinguishable in both lithostratigraphic and metamorphic criteria (Treloar *et al.* 1988*b*) are two of a number of such nappes stacked late in the southward shearing phase of collision.

Cover metasediments outcrop to the west and south of Besham as parallel inliers imbricated within the basement gneiss and east of Besham, around Banna, they occur as dominantly calcareous rocks in a thrust sheet in the immediate footwall to the MMT. Seven kilometres west of Besham, a slice of cover metasediments rests with clear unconformity on the gneissic-granite basement of the Besham Group. The contact is marked by a conglomerate that contains clasts of pegmatite, foliated granite gneiss, psammite and pelite. This slice is the best for determining a cover stratigraphy; the conglomerate is overlain by psammite, then graphitic pelites, which become increasingly calcareous towards the top of the sequence, eventually replaced by marbles. Intrusive amphibolites occur within the sediments. In the west, near Swat, there is another group of cover sediments, the Alpurai Schists with a more pelitic stratigraphy than the Besham cover sequences.

## Structure

A polyphase deformation history (table 1) affects the basement gneisses and cover metasediments. There is an early schistosity (S1) forming event which is sub-parallel to bedding. This tectonic fabric is crenulated by second phase structures on a mesoscopic and

TABLE 1. DEFORMATION CHRONOLOGY FOR THE INTERIOR ZONES OF THE NORTHERN MARGIN OF THE INDIAN PLATE

*Structures that post-date the main southward obduction of Kohistan*

| | |
|---|---|
| post-D3 | N–S striking brittle steep reverse faults. E side up. |
| D3 | NW- and NNW-verging backfolds and backthrusts which locally breach the MMT. Large-scale NNE-plunging WNW-verging (syntaxial) upright folds. These structures are probably coeval with the thrusts in the external zone. |
| pre-D3 | Top-side E- or NE-directed shears that re-activated D2*a* surfaces. |

*Structures generated in the footwall of the MMT during southward obduction of Kohistan*

| | |
|---|---|
| D2*a* | Thrusts with southerly transport which stack and internally imbricate crustal scale nappes. |
| D2 | Crenulations and folds, through S-facing, although strongly sheath-like in the N. Probably a diachronous continuation of D1. |
| D1 | Main fabric forming event related to southward overthrusting of Kohistan over India. Ductile blastomylonites, thickened by small-scale thrust-related folds in the N. Becoming more brittle to the S, where the fabric is increasingly typical of a shortening rather than shearing deformation, with shear strains increasingly accommodated in narrow mylonite zones. |

macroscopic scale, deforming lithological layering and S1 fabric into tight folds whose axes are curvilinear, varying in plunge from 60° to the SW to 30° to the NE (Coward *et al.* 1982). These sheath-like folds and the mineral lineations that plunge to the N, suggest S-directed movements during development of the F2 structures (Treloar *et al.* 1988*a*). The second deformation phase culminated in, or was followed by, a discrete SSE-directed thrusting event that stacked the crustal nappes along shears such as the Thakot Shear and also internally imbricated them.

Fabric porphyroblast relations show that the main metamorphism was largely synchronous with the early stages of deformation. A metamorphic map of the internal zones, from Treloar *et al.* (1988*b*) is shown in figure 5*b*. There are metamorphic breaks across the shears that stack the crustal nappes and within each nappe there is an overall upward increase in metamorphic grade. Within the highest grade rocks of the Besham nappe, the Alpurai schists, the metamorphic peak ($600 \pm 50$ °C at $9 \pm 2$ kbar†) was synchronous with the D1 ductile shearing event. Thermobarometry and garnet inclusion studies (Treloar *et al.* 1988*a*) imply that this metamorphism accompanied a steep pressure increase, interpreted as involving subduction and thickening of the leading edge of the Indian Plate beneath Kohistan. The early cessation of the metamorphism here reflects a rapid crustal rebound. Within the Hazara nappe (figure 5), however, metamorphism was at lower pressures (about 5–8 kbar) (Treloar *et al.* 1988*b*) and continued for longer, until after development of D2 crenulations. The metamorphism of this zone is more typically characteristic of thermal relaxation following crustal thickening.

West of Besham the lowest-grade rocks occur in the core of the late large Besham antiform and the grade generally increases upwards on a gross scale, suggesting some form of inverted metamorphism (cf. Le Fort 1986). There are sharp metamorphic breaks across the late thrusts that imbricate basement and cover slices west of Besham although there is no direct evidence as to whether metamorphic grade within individual sheets increased upwards or not. As the rocks immediately beneath the MMT near Swat contain low-temperature blueschist assemblages, this suggests that the uppermost part of the thrust stack was cold. Therefore, we suggest a model for the inverted metamorphism in the Besham area including the following.

(i) Overthrusting of the Indian Plate by the Kohistan arc with a zone of blueschists preserved locally beneath the footwall to the suture. This deformation presumably produced the first fabric, possibly locally stacking the basement-cover sequence and metamorphosing them to give a normal metamorphic gradient. The original metamorphic gradient presumably decreased towards the south, away from the overthrust island arc.

(ii) This package was then re-imbricated during or following the second phase of deformation, slicing and restacking the metamorphic sequence, placing originally deeply buried high-grade rocks on top of more shallowly buried low-grade rocks.

Within the other crustal nappes, such as the Hazara and Kaghan Valley nappes, metamorphic rocks were also imbricated by late thrusts that have higher-grade rocks on their hanging walls than on their footwalls. This late imbrication gives an overall sense of metamorphic inversion resulting from a post-metamorphic disruption of the metamorphic pile, rather than from an originally inverted thermal gradient. This is a fundamentally different model to that proposed for the metamorphic inversion along the MCT in India and Nepal, where a hot slab is interpreted as having been emplaced above colder rocks (Le Fort 1986), with the immediate imposition of a saw-tooth geothermal profile, subsequently modified by a downward heat flux.

A map of the main shears and ductile thrusts is shown in figure 5, based on detailed mapping in the Besham–Swat–Mansehra region, plus reconnaissance mapping to the east. The N–S trending Thakot shear zone either slices through, or is joined by, E–W trending shears in the Hazara basement nappe. It can be interpreted as the western lateral ramp to the Hazara nappe, the Balakot Shear Zone (BSZ) forming the eastern lateral ramp boundary. This Balakot shear is the same structure as the mylonite belt mapped NW of the Hazara syntaxis by Bossart

† 1 kbar = $10^8$ Pa.

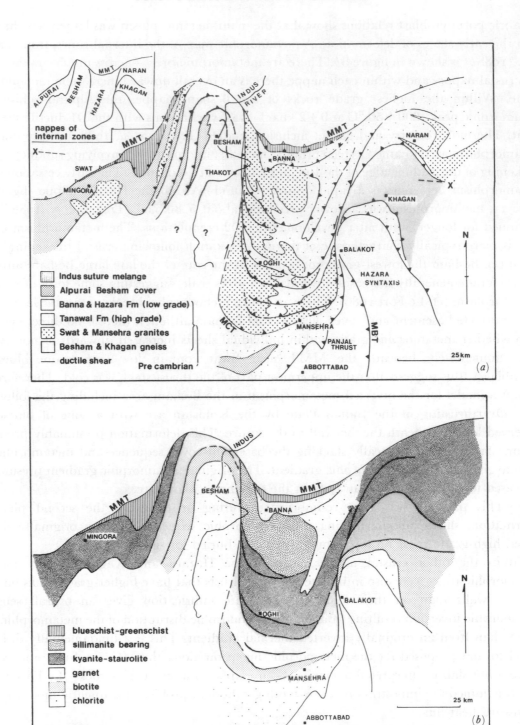

FIGURE 5. (a) Map of the internal zones in the Besham–Mansehra region, showing the distribution of metasediments and gneisses. Tentative correlations are made between metasediments and granites of the Naran nappe and those of the Hazara nappe and between gneisses of the Khazan and Besham nappes. (b) Simplified map showing variations in metamorphic grade (after Treloar et al. 1988a, b).

*et al.* (1984). If the imbricate ductile thrusts within the Hazara nappes, such as the Oghi shear (figure 5) (Coward *et al.* 1982), are the same age as the major shears that bound and separate the nappes, they may be modelled as hanging wall splays off a major ductile thrust system. Figure 6 shows an E–W cross section through the thrust stack roughly perpendicular to the thrust transport. The trend of the lateral ramps suggests a N–S overthrust direction.

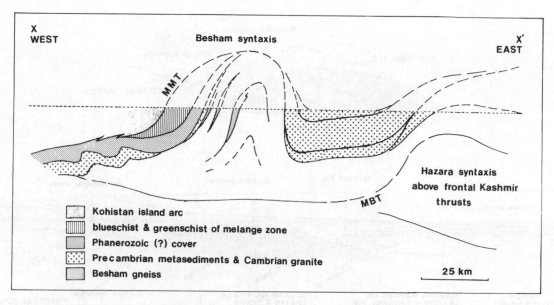

FIGURE 6. East–west simplified section through the main thrust packages of the internal zone; for section line see figure 5*a*).

This relatively straightforward package of ductile shears in metamorphic rocks is modified by at least three later tectonic events. The Banna sediments form a low-grade package above the Hazara nappe. From the sense of the small-scale folds and thrusts and the trend of mineral lineations, they would appear to have been transported from the south as a backthrust sheet. Figure 7 shows our interpretation of the Banna sediments thrust as a wedge between the Hazara nappe and the MMT.

The gneisses and cover sediments near Besham show several shear zones, locally occurring as 5–10 m wide zones of low-grade fault rock, with kinematic movement indicators which suggest a displacement of top to the east or northeast. These shears are folded round the later Besham antiform, so that in the west they are upward and eastward verging, in the east they are downward verging. They may be related to an eastward-directed low-angle thrust, subsequently folded round the Besham antiform or to a low-angle extensional fault, similar to the NE-directed extensional fault in Kashmir (Herren 1987), and generated by the collapse of the more ductile thickened crust.

The Besham antiform is part of a series of N–W trending upright to westward verging folds which occur south of the MMT between the Indus River and the Afghanistan border (figures 5 and 8). North of Besham, these upright structures fold the MMT into a syntaxial zone similar to that of Nanga Parbat. To the west, however, the MMT appears planar and hence these late structures must decouple up onto the MMT. Their origin is unknown (cf. Coward *et al.* 1986).

They may be related to deformation around the western tip of the main Himalayan thrust belt, that is, they may be unrelated to the Kohistan thrust sequence. Alternatively, they may be oblique back thrusts, developed during Kohistan thrusting.

Figure 7*b* shows a sketch restoration of the rocks of the internal zone. It suggests that the high-grade rocks of the Hazara nappes formed originally the northernmost part of the Indian Plate in southern Kohistan.

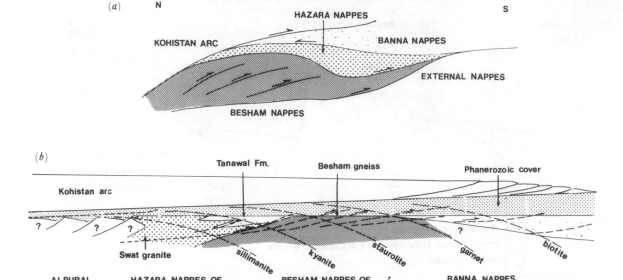

FIGURE 7. (*a*) Simplified cross section through the internal zone to show the distribution of the major nappe packages. (*b*) Simplified restored cross section to show the original distribution of the nappe packages; see text for discussion.

The early Himalayan metamorphism (S1 to post-S1) is assumed to be related to crustal thickening during the initial stages of overthrusting by the Kohistan arc. That the arc did overthrust the Indian Plate a considerable distance, in excess of 100 km, is shown by the Nanga Parbat syntaxis (figure 1), where the suture is folded and thrust by structures in the underlying Indian Plate and then eroded to leave a long thin half-window.

The Hazara nappe package shows Precambrian to Cambrian deformation and meta-morphism in the Tanawal sediments, largely overprinted by Himalayan effects and intrusion of the Swat–Mansehra granites. Similar high-grade metasediments and coarse porphyritic granites occur in NE Kohistan and form much of the Indian Plate gneisses of the Nanga Parbat syntaxis. The basement metamorphism probably represents part of the widespread late Precambrian–early Palaeozoic tectonic event that affected much of Gondwanaland, i.e. it may represent the local equivalent of the 'Pan African' event. This tectonic event is absent over much of the north Indian craton, occurring mainly in the Himalayan thrust sheets (Le Fort *et al.* 1980). A late Precambrian–early Palaeozoic event is also recognized in the basement rocks of the Lhasa Block, north of the Indus Suture (Chang Chengfa *et al.* 1986).

The Besham thrust package originated south of the Hazara package. The origin of the Besham gneisses is unknown, although they may be basement to the Tanawal Formation to the north and Hazara slates to the south. The Banna nappe is considered to be the originally southernmost of the internal thrust sheets, back-thrust as a wedge between the Hazara nappe package and the MMT. This back-thrust and the upright to west-verging folds at Besham probably developed to maintain critical taper to the overall thrust wedge as it over-rode the Indian Plate (figure 8). The Himalayan leucogranites are confined to the Hazara and Besham packages and probably formed by melting of the thickened crust at depth, and were then intruded after the F1/F2 deformation.

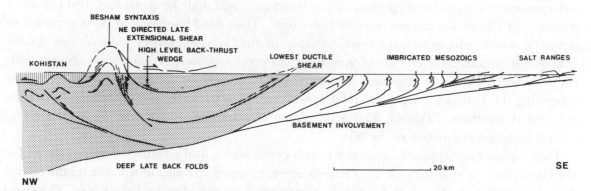

FIGURE 8. Simplified section to show the distribution of late shears, thrusts and folds near Besham and the relation of internal (stippled) and external (blank) zones. Foreland basin sediments shown by fine stipple.

## THE EXTERNAL THRUST ZONES

### Stratigraphy

The Precambrian mudstones and greywackes of the external zones show little of the late Precambrian–early Palaeozoic tectonics of the internal zones. Near Abbottabad (figure 1), the older sediments were tilted before deposition of the Cambrian(?) Abbottabad Formation. However, only a Himalayan-age cleavage has been recognized in this area, common to the Precambrian and Phanerozoic sediments. The exact age of the Abbottabad Formation is uncertain; a Cambrian age is suggested on the basis of acretarchs from the upper limestones (Latif 1974). In the frontal part of the thrust zone, south of Islamabad, reviewed by Butler *et al.* (1987), Eocambrian–Lower Cambrian rocks form the Salt Range Formation, overlain by Lower to Middle Cambrian clastic sediments (Schindewolf & Seilacher 1955; Teichert 1964; Butler *et al.* 1987). The Salt Range Formation is a major evaporite-bearing sequence, locally over 1 km thick, of red marls, gypsum and dolomite passing up into thick beds of halite with thinner marls and some poor-grade oil shales. It represents a thick series of playa and distal alluvial fan evaporites, with palaeocurrent directions indicating transport from the south (Asrarullah 1967). The relation between the Abbottabad Formation and the Cambrian rocks of the Salt Ranges is unknown; no Cambrian rocks occur in the intervening thrust sheets and the two regions must have been several hundred kilometres apart before Himalayan-age thrusting.

In northern Pakistan the region formed a high throughout most of the Palaeozoic and in the Salt Ranges, Permian sediments overlie the Cambrian formations with only a slight angular unconformity. Permian sediments comprise a basal group of tillites, with diamicrites, grading upwards into massive sandstones. These presumably reflect the widespread Gondwanaland

Permian glacial event. The overlying Permian sediments record a change to warmer conditions with a good marine fossil fauna and upper limestones and dolomites. Elsewhere Mesozoic carbonates rest directly on the Precambrian or Cambrian sediments. They are mainly shallow marine to non-marine carbonates, essentially dolomites and dolomitic limestones, with minor shales and sandstones (Shah 1977). There were phases of emergence with local disconformities, notably at the end of Triassic times and in the Upper Jurassic. The Lower Cretaceous consists of glauconitic and iron-rich sandstones with probable non-sequences, indicative of a change to a deeper-water environment, passing up into sandstone and shale horizons and then to a thick sequence of Upper Cretaceous micritic limestones. Thus the Mesozoic can be interpreted in terms of several phases of subsidence, notably in the Triassic, Lower Jurassic and Lower Cretaceous, producing over 1 km of essentially platform carbonates. We interpret these as post-rift sediments, probably deposited during thermal subsidence following phases of rifting producing the Tethyan margin. No thick syn-rift sediment packages and graben-fill have been detected in northern Pakistan. The main region of extension was on the Tethyan margin, several hundred kilometres to the NW.

There was a major phase of emergence with gentle tilting and possibly folding at the end of the Cretaceous, so that the overlying Palaeocene carbonates rest unconformably on the eroded Mesozoic sediments. Figure 9 shows a simplified sub-crop map for the Palaeocene. The trend of the gentle end-Cretaceous folds is N–S to NE–SW, highly oblique to the trend of Himalayan-age thrusts, but approximately parallel to the Tethyan margin of NW Pakistan. This phase of basin inversion coincides with the commencement of rapid spreading of the Indian Ocean as India moved northwards away from the rest of Gondwanaland. It also coincides with the emplacement of ophiolites on to the northern margin of the Indian Plate, in Zanskar (Searle 1983, 1986) and on to the Arabian Plate in Oman (Searle 1985).

The lowest Palaeogene sediments are local deltaic sandstones and shales with some coals, overlain by a thick sequence of foraminiferal limestones, recording continued subsidence. During the Eocene there was a gradual influx of red clastic sediments. Evaporites occur in the Eocene of the Kohat region and in the northern part of the Hazara syntaxis there is an interleaving of molassic deposits and Eocene limestones (Bossart & Ottiger 1988) recording the onset of Himalayan-derived sedimentation.

The transgressive package of molassic sediments has been divided into a lower Rawalpindi Group and an upper Siwalik Group. In general, the sediments coarsen upwards, so that the Siwaliks have a much greater sandstone content with conglomerates. The groups thin towards the SW; in the northern Potwar the Rawalpindi Group attains a thickness of over 3 km, but in the SW Salt Ranges it is only 200 m thick. Similarly, the Siwaliks thin from about 4 km in the east, to zero south of the Salt Ranges, where basement rocks are uplifted along the Kirana Hills. This change in sediment thickness is also reflected in the Bouguer gravity-anomaly map (figure 10), which shows contours trending WNW–ESE. The Kirana Hills coincide with the ridge of positive Bouguer anomalies and may represent a low amplitude peripheral bulge at the edge of the main Himalayan flexural basin, though they are also the site of a pre-existing basement high during Mesozoic sedimentation.

Note that this flexural basin is oblique to the trends of thrusts in northern Pakistan and hence causes important lateral changes in thrust geometry.

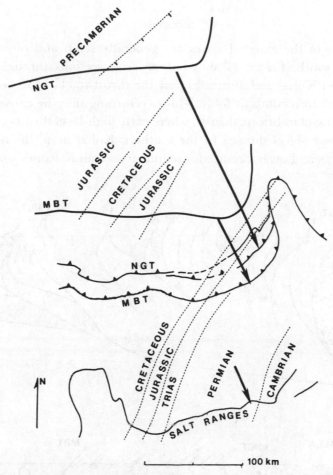

FIGURE 9. Sub-crop map of the Palaeocene, restored for Himalayan thrusts following the model of Coward and Butler (1985). NGT, Nathia Gali Thrust; MBT, Main Boundary Thrust. Data from Latif (1977). Stratigraphic data are compiled from Gee (1980), Shah (1977), Meissner *et al.* (1974) and Yeats & Hussain (1987).

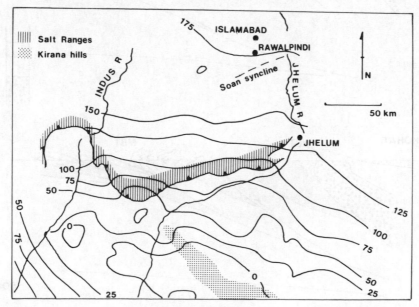

FIGURE 10. Bouguer gravity map of northern Pakistan (after Farah *et al.* 1977). The units of the contours are milligal (1 mGal = $10^{-3}$ cm s$^{-2}$).

*Structure*

The thrust sheets of the external zones are generally steep and often overturned to dip moderately to the south. Figure 11 shows three cross sections through the frontal ranges between Murree and Kohat and illustrates how the thrust stacks have been reorientated into north-facing inclined to recumbent folds. This overturning may be caused by the following.

(i) Back-steepening of imbricate thrusts, where early high-level thrusts are back-tilted by the accretion of later lower-level thrusts. In the southern Kohat area, the development of local duplex zones in Jurassic–Lower Cretaceous sediments certainly causes some back-steepening.

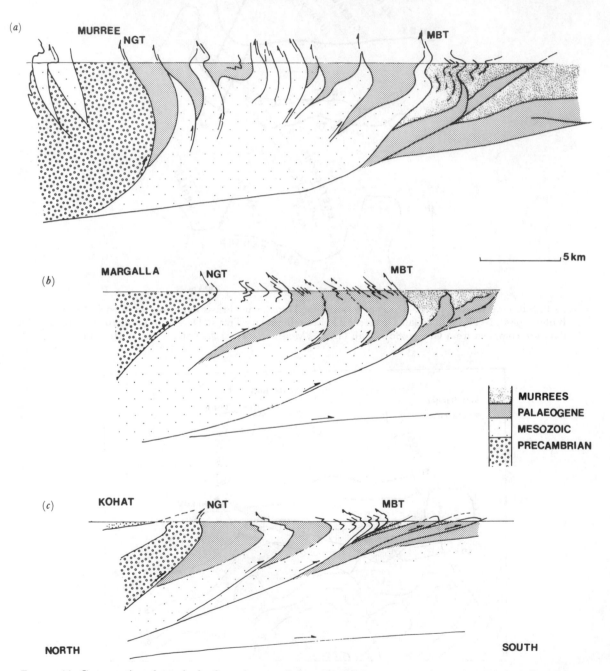

FIGURE 11. Cross section through the frontal parts of the external thrust zones; for section lines see figure 1.

(ii) Refolding by a later phase of back-folds, developed above the later low-level thrusts. This is considered to be the principle mechanism and produces several orders of folding; macroscopic parasitic folds occur on the overturned limb of the major fold.

This major back-folding event, close to the thrust front, is probably related to the change in critical thrust wedge taper as the basal detachment climbs into the sands and conglomerates of the Rawalpindi–Siwalik Groups.

South of Kohat, later thrusts form above a detachment in Eocene evaporites (figure 12). Here the fault wedge had a much lower basal shear strength. The thrusts show less evidence of back-steepening; many dip at moderate angles to the north and many of the folds can be considered as simple ramp anticlines. There are several zones of south-dipping normal faults along the Kohat–Margalla Hills north of this zone, suggesting that there may have been changes in wedge shape by extensional collapse of the relatively steep mountain front. Further south, near the Surghar Ranges, the southern Kohat thrusts also become back-steepened, suggesting local loss of easy basal slip in the Eocene sediments.

During the latest movements, slip has transferred down into Cambrian salt. Throughout much of the western Potwar Plateau displacement occurred by easy slip beneath the previously formed high level thrusts, to form thrust ramps at the Surghar and Salt Ranges, where the easy slip was disturbed by salt diapirism, local loss of salt, or earlier normal faults (Baker *et al.* 1988). The geometry of the Salt Range décollement is described at length by Butler *et al.* (1987). The progressive changes in style of thrust front structures and their variations along strike seem to record the changes in basal shear strength of the thrust wedge as it climbs from Mesozoic limestones to Cainozoic molasse deposits and subsequently locally back down to Eocene and then to Cambrian easy-slip horizons by major footwall collapse (figure 12).

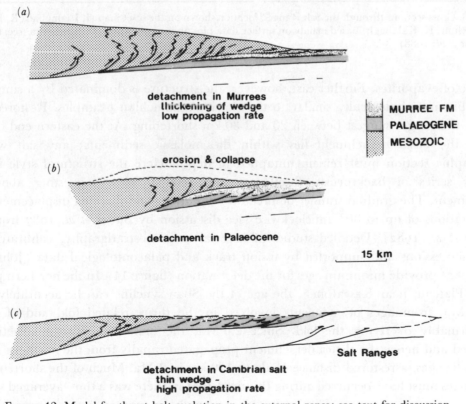

FIGURE 12. Model for thrust belt evolution in the external zones; see text for discussion.

In plan form, the Salt Ranges have a markedly irregular pattern (figure 1) and show pronounced change in structural style along strike. In the west, on the eastern side of the Surghar re-entrant, thrusts and folds detach within, or at the base of, the Salt Range Formation evaporites, so that a full stratigraphic section is involved in the deformation. The westernmost section (figure 13) shows a simple fold train, underlain by a greatly thickened

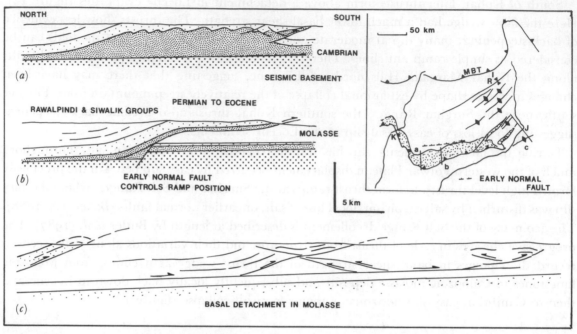

FIGURE 13. Cross sections through the Salt Ranges, location shown on the inset map (I, Islamabad; R, Rawalpindi; J, Jhelum; K, Kalabagh) based mainly on surface data plus an interpretation of seismic data (see, for example, Baker *et al.* 1988).

sequence of evaporites. Further east, however, the structure is dominated by a simple ramp-type uplift, morphologically similar to the classic Appalachian examples. Restored sections through the ranges suggest between 20 and 30 km shortening. At the eastern end of the Salt Ranges the basal detachment lies within the molassic sediments; any salt within the stratigraphic section must remain untapped at depth. Here the structural style involves a complex series of backthrusts, forethrusts and related folds, suggesting about 30 km displacement. The gradual transfer of foreland- and hinterland-directed displacements caused local rotations of up to 30° anticlockwise (see discussion by Butler *et al.* 1987 from data of Opdyke *et al.* 1982). Detailed studies of reversal magnetostratigraphy, calibrated against known successions and supported by fission track and palaeontological data (Johnson *et al.* 1979, 1982) provide minimum ages for the deformation (figure 14). In the northern part of the Potwar Plateau, near Rawalpindi, the age of the Soan syncline can be accurately dated at 2.1–1.9 Ma, from ages of sediments involved in the thrust-related fold and of sediments unconformably overlying the fold. Since 2.0 Ma, the main Salt Range structures have developed and hence the thrust detachment propagated rapidly from the northern Potwar to the Salt Ranges, a restored distance of about 120 km in 2 Ma. Much of the shortening in the Salt Ranges must have occurred during this time, that is, there was a time-averaged shortening

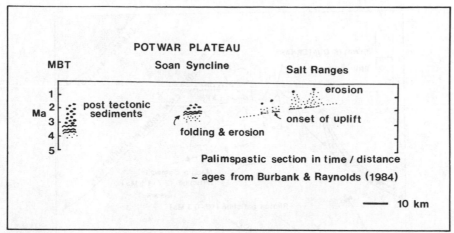

FIGURE 14. Schematic profile through the Potwar Plateau–Salt Ranges, to show the timing of folding (coarse wavy lines), local uplift (fine wavy lines), coarse sedimentation at final uplift and erosion (circles); data from Burbank & Raynolds (1984).

rate of 1.0–1.5 cm a$^{-1}$. This matches the migration rate of the foredeep basin in India (Lyon-Caen & Molnar 1985), although direct comparisons are unwise in view of the uncertainties in flexural rigidity, local morphology and crustal structures in these areas.

### INTERFERENCES OF HIMALAYAN AND KOHISTAN MOVEMENTS

Northern Pakistan shows the effects of interference between two thrust transport directions, i.e. the S–SSE directed transport of the Kohistan system and the SW–W directed thrusts of the western Himalayan arc. The thickness variations of the Rawalpindi and Siwalik Groups and the Bouguer anomaly map show that the foreland basin south of the Salt Ranges has been controlled by flexure related to thickening along the NW–SE trending Himalayan arc and not in the Kohistan mountains.

The Kohistan thrust direction can be obtained from the maps of cut-offs and lateral ramps as well as lineation data and axes of sheath folds in the internal zones (figure 16). The lateral ramps that bound the Salt Ranges and Surghar Ranges suggest a clear 160° direction for thrust transport, although there is evidence for Recent anticlockwise rotation at the eastern lateral tips to the larger thrusts in the Salt Ranges (Opdyke et al. 1982; Butler et al. 1987).

At the eastern end of the Salt Ranges there is interference between Himalayan and Kohistan thrust movements. The most outlying structure of the Himalayan mountains is the Mangla–Samwali anticline (figure 15), which deforms alluvium with reversal stratigraphies indicative of deposition between 2.7 and 1.5 Ma (Johnson et al. 1979). This anticline immediately predates the eastern folds of the Salt Range and probably inhibited the lateral propagation of the Salt Range detachment. Thus the lateral termination of the Salt Range system may reflect either the proximity of the main Himalayan belts or the associated flexural depression and its thicker pile of molassic sediments to the NE. This lateral pinning of Salt Range structures may cause the local obliquity of the thrusts in the eastern Salt Ranges.

Slickensides and other shear criteria in the fault rocks of the Kohat, Kala Chitta Ranges and Margalla Hills (figure 1) also suggest a dominant SSE thrust direction, although in the east,

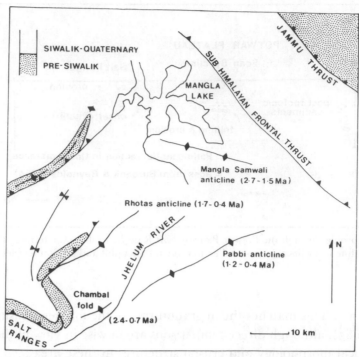

FIGURE 15. Simplified map of the distribution of folds and thrusts at the eastern termination of the Salt Range in the Jhelam district (see figure 1 for location), modified after Gee (1980) and Johnson *et al.* (1979), illustrating the ages of folding for various structures determined by magnetic reversal stratigraphy in the Siwalik Group of molasse.

near Murree, some of the thrusts carry large folds with steeply plunging but curvilinear hinges, which suggest a strike-slip movement (figure 16). These could be SW-directed folds and thrusts of the Himalayan system, tilted to give an apparent strike-slip attitude during subsequent SSE-directed thrusting. The Hazara syntaxis is an uplifted zone of Palaeogene sediments beneath the Hazara thrust system (Bossart *et al.* 1984). The trend of the syntaxis suggests it originated from a complex strain pattern and uplift in SW-directed structures at depth (see also Seeber *et al.* 1981; Coward 1983), though in detail the bedding-cleavage intersection directions and incremental strain indicators show that the uplift has been variable and connected with pronounced rotation (Bossart *et al.* 1984).

In the internal thrust zones of N Pakistan, the transport direction may be estimated from lineation data (figure 16). In the Mansehra–Besham region and NE of Nanga Parbat the majority of mineral lineations and hinges of large scale sheath folds have a NW–SE trend (Coward *et al.* 1986, 1987). North of Besham and near Swat, however, lineations are more variable; figure 17 shows the range in the Besham–Alpurai/Swat area, from N45E to N45W. Perhaps a more accurate indicator of thrust transport is given by the trends of major lateral structures of the Thakot and Balakot shears (figures 5 and 6), indicating N–S transport.

The Nanga Parbat syntaxis is a zone of recent uplift as shown by K–Ar cooling ages (Coward *et al.* 1986) and fission track data (Zeitler *et al.* 1982; Zeitler 1985). It is dominantly a large antiformal fold, buckling the MMT and underlying sheared gneisses of the Indian Plate. A structural cross section is given in Coward *et al.* (1986). The gneisses have been intensely re-deformed during the uplift and two dominant shear senses can be recognized: of ductile thrust

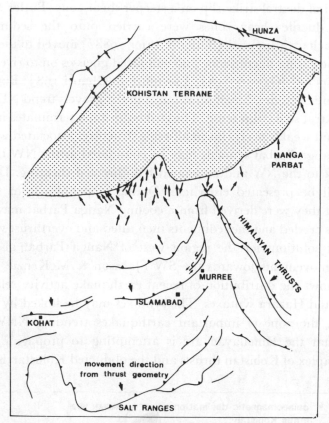

FIGURE 16. Simplified map of Kohistan showing the orientations of mineral lineations and early fold hinges for the internal zones of the Indian Plate, and from the Himalayan age shears and thrusts of the Asian Plate to the north. (Data on Asian Plate from Coward *et al.* 1986; Pudsey *et al.* 1986.)

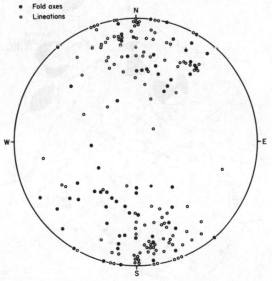

FIGURE 17. Stereoplot of stretching lineations and fold axes from the imbricated sequence west of Besham (MPW in Treloar *et al.* 1988a).

sense to the NW and of dextral strike-slip on a steeper shear zone (Butler & Prior 1988 b). The amphibolite facies ductile shear zones were carried onto the sediments by continued movements. The Liachar thrust zone (Butler & Prior 1988 b) moved to the NW, parallel to the shear direction in the earlier ductile zone and emplaced gneisses on to river gravels and glacial deposits (50 000 years BP) (Gansser 1981; Lawrence & Ghauri 1983; Butler & Prior 1988 b). It has historical seismic activity. Small-scale faults in the gravels trend NE–SW and have both thrust sense and dextral strike-slip sense. The strike-slip zone terminates in zones of shattering towards the south that weakly overprint small-scale structures associated with the NW-directed thrust. Thus, the Nanga Parbat syntaxis may be considered as the NW tip of the Himalayan arc with some uplift to the NW and associated SW-directed shearing. Detailed discussion of these structures will be presented elsewhere. Dating of pebbles from the deformed Indus gravels, suggest that they were derived from a cooling Nanga Parbat mass at only 2 Ma, that is Nanga Parbat was eroded and the sediments then tilted and overthrust within the past 2 Ma.

Earthquake fault solutions for the region west of Nanga Parbat also suggest dominant present-day thrust movements towards the SW (Jackson & McKenzie 1984; Coward *et al.* 1987). Figure 18 shows the distribution of recent earthquake activity related to the uplifts of the Nanga Parbat and Hazara syntaxes. The syntaxes may be linked by a NE-trending zone of uplift. However, the zone of important earthquakes trending NNW from the Hazara syntaxis suggests that the Himalayan arc is attempting to propagate along strike but is hindered by the complex of Kohistan thrusts and the obducted Kohistan island arc. The initial

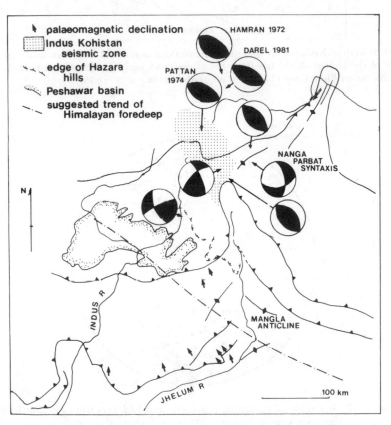

FIGURE 18. Distribution of the main neotectonic structures of the western Himalayan arc.

age of the Nanga Parbat and Hazara syntaxes is unknown. However, the later thrust movements in the southern part of the Hazara syntaxis uplifted the Mangla Samwali hills 2.7–1.5 Ma ago and presumably uplifted the Murree to Nathia Gali hills to their present altitude of over 2 km at this time (see figure 18). Some of this uplift was transferred on to the Nanga Parbat syntaxis causing overthrusting of Indus sediments. This late uplift and crustal thickening presumably modified the foreland basin. The Peshawar Basin (figure 18) may be partially a flexural basin ahead of the Himalayan thrusts, superimposed on and depressing the earlier Hazara thrusts. Sediments in the Peshawar basin date from about 2.8 Ma to the present day, fitting closely the timing of uplift in the Mangla Samwali hills and Nanga Parbat.

The tectonics of northern Pakistan therefore, cannot be considered accurately in simple two-dimensional sections. Accurate restorations require a three-dimensional thrust analysis. Furthermore, any modelling of lithospheric flexure (cf. Duroy 1986) needs to consider three-dimensional data before values for the effective elastic thickness or rates of shortening from the migration of the foreland basin can be calculated. Duroy (1986) and Duroy et al. (1988) consider the flexure in terms of a NNW–SSE section, perpendicular to the Kohistan thrusting. However, as argued above, the flexural basin is probably largely developed by the load of the main Himalayan arc, but to estimate effective elastic parameters for Northern Pakistan, the loads of Kohistan and the Himalayas need to be taken into account.

## Conclusions

1. Large displacements of Kohistan relative to the Indian Plate took place on detachments within the upper part of the crust. Estimates of about 470 km have been made for shortening within the non-metamorphosed Phanerozoic cover in the external zone, and at least an extra 150 km displacement occurred by shear in the metamorphosed cover sequence in the Nanga Parbat syntaxis. Indian lower crust, therefore, must have been subducted northwards beneath Kohistan. The continental crust beneath the Indian Plate is on average 35–40 km thick (Chaudhury 1965) but beneath Nanga Parbat it may be over 70 km thick (Kaila 1981). Much of this thickening may occur by underplating of Indian crust (see also Malinconico 1988). Certainly the Kohistan arc remained relatively undeformed during Himalayan events, locally shortening by SE-directed shears near the NW margin (Pudsey et al. 1986; Searle et al. 1987). However, it is unlikely that lower Indian crust extends beneath Kohistan as a rigid sheet; more likely it has been internally folded and imbricated, similar to the gneisses in the Besham area (figure 19).

2. Much of the thrusting involved thin-skinned tectonics in a simple, almost layer-cake, stratigraphy. Very little of the rift margin has been preserved and Tethyan sediments consists of shelf carbonates with several phases of Mesozoic subsidence and emergence. A minor phase of inversion and erosion occurred at the end of the Cretaceous, so that Palaeocene sediments rest unconformably on a range of Phanerozoic rocks.

3. The northernmost Tethyan sediments form the cover sequence to the basement rocks in the internal zones. The Nanga Parbat and Hazara nappe units are the more distal, the Besham nappes more proximal. Collision involved subduction of the basement gneisses and some cover rocks beneath Kohistan, associated with locally high-grade metamorphism. Generally, the more distal rocks of the Hazara nappe show a higher metamorphic grade than those of the

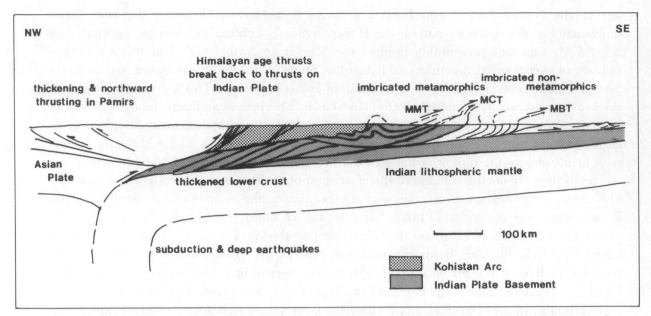

FIGURE 19. Simplified section through Kohistan and Pamirs showing how the Indian continental crust must be subducted partially beneath the Asian Plate, hence causing some thickening and uplift. Some Himalayan age deformation of the Asian Plate occurs by SE-directed thrusts, which are break-back to the main deformation on the Indian Plate.

Besham nappes, suggesting deeper subduction. However, there may be east–west lateral variations in metamorphic grade associated with lateral changes in thrust structure and associated crustal thickening. This metamorphic sequence was subsequently imbricated during later stages of collision, so that high-grade rocks were stacked above rocks preserving lower-grade assemblages, producing a form of metamorphic inversion.

4. The thrust package has been modified by several phases of back-folding, back-thrusting and extensional shears, probably largely related to the bulk strain necessary to maintain critical wedge shape, as the internal zone of nappes was thrust onto cover sediments of the Indian Plate. Relatively late structures include those of the Besham syntaxis and associated folds to the west, some of which detach onto the MMT and may link with the Himalayan thrusts which affect NW Kohistan (figure 20).

5. Within the external zones there are complex variations in thrust geometry and dramatic zones of back-steepening and large-scale back-folds that also are probably related to bulk strains maintaining wedge shape. Eocene salt horizons in the Kohat area and Cambrian evaporites beneath the Potwar Plateau and Salt Ranges have allowed late easy slip. The basal detachment has propagated rapidly through the salt, at a time-averaged rate of over 6 cm a$^{-1}$ and thrust movement has been possible even though the critical taper of the thrust zone was low.

6. Time averaged displacement rates for the Potwar Plateau–Salt Ranges are in the order of 1–1.5 cm a$^{-1}$, comparable with those estimated for the Indian foothill thrusts (Lyon-Caen & Molnar 1984). A similar time-averaged displacement rate can be estimated for Kohistan tectonics as a whole, with over 600 km shortening in the 50 Ma since collision. At present, the Indian Plate is moving northwards at a rate of between 3.5 cm a$^{-1}$ in the west and 5.0 cm a$^{-1}$ in the east, relative to a fixed Asian Plate (cf. Molnar & Tapponnier 1975). The difference

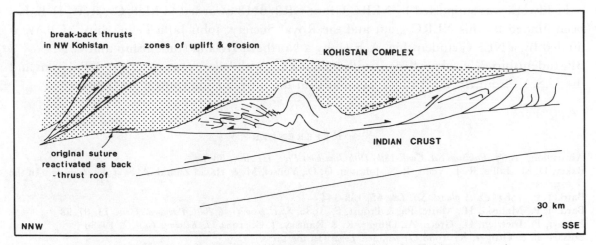

FIGURE 20. Suggested links between the break-back thrusts in the Kohistan arc and
the back-thrusts and back-folds at Besham.

in shortening rate between that measured from field studies and that estimated from plate
reconstruction must be accommodated by extra strains north of the MMT, within the Pamirs,
Tibet and west China–central Asia.

7. The Himalayas show a range in overthrust directions, particularly around the western
Himalayan arc. Reasons for these variations, whether they are caused by rotation around
lateral tips to the thrust system or to spreading of thickened Tibetan–Himalayan lithosphere,
are discussed by Coward et al. (1987). The Kohistan thrust system involved tectonic transport
to the S or SSE as determined from minor structures and larger-scale thrust fault geometries.
There is major interference between the western Himalayan and Kohistan thrust directions,
whereby both thrust systems hinder lateral propagation of the other set. Hence accurate
restorations and models of lithospheric flexing require more detailed three-dimensional surveys
of the region. Such work is in progress.

8. The most recent Kohistan thrust movements involved above 30 km displacement on the
Salt Range thrusts. These structures terminate laterally at the frontal ranges of the Himalayan
arc, at the Mangla Samwali hills, formed at about 2.7–1.5 Ma. The zone of Himalayan uplift,
folding and thickening can be traced northwards, to the west of the Hazara syntaxis, to form
a zone of active seismicity (the Indus–Kohistan Seismic Zone of Seeber et al. (1981)). This
probably links with uplift of the Nanga Parbat mountains, where paired strike-slip and thrust
movements on the western margin of the massif affect Recent river gravels and appear to have
stepped southwestward with time. This probably reflects the SW migration of the lateral tip
to the Himalayan system. The foreland basin seems to be related to crustal thickening in the
Himalayan arc rather than in Kohistan, where younger Cainozoic deformation is thin skinned.
Hence the foreland basin trends WNW, or NW, oblique to Kohistan structures and the
Peshawar Basin may be a foreland depression superimposed on top of Hazara thrusts by the
excessive crustal thickening in the western Himalayan arc, in particular in the Nanga Parbat
region. Presumably, only the presence of salt beneath the frontal Kohistan thrusts has allowed
continued movement without the necessary build up of wedge shape.

Fieldwork was supported by NERC Grant GR3/6113 awarded to M. P. C.; R. W. H. B. has been funded by this NERC grant and the Royal Society, John Jaffa Fellowship. M. P. W. is funded by a NERC studentship, A. F. C. by a Northern Ireland studentship and C. N. I. by a BP studentship. We acknowledge fruitful discussions with fellow researchers from Pakistan and the U.S.A., from the institutions mentioned in the introduction, plus many more.

## REFERENCES

Asrarullah, 1967 *Pakistan Sci. Conf. 18th/19th Jamshoro Proc. III Abs.*, F3–F4

Baker, D. M., Lillie, R. J., Yeats, R. S., Johnson, G. D., Yousef, M. & Havid Zamid, A. S. 1988 *Geology.* (In the press.)

Bard, J. P. 1983 *Earth planet. Sci. Lett.* **65**, 133–144.

Bard, J. P., Maluski, H., Matte, Ph. & Proust, F. 1980 *Spec. Issue Geol. Bull. Peshawar Univ.* **13**, 87–93.

Bossart, P., Dietrich, D., Greco, A., Ottiger, R. & Ramsay, J. G. 1984 *J. Kashmir Geol.* **2**, 19–36.

Bossart, P. & Ottiger, R. 1988 *J. geol. Soc. Lond.* (In the press.)

Burbank, D. W. 1983 *Earth planet Sci. Lett.* **64**, 77–92.

Burbank, D. W. & Raynolds, R. G. H. 1984 *Nature, Lond.* **311**, 114–118.

Burg, J.-P. & Chen, G.-M. 1984 *Nature, Lond.* **311**, 219–223.

Butler, R. W. H. 1987 *Terra Cogn.* **7**, 152.

Butler, R. W. H. & Coward, M. P. 1988 *NATO ASI Volume.* (In the press.)

Butler, R. W. H., Coward, M. P., Harwood, G. M. & Knipe, R. J. 1987 *Dynamical geology of salt and related structures*, pp. 339–418. London: Academic Press.

Butler, R. W. H. & Prior, D. 1988*a* *Geol. Rdsch.* (In the press.)

Butler, R. W. H. & Prior, D. 1988*b* *Nature, Lond.* (In the press.)

Calkins, J. A., Oldfield, T. Q., Abdullah, S. K. M. & Tayyat Ali, S. 1975 *Prof. Pap. U.S. geol. Surv.*, 716C.

Chang Chengfa *et al.* 1986 *Nature, Lond.* **323**, 501–507.

Chaudhury, H. M. 1965 *Indian J. met. Geophys.* **17**, 385–394.

Coward, M. P. 1982 *J. struct. Geol.* **4**, 181–197.

Coward, M. P. 1983 *J. struct. Geol.* **5**, 113–123.

Coward, M. P. & Butler, R. W. H. 1985 *Geology* **13**, 415–420.

Coward, M. P., Butler, R. W. H., Asif Khan, M. & Knipe, R. J. 1987 *J. geol. Soc. Lond.* **144**, 377–391.

Coward, M. P., Jan, M. Q., Rex, D., Tarney, J., Thirlwell, M. & Windley, B. F. 1982 *J. geol. Soc. Lond.* **139**, 299–308.

Coward, M. P., Windley, B. F., Broughton, R., Luft, I. W., Petterson, M., Pudsey, C., Rex, D. & Asif Khan, M. 1986 In *Collision tectonics* (ed. M. P. Coward & A. C. Ries), Spec. Publ. Geol. Soc. Lond. vol. 19, pp. 203–219.

Davis, D. M. & Engelder, T. 1985 *Tectonophysics* **119**, 67–88.

Davis, D. M., Suppe, J. & Dahlen, F. A. 1983 *J. geophys. Res.* **88**, 1153–1172.

Duroy, Y. 1986 M.Sc. thesis, Oregon State University, Corvallis. (**76** pages.)

Duroy, Y., Farah, A., Lillie, R. J. & Malinconico, L. L. 1988 *Geol. Soc. Am. Spec. Publ.* (In the press.)

Farah, A., Mirza, M. A., Ahmad, M. A. & Butt, M. H. 1977 *Bull. geol. Soc. Am.* **88**, 1147–1155.

Gansser, A. 1964 *Geology of the Himalayas.* (289 pages.) London: Wiley Interscience.

Gansser, A. 1981 *Zagia, Hindu Kush, Himalaya, geodynamic evolution* (ed. H. K. Gupta & F. M. Delaney), Am. Geophys. Union, Geodynamics Series, vol. 3, pp. 111–121.

Gee, E. R. 1980 *Salt range series*, Pakistan Geological Maps at 1: 50,000, 6 sheets. U.K.: Directorate Overseas Surveys.

Herren, E. 1987 *Geology* **15**, 409–413.

Jackson, J. & McKenzie, D. 1984 *Geophys. Jl R. astr. Soc.* **77**, 185–264.

Johnson, G. D., Johnson, N. M., Opdyke, N. D. & Tahirkheli, R. A. K. 1979 *Geodynamics of Pakistan* (ed. A. Farah & K. DeJong), pp. 149–165. Quetta: Geol. Surv. Pakistan.

Johnson, G. D., Zeitler, P., Naeser, C. W., Johnson, N. M., Summers, D. M., Frost, C. D., Opdyke, N. D. & Tahirkheli, R. A. K. 1982 *Palaeogeogr. Palaeoclimatol. Palaeoecol.* **37**, 63–93.

Kaila, K. L. 1981 *Zagros, Hindu Kush, Himalaya, geodynamic evolution* (ed. H. K. Gupta & F. M. Delaney), Am. Geophys. Union, Geodynamics series, vol. 3, pp. 272–293.

Kazmi, A. H., Lawrence, R. D., Anwar, J., Snee, L. W. & Hussain, S. S. 1986 *Econ. Geol.* **81**, 2022–2028.

Kazmi, A. H., Lawrence, R. D., Dawood, H., Snee, L. W. & Hussain, S. S. 1984 *Geol. Bull. Peshawar Univ.* **17**, 127–144.

Latif, M. A. 1974 *Geol. Bull. Punjab Univ.* **10**, 1–20.

Lawrence, R. D. & Ghauri, A. A. K. 1983 *Geol. Bull. Peshawar Univ.* **16**, 185–186.

Le Fort, P. 1986 In *Collision tectonics* (ed. M. P. Coward & A. C. Ries), Spec. Publ. Geol. Soc. Lond. vol. 19, pp. 159–172.

Le Fort, P., Debon, F. & Sonet, J. 1980 *Geol. Bull. Peshawar Univ.* **13**, 51–62.

Lyon-Caen, H. & Molnar, P. 1985 *Tectonics* **4**, 513–538.

Malinconico, L. 1988 *Bull. geol. Soc. Am.* (In the press.)

Maluski, H. & Matte, Ph. 1984 *Tectonics* **3**, 1–18.

Meissner, C. R., Master, J. M., Rashil, M. A. & Hussain, M. 1974 *Prof. Pap. U.S. geol. Surv.*, **716D** (30 pages.)

Molnar, P. & Tapponnier, P. 1975 *Science, Wash.* **189**, 419–426.

Opdyke, N. D., Johnson, N. M., Johnson, G. D., Lindsay, E. H. & Tahirkheli, R. A. K. 1982 *Palaeogeogr. Palaeoclimatol. Palaeoecol.* **37**, 1–15.

Patriat, P. & Achache, J. 1984 *Nature, Lond.* **311**, 615–621.

Pilgrim, G. E. 1910 *Rec. geol. Surv. India* **40**, 185–205.

Platt, J. 1986 *Bull. geol. Soc. Am.* **97**, 1037–1053.

Pudsey, C. J., Coward, M. P., Luff, I. W., Shackleton, R. M., Windley, B. F. & Jan, M. Q. 1985 *Trans. R. Soc. Edinb.* **76**, 463–479.

Royden, L. & Burchfiel, C. 1987 *Continental Extensional Tectonics* (ed. M. P. Coward, J. F. Dewey & P. Hancock), Spec. Publ. Geol. Soc. Lond. vol. **28**, pp. 611–619.

Scharer, V., Hamet, J. & Allègre, C. J. 1984 *Earth planet. Sci. Lett.* **67**, 327–339.

Schindewolf, D. H. & Seilacher, A. 1955 *Abh. math.-naturw. Kl. Akad. Wiss. Mainz* **10**, 446.

Searle, M. P. 1983 *Trans. R. Soc. Edinb.* **73**, 205–219.

Searle, M. P. 1985 *J. struct. Geol.* **7**, 129–144.

Searle, M. P. 1986 *J. struct. Geol.* **8**, 923–936.

Searle, M. P., Windley, B. F., Coward, M. P., Cooper, D. J. W., Rex, A. J., Rex, D., Li Tingdong, Xiao Xuchang, Jan, M. Q., Thakur, V. C. & Kumar, S. 1987 *Bull. geol. Soc. Am.* **98**, 678–701.

Seeber, L., Armbruster, J. & Quittmeyer, R. C. 1981 *Zagros, Hindu Kush, Himalaya, geodynamic evolution* (ed. H. K. Gupta & F. M. Delaney), Am. Geophys. Union, Geodynamic Series, vol. 3, pp. 215–242.

Shah, S. H. I. 1977 *Mem. geol. Surv. Pakistan* **12**. (138 pages.)

Tahirkheli, R. A. K. & Jan, M. Q. 1979 *Spec. Issue Geol. Bull Peshawar Univ.* **11**, 1–30.

Teichert, C. 1964 *Rec. geol. Surv. Pakistan* **11**, 1–2.

Treloar, P. J., Coward, M. P., Williams, M. P. & Asif Khan, M. 1988*a Bull. geol. Soc. Am.* (In the press.)

Treloar, P. J., Broughton, R. D., Coward, M. P., Williams, M. P. & Windley, B. F. 1988*b J. Metal Geol.* (In the press.)

Windley, B. F. 1983 *J. geol. Soc. Lond.* **140**, 849–866.

Yeats, R. S. & Hussain, A. 1987 *Bull. geol. Soc. Am.* **99**, 161–176.

Yeats, R. S., Khan, S. H. & Akhtar, M. 1984 *Bull. geol. Soc. Am.* **95**, 958–966.

Zeitler, P. K. 1985 *Tectonics* **4**, 127–151.

Zeitler, P. K., Johnson, N. M., Naeser, C. W. & Tahirkheli, R. A. K. 1982 *Nature, Lond.* **298**, 255–257.

## Discussion

A. BARNICOAT (*Department of Geology, University College of Wales, U.K.*). The metamorphism seen imbricated within the thrust sheets in the Besham area is of Barrovian type, and hence presumably caused by thermal relaxation following crustal thickening. Would one of the authors care to speculate on the nature and geometry of that thickening?

P. J. TRELOAR. We welcome Dr Barnicoat's offer to speculate about metamorphism and crustal thickening in North Pakistan, but would rather follow our established practice and confine ourselves to a description of the facts and their implications. Himalayan age metamorphism at Besham accompanied a rapid pressure increase with the peak synchronous with the main D1 ductile shearing event. We equate this syn-tectonic metamorphism to thickening of the Indian Plate during its subduction underneath the over-riding Kohistan complex, rather than to metamorphism during thermal relaxation after crustal thickening. The imbrication of the metamorphic pile during thrusting late in the period of southward shearing in the footwall of the MMT generated an overall 'inverted' metamorphic profile in that it stacked higher-grade rocks on top of lower-grade ones. Such a tectonic inversion is the norm in the Indian Plate rocks south of the MMT in North Pakistan where the late thrusting stacked a number of internally imbricated metamorphic blocks on top of each other. Unfortunately, either individual slices are too thin, or errors on the thermobarometry too large, to state whether or not the metamorphism was originally the right way up (temperature increasing with depth), although over the region as a whole that is what we would expect to have been the case. What is noticeable is that metamorphism was diachronous with the syn-D1 metamorphic peak at Besham pre-dating the syn- to post-D2 peak in the Hazara region where textural relations imply a more normal post or late thickening thermal relaxation. Perhaps part of the reason for the difference is that the rocks at Besham actually underwent some subduction under Kohistan whereas those in Hazara saw nothing more than normal orogenic thickening in the footwall of the MMT.

*Phil. Trans. R. Soc. Lond.* A **326**, 117–150 (1988)    [ 117 ]

*Printed in Great Britain*

# Collision tectonics of the Ladakh–Zanskar Himalaya

By M. P. Searle, D. J. W. Cooper and A. J. Rex

*Department of Geology, University of Leicester, Leicester LE1 7RH, U.K.*

[Plates 1–4]

The collision of the Indian Plate with the Karakoram–Lhasa Blocks and the closing of Neo-Tethys along the Indus Suture Zone (ISZ) is well constrained by sedimentologic, structural and palaeomagnetic data at *ca.* 50 Ma. Pre-collision high *P*-low *T* blueschist facies metamorphism in the ISZ is related to subduction of Tethyan oceanic crust northwards beneath the Jurassic–early Cretaceous Dras island arc. The Spontang ophiolite was obducted southwestwards onto the Zanskar shelf before the Eocene closure (D1). The youngest marine sediments on the Zanskar shelf and along the ISZ are Lower Eocene, after which continental molasse deposition occurred.

After ocean closure, thrusting followed a SW-directed piggy-back sequence (D2). This has been modified by late-stage breakback thrusts, overturned thrusts and extensional normal faulting associated with culmination collapse and underplating. The ISZ and northern Zanskar shelf sequence are affected by late Tertiary N-directed backthrusting (D3), which also affects the Indus molasse. A 50 km wide 'pop-up' zone with divergent thrust vergence was developed across the Zanskar Range. Balanced and restored cross sections indicate a minimum of 150 km of shortening across the Zanskar shelf and ISZ.

Post-collision crustal thickening by thrust stacking resulted in widespread Barrovian metamorphism in the High Himalaya that reached a thermal climax during Oligocene–Miocene times. Garnet–biotite–muscovite ± tourmaline granites were generated by intracrustal partial melting during the Miocene within the Central Crystalline Complex. Their emplacement on the hangingwall of localized ductile shear zones was associated with SW-directed thrusting along the Main Central Thrust (MCT) zone and concomitant culmination collapse normal faulting along the Zanskar Shear Zone (ZSZ) at the top of the slab. Metamorphic isograds have become inverted by post-metamorphic SW-verging recumbent folding and thrusting along the base of the High Himalayan slab. Along the top of the slab, isograds are the right way up but are structurally and thermally telescoped by normal faulting along the ZSZ.

## 1. Introduction

It is now widely accepted that the collision between the Indian Plate and the collage of previously sutured micro-continental plates of Central Asia occurred during the mid- to late Eocene, at approximately 50–45 Ma (see, for example, Allègre *et al.* 1984; Searle *et al.* 1987). The timing of terminal collision of the two plates is deduced from (i) the ending of marine sedimentation in the Indus Suture Zone (ISZ), (ii) the beginning of continental molasse sedimentation along the suture zone, (iii) the ending of Andean-type calc-alkaline magmatism along the Trans-Himalayan (Ladakh–Kohistan–Gangdese) batholith and (iv) the initiation of the major collision-related thrust systems in the Himalayan Ranges.

Palaeomagnetic data indicate that over 2000–2500 km of southern (Neo-) Tethys separated the Indian and Central Asian Plates during the late Cretaceous, with initial collision at about 55 Ma (Klootwijk 1979) and terminal collision at about 40 Ma (Molnar & Tapponnier 1975; Klootwijk & Radhakrishnamurthy 1981). The northward relative motion of the Indian Plate decreased threefold at 40 Ma from average rates of $14.9 \pm 4.5$ cm a$^{-1}$ to $5.2 \pm 0.8$ cm a$^{-1}$ (Pierce 1978). Seafloor spreading rates in the Indian Ocean also decreased drastically at anomaly 22, which corresponds in time to *ca.* 50 Ma (Sclater & Fisher 1974; Johnson *et al.* 1976). Major directional shifts in the relative motion of the Indian Plate at anomalies 22 and 21 (50–48 Ma) are also thought to indicate the onset of collision (Patriat & Achache 1984). Another major shift occurred at anomaly 13 (36 Ma) after which India resumed stable northwards convergence with a constant rate of 5 cm a$^{-1}$ (Patriat & Achache 1984).

The post-collision crustal shortening in the Indian Plate can only be deduced from the restoration of balanced cross sections across the Himalaya. It is widely recognized that in

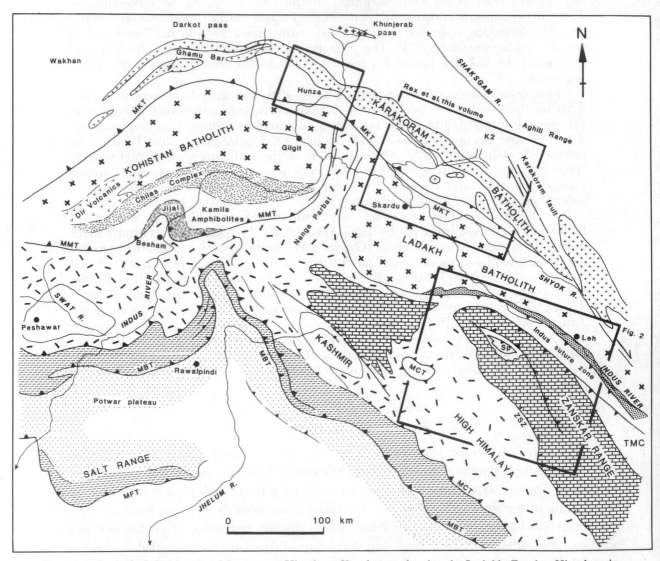

FIGURE 1. Geological sketch map of the western Himalaya–Karakoram showing the Ladakh–Zanskar Himalaya in box and central Karakoram described in the accompanying paper (Rex *et al.* this symposium). Abbreviations: MKT, Main Karakoram Thrust; MMT, Main Mantle Thrust; MCT, Main Central Thrust; MBT, Main Boundary Thrust; MFT, Main Frontal Thrust; Sp, Spontang ophiolite.

general terms, southward-propagating thrust stacking had occurred across the Himalaya since the mid-Eocene collision and suturing (see, for example, Molnar 1984; Mattauer 1986; Searle *et al.* 1987).

The climax of crustal shortening and thrust stacking occurred during the mid-Tertiary in the High Himalaya and Zanskar Ranges, with thrusts propagating southwestwards from the Main Central Thrust (MCT) system to the Lesser Himalaya and the late Tertiary Main Boundary Thrust (MBT) system (figure 1). The youngest thrusts occur along the southern boundary of the Himalaya, in the Siwalik molasse deposits of the Indian foreland basin, where the Main Frontal Thrust (MFT) terminates at tip lines below the Indo–Gangetic plains.

Four major tectonic zones constitute the Ladakh–Zanskar Himalaya. These are, from north to south, the Ladakh (Transhimalayan) batholith, the Indus Suture Zone, the Tibetan-Tethys (Zanskar shelf) Zone, and the High Himalaya. A generalized geological map depicting these

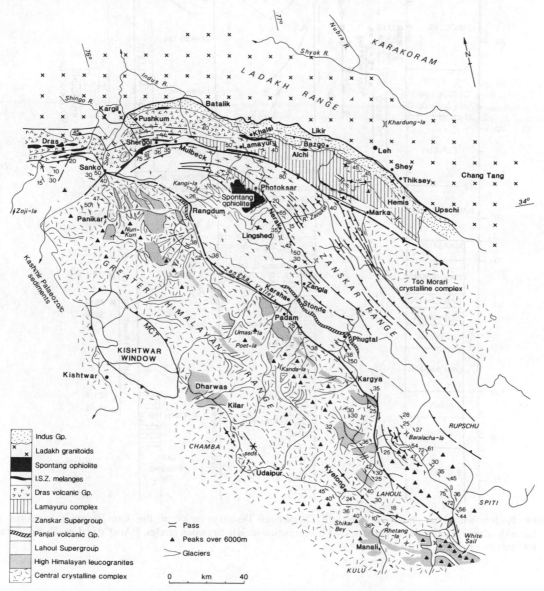

FIGURE 2. Geological map of Ladakh–Zanskar (after Searle 1983, 1986).

zones is shown in figure 2. This paper reviews and summarizes the geology of these four zones in the western Himalaya and discusses the timing of deformation, the mechanism of crustal thickening, the amounts of crustal shortening and models for collision tectonics in the Himalaya. Figure 3 shows a late Cretaceous and Tertiary time chart for the Ladakh–Zanskar Himalaya (after Searle 1983, 1986) updated with all radiometric ages and stratigraphic time spans for the four tectonic zones.

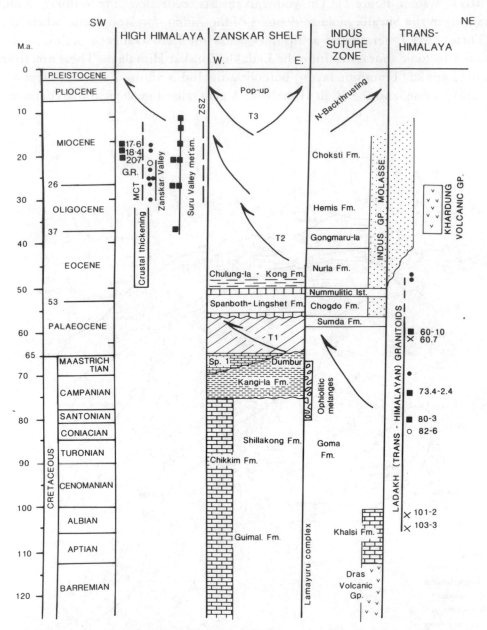

FIGURE 3. Revised time chart for the late Cretaceous–Tertiary rocks of the four tectonic zones in the Ladakh–Zanskar Himalaya. Crosses, U–Pb ages; squares, Rb–Sr; open circles, $^{39}$Ar–$^{40}$Ar; dots, K-Ar. See text for sources of data and discussion.

## 2. STRUCTURAL FRAMEWORK

Searle (1983, 1986) defined three major stages of deformation in the Western Himalaya: (1) a pre-collision ophiolite obduction phase (T1: 75–60 Ma), (2) a continental collision stage (T2: 45–25 Ma) and (3) a post-collision stage (T3: 15–0 Ma). The timing of motion on thrusts was determined by combining detailed structural mapping with stratigraphy and fault geometry (folded thrusts, truncated folds, truncated thrusts, etc.). In the High Himalaya, offsets were estimated by determining the differences in $P$–$T$ conditions in the footwall and hangingwall of shear zones.

Determination of the sequence of thrusting is critical to the understanding of the geology of Ladakh–Zanskar. Therefore it is necessary at this point to define four major types of thrusts. Piggy-back thrusts propagate in-sequence towards the foreland (India) and preserve intact stratigraphy in the footwall. They cut up-section in the transport direction and place older rocks over younger (see Butler (1987) for a review). Out-of-sequence thrusts develop in the hangingwall of earlier thrusts and may eliminate some stratigraphic section in the footwall. They may truncate folds and thrusts in the footwall and place younger rocks on to older. Breaching thrusts are thrusts that cut through the roof thrust of a duplex and may cause small-scale reversals of stacking order. Breakback thrusts cut through a previously assembled stack of thrust sheets and place originally lower, younger thrust sheets over originally higher, older thrust sheets. They may truncate earlier folds and thrusts in the footwall and therefore cross sections must be sequentially restored in a reverse time sequence.

## 3. LADAKH BATHOLITH

The Ladakh batholith, a part of the 2000 km long Transhimalayan batholith, is situated to the north of the ISZ in the Ladakh Ranges (figure 2). The Transhimalaya belt continues westward to include the Kohistan batholith in northern Pakistan (Petterson & Windley 1985) and eastward to include the Gangdese batholith in south Tibet (Allègre *et al.* 1984; Tapponnier *et al.* 1981). The Ladakh batholith ranges in composition from olivine–norite to leucogranites with grandiorites dominating (Thakur 1981; Honegger *et al.* 1982). Olivine–orthopyroxene-bearing norites, gabbros, diorites, granodiorites, granites and leucogranites are all represented and collectively represent a continental, subduction-related batholith (Honegger *et al.* 1982). The presence of granitic gneisses and metasedimentary sequences in the Ladakh Range provides evidence for the existence of precursory continental crust (Honegger & Raz 1985). Furthermore, a substantial component of inherited lead in zircons extracted from samples of the batholith implies involvement of continental crust during petrogenesis (Schärer *et al.* 1984). Initial $^{87}Sr/^{86}Sr$ ratios of around 0.704 (Schärer *et al.* 1984) and 0.705 (Honegger *et al.* 1982) are unenriched, inferring a primary mantle derivation of the magmas. The amount of isotopic enrichment due to the envisaged crustal component remains to be quantified.

Available geochronological determinations indicate that the Ladakh batholith is composite and dominated by pre-collisional magmatic events. U–Pb zircon ages of $103\pm3$ Ma (Honegger *et al.* 1982) and $101\pm2$ Ma (Schärer *et al.* 1984) from samples collected near Kargil refer to important mid-Cretaceous magmatism. A second U–Pb determination of monazite/allanite from a biotite–granite near Leh gave an age of $60.7\pm0.4$ Ma, whereas a Rb–Sr isochron age of $73\pm2.4$ Ma (Schärer *et al.* 1984) and a $^{39}Ar/^{40}Ar$ age of $82\pm6$ Ma (Schärer & Allègre 1982) may indicate a continuity of magmatic activity for some 40 Ma, finally terminating with continent–continent collision (figure 3). However, the latter two age determinations may also be interpreted as cooling events on the argument that these isotopic systems have lower blocking temperatures (Schärer *et al.* 1984). Likewise, a span of K–Ar mineral ages of 50–40 Ma is considered to reflect a cooling period associated with uplift.

The termination of granitic magmatism in the Ladakh batholith is contemporaneous with the closing of Neo-Tethys along the ISZ and the change from marine to continental sedimentation. Clasts of granites, andesites and rhyolites are particularly widespread in the conglomerates of the Chogdo, Nurla and Hemis Formations of the Indus Group molasse, which unconformably overlies the batholith. Volcanic rocks overlying the Ladakh batholith include andesites and rhyolites (Srimal *et al.* 1987). Late-stage garnet–muscovite leucogranitic dykes intrude the granite and may reflect a final phase of intracrustal melting after the 50 Ma collision along the ISZ.

Granitoids of the Ladakh batholith intrude the Dras island arc volcanics and rocks of the suture zone around Kargil. This provides strong evidence that the Dras island arc was accreted to the Karakoram–Lhasa Blocks by the mid-Cretaceous and that the Shyok (Northern) Suture between the Ladakh and Karakoram terranes closed in the mid-Cretaceous (Coward *et al.* 1986; Pudsey 1986; Searle *et al.* 1988) and not in the late Tertiary (Thakur 1981, 1987).

The Ladakh batholith has undergone an unknown amount of post-collision shortening evidenced by thrusting within the granitoids. Near Kargil, a number of late stage shear zones with mylonitized amphibolites and greenschists cut intrusive rocks of the batholith.

## 4. INDUS SUTURE ZONE

The Indus Suture Zone (ISZ) defines the zone of the collision between the Indian Plate and the Karakoram–Lhasa Block to the north (Gansser 1964, 1977; Allègre *et al.* 1984; Searle *et al.* 1987) and can be traced for over 2000 km from Kohistan and Ladakh in the east right across southern Tibet to the NE Frontier region of India–Burma. In Tibet it is commonly referred to as the Yarlung–Tsangpo Suture; in the western Himalaya it is the Indus Suture Zone. To the west of Ladakh, the ISZ is folded around the giant Nanga Parbat fold, a 35 km wavelength, upright anticlinorium that exposes Indian Plate gneisses equivalent to the Central Crystalline Complex in its core (figure 1). The ISZ has been offset by the late Tertiary culmination of Nanga Parbat and downfaulted west of Nanga Parbat along the Rakhiot Fault zone. It continues westwards as the Main Mantle Thrust (MMT) across southern Kohistan (Tahirkheli & Jan 1979).

In Ladakh, the Indus Suture is bounded to the south by the backthrust shelf sediments of the Zanskar Supergroup and the Tso Morari crystalline complex, and to the north by the Ladakh batholith. The geology of the ISZ has been described by Gupta & Kumar (1975), Shah *et al.* (1976), Andrieux *et al.* (1981), Fuchs (1977, 1979), Bassoullet *et al.* (1978, 1981), Frank *et al.* (1977), Baud *et al.* (1982), Brookfield & Andrews-Speed (1984*a*,*b*), Thakur (1981, 1987) and Searle (1983, 1986).

The Indus Suture Zone in Ladakh consists essentially of three major linear thrust belts, the Lamayuru complex, the Nindam–Dras Volcanic Group and the Indus Group molasse. They are separated by major fault zones or ophiolitic melange belts (figure 7*a*, plate 1). Figure 4 is a palaeogeographic reconstruction of the Neo-Tethyan basin showing the pre-Eocene stratigraphy of the rocks now preserved in the ISZ in Ladakh.

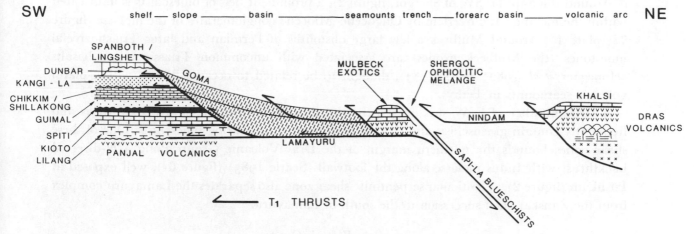

FIGURE 4. Palaeogeographic reconstruction of the Ladakh Tethys showing stratigraphic units along the Indus Suture Zone in Ladakh.

### (a) Lamayuru complex

This complex is the passive-margin deep-water sediments, consisting mainly of shales with indurated sandstones and limestones, that show evidence of distal turbidity current deposition (Frank et al. 1977; Fuchs 1977, 1979, 1982). The Lamayuru complex rocks are now preserved as a series of SW-dipping inverted imbricate slices, the early SW-directed thrusts having been rotated during the late Tertiary phase of N-directed backthrusting. In the Indus Suture Zone the Lamayuru complex has been dated at late Triassic to (?)mid-Jurassic (Fuchs 1977; Bassoullet et al. 1981). The highest preserved stratigraphic units are thick-bedded limestone conglomerates and oolitic grainstones that crop out as dismembered sheets between Prinkiti-la (Lamayuru gompa) and the Namika-la east of Mulbeck (figure 2), and record a period of platform margin collapse.

On the Zanskar shelf, a sequence of slates of possible Cretaceous age (Goma Formation) (Colchen et al. 1986) crops out in a thrust slice beneath the Spontang ophiolite. These rocks are part of the Lamayuru complex (Bassoullet et al. 1978, 1981) and were emplaced along with the overlying Dras volcanic, Spontang ophiolite and mélange units onto the Zanskar shelf from the Indus Suture before ocean closure.

### (b) Ophiolitic mélanges

Narrow zones of ophiolitic mélange bound both northern and southern margins of the Dras volcanic and Lamayuru complex rocks along the ISZ. The widest zone separates the Lamayuru complex from the Nindam Formation–Dras Volcanic Group (the Shergol ophiolitic mélange of Thakur (1981) and Searle (1983)). Blocks consist of MORB–tholeiitic basalts, alkali basalts and amygdaloidal pillow lavas, agglomerates, gabbros, rodingites, dunites, harzburgites, radiolarian cherts and deep-sea sediments.

Pelagic limestones intercalated within the mélange in the Mulbeck–Pushkum area (figure 2) have yielded Campanian–Maastrichtian Foraminifera (Colchen et al. 1986). The late Cretaceous age of the ophiolitic mélange corresponds to the proposed timing of obduction of the Spontang ophiolite on to the Zanskar shelf (Searle (1986, 1987; see also following section).

Around the Sapi-la, SW of Shergol (figure 2), a prominent belt of blueschists is imbricated within the mélange and overlain by Oligocene–Miocene conglomerates of the molasse (figure 7b, plate 1). Around Mulbeck a few large olistoliths of Permian and later Triassic reefal limestones (the Mulbeck exotics) are associated with uncommon Triassic alkali basalts (Honegger et al. 1982; Searle 1983), thought to be related to oceanic island (within-plate) volcanic seamounts in Tethys.

A large number of E–W striking serpentinite shear zones cut the suture zone rocks and frequently contain greenschists and crossite or glaucophane-bearing blueschists. One of these shear zones bounds the northern margin of the Dras Volcanic Group along a N-directed backthrust with Indus molasse along the footwall (Searle 1983) (figure 6e), well exposed at Pushkum (figure 2). A lenticular serpentinite shear zone also separates the Lamayuru complex from the Zanskar shelf succession to the south of Lamayuru.

## (c) Dras Volcanic Group

A thick sequence of clinopyroxene, hornblende and plagioclase–phyric basalts, andesites and dacites together with intercalated volcaniclastics and minor amounts of tholeiitic pillow lavas make up the Dras Volcanic Group in western Ladakh (Wadia 1937; Thakur 1981; Honegger et al. 1982; Searle 1983). They represent the lateral equivalents to the Chalt volcanics in the upper part of the Kohistan island arc sequence in Pakistan (Coward et al. 1986), but are not present anywhere along the ISZ east of Ladakh. Uncommon radiolarian cherts interbedded in the Dras volcanic rocks west of Dras (figure 2) have yielded Callovian–Tithonian ages (Honegger et al. 1982). In central Ladakh, Orbitolina-bearing limestones of Albian–Aptian age (Khalsi limestone, figure 4) stratigraphically overlie the Dras volcanic rocks.

Basalts of the Dras Volcanic Group are enriched in K, Ba, Rb, Sr and the light rare earth elements (LREEs) and depleted in the high field strength (HFS) elements Ti, Zr, Y, Nb and the HREEs relative to MORB (Honegger et al. 1982; Dietrich et al. 1983; Radhakrishna et al. 1984). Their geochemical signature may be compared with modern-day island-arc tholeiites and primitive calc-alkaline volcanic rocks (Pearce & Cann 1973).

Around Kargil, olivine–orthopyroxene-bearing gabbro–norites may represent relic magma chamber cumulates associated with the Dras volcanics. They are similar to the volumetrically abundant granulite–facies rocks of the Chilas complex at the base of the Kohistan island arc in Pakistan (Tahirkheli & Jan 1979; Coward et al. 1986).

In the Suru Valley, large amounts of volcaniclastic and some limestone blocks are included in the Dras Volcanic Group. In the eastern part of the ISZ in Ladakh, the Nindam Formation is a thick volcaniclastic sedimentary succession (Bassoullet et al. 1978, 1981; Colchen et al. 1986). The Dras volcanic rocks are interpreted as representing Jurassic–Lower Cretaceous island-arc volcanism above a N-dipping subduction zone within Tethys.

The Dras Volcanic Group is associated with a volcano-sedimentary succession, the Nindam Formation (Bassoullet et al. 1978). To the south and west of Rusi-la, agglomerates and tuffs with subordinate intrusive basalt are interbedded with shales and some bedded limestone that have yielded Cenomanian Foraminifera (I. Reuber, personal communication 1987). East of Trezpone, these limestones are compatible with a slope facies, containing conglomerates and graded beds of probable turbidite origin.

To the east, near Lamayuru, thick-bedded, graded volcaniclastic sandstones are interpreted as the products of high-density turbidity current deposition, and are interbedded with finer-grained low density turbidity current graded sands, silts and shales. Diagenetic remobilization of silica has resulted in the silicification of shales. Sediments of the Nindam Formation are interpreted as being derived from the Dras volcanic arc (Fuchs 1977; Thakur 1981).

### (d) Indus Group molasse

A continental clastic sequence approximately 2000 m in thickness, comprising alluvial fan, braided stream and fluvio-lacustrine sediments, constitutes the Indus Group (Brookfield & Andrews-Speed 1984a, b; Van Haver 1984). Clasts in the conglomerates were derived mainly from the uplifted and eroded Ladakh batholith to the north, but also from the suture zone itself (cherts, limestones, serpentinised peridotites) and from the Zanskar shelf carbonates to the south. Palaeocurrents show dominantly east to west basin axis-parallel sediment transport paths (Pickering et al. 1987). Near Kargil the Indus molasse unconformably rests on the Ladakh granitoids along the northern margin of the ISZ (figure 7c, plate 1).

The collision of India with the Karakoram and Lhasa Blocks and the closing of Tethys at 50 Ma marks the initiation of Indus Group molasse sedimentation (Searle et al. 1987). The Chogdo Formation at the base of the molasse rests stratigraphically above late Palaeocene shallow marine Sumda Formation limestones (figure 3). A Nummulites-bearing limestone interval (mid-upper Ilerdian–Cusinien (Van Haver 1984) marks the final marine incursion, and the ages of subsequent molasse deposits are poorly constrained, although they probably extend up into the Miocene (Tewari & Sharma 1972). The structure of the Indus molasse is dominated by late Tertiary N-directed backthrusting.

Figure 5 is a balanced cross section across the Indus Suture Zone along the Zanskar River section. Restoration of the late stage (T3) thrusts along this section (figure 6) indicates a minimum post-middle Miocene shortening of 36 km and an original minimum width of the Indus molasse basin of 60 km.

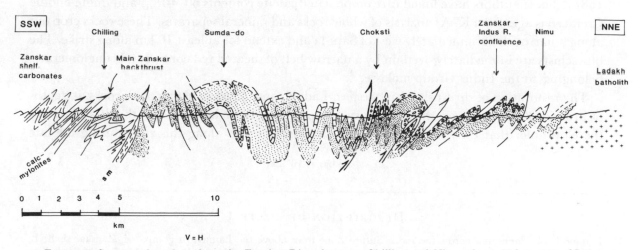

FIGURE 5. Structural section along the Zanskar River between Chilling and Nimu about 40 km west of Leh.

### (e) Subduction-related metamorphism

Imbricated slices of blueschists occur in the Shergol ophiolitic mélange between the Dras volcanic thrust sheets to the north and the Lamayuru complex thrust sheets to the south. They are associated with a complex series of metabasic, volcaniclastic and sedimentary (dominantly

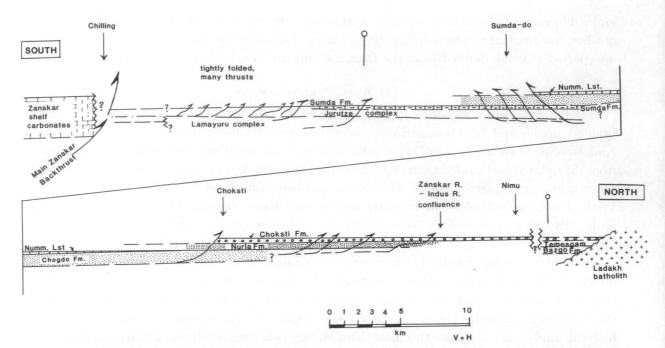

FIGURE 6. Restored section of the Indus Group along the Zanskar River shown in figure 5. The section has been restored sequentially along the Sumda Formation, the Nummulitic limestone and the Choksti Formation. The Jurutze complex is from data of Brookfield & Andrews-Speed (1984a).

red chert with some oolitic limestones) rocks and have a characteristic mineral assemblage: glaucophane–crossite–lawsonite–albite–phengite–chlorite–garnet–stilpnomelane–sphene, which indicate temperatures of formation of 400–450 °C at 10–12 kbar† (Honegger et al. 1985). These authors have found rare omphacite (jadeite contents 60–40%) and quote middle Cretaceous ages from K–Ar analysis of whole rocks and mineral separates. These rocks crop out along a narrow belt immediately west of Sapi-la and extend for at least 10 km along strike. The blueschists are immediately overlain by a narrow belt of sheared red continental conglomerates belonging to the Indus Group molasse.

The restoration of thrust sheets in western Ladakh and Zanskar places these rocks at depths of around 30 km between the Lamayuru complex and the Nindam–Dras thrust sheets. We therefore infer that the site of major oceanic subduction was along this zone during the late

† 1 kbar = $10^8$ Pa.

DESCRIPTION OF PLATE 1

FIGURE 7. (a) Panorama across the Indus Suture Zone from above the Lamayuru gompa. Z, Zanskar shelf; L, Lamayuru complex; O, ophiolitic mélange; N, Nimdam Formation; M, Indus molasse; Lb, Ladakh batholith. All units dip towards S and are backthrust towards N. (b) The Shergol ophiolitic mélange with glaucophane-bearing blueschists (B), overlain by Indus molasse (M), north of Sapi-la. This belt of mélange separates the Lamayuru complex (L) from the Dras Volcanic Group (DV). (c) Indus molasse (IM) unconformably overlying the Ladakh batholith (LG) and overthrusted by the Dras Volcanic Group (DV) along a N-directed backthrust. North of Kargil. (d) Upright anticline in the Indus molasse along the Zanskar River section. The lower molasse unit, the Chogdo Formation (Ch) is overlain by the marine Nummulitic limestone (Numm) and the continental molasse of the Nurla Formation (N). (e) Conglomerates of the Hemis Formation in the Indus molasse. Light-coloured pebbles are Ladakh granites, andesites and dacites, dark pebbles are serpentinites and red cherts.

FIGURE 7. For description see opposite.

FIGURE 8. Zanskar Supergroup stratigraphy exposed at (*a*) Rangdum and (*b*) Zangla. Pv, Panjal Volcanic Group; TK, Tamba Kurkur; H, Hanse; Z, Zozar; T, Tsatsa; K, Kioto; L, Laptal; S, Spiti; G, Guimal; Sh, Shillakong; KL, Kanji-la Formations.

*Phil. Trans. R. Soc. Lond. A, volume* 326

*Searle et al., plate* 3

FIGURE 15. (*a*) Aerial view looking east across the northern part of the Zanskar shelf showing the northern Zanskar 'pop-up' structure with N-facing folds in the N and S facing folds in the south. (*b*) SW-facing isoclinal fold in Spanboth Formation (S) marine limestones and Chulung-la Formation (C) continental slates, N of Dibling. (*c*) SW-facing syncline in Kanji-la Formation (KL) and Spanboth Formation (S) limestones, near Kanji-la, western Zanskar. (*d*) Tight to isoclinal SW-verging folding and imbricate thrusting in Lilang group limestones, Zanskar River, N of Zangla. (*e*) Giant sheath fold in the Mesozoic shelf carbonates of the northern Zanskar unit. View S from Rubering-la, eastern Zanskar. (*f*) Isoclinal fold in Zozar Formation limestones with flat-lying axial plane, in the western Zanskar zone of lateral spreading, unnamed southern tributary of Wakka Chu, near Shergol.

FIGURE 16. For description see opposite.

Cretaceous and Palaeocene (figure 4) with the oceanic crustal slab subducting to the north beneath the Dras island arc. Blueschists are also present on the hangingwall of the MMT in Kohistan (Shams 1980; Shams *et al.* 1980) and are dated at $80 \pm 5$ Ma ($^{39}$Ar–$^{40}$Ar) by Maluski & Matte (1984). Bard (1983) has related the high-pressure metamorphism in Kohistan to the southward obduction of the Kohistan island arc sequence during the late Cretaceous.

In eastern Ladakh, Virdi *et al.* (1977) and Jan (1985) have described glaucophane–epidote–garnet–rutile–stilpnomelane–actinolite–quartz–albite assemblages with uncommon lawsonite in metabasic rocks in the ISZ that are associated with the Nidar ophiolite complex (Thakur 1981).

### (f) Structural evolution

The structure of the Indus Suture Zone is complex and involves pre-collision oceanic thrusting associated with obduction of the Dras–Kohistan arc, and obduction of the Spontang ophiolite and associated sub-ophiolitic sheets. Additionally, continuous deformation from closure of the suture (50 Ma) to the present has been incurred. The early phases of thrusting are difficult to recognize because of subsequent extreme deformation during and after closure.

The presence of glaucophane, lawsonite and (?)omphacite in the Sapi-la blueschists indicate that they were formed at minimum pressures of around 10 kbar, corresponding to depths of around 30 km in an oceanic subduction environment (figure 4). Rapid thrust exhumation of these rocks is implied for them to retain their blueschist mineralogy without being thermally reset by decreasing $P$–$T$. Their incorporation into the ophiolitic mélange along with unmetamorphosed sedimentary and volcanic blocks must have been a shallow-level process. Late Cretaceous thrusting of the Lamayuru complex and Dras arc is also implied by the Santonian–Maastrichtian age on the ISZ ophiolitic mélanges and the ages of the blueschists in Ladakh and Kohistan (figure 3).

A major phase of SW-directed thrusting occurred along the north Indian continental margin during the Santonian to Palaeocene (80–60 Ma). Thrusts in the Lamayuru complex are not continuous upwards into the late Palaeocene-Eocene rocks. Two marine transgressions during the Upper Palaeocene (Sumda Formation) and lower Eocene (*Nummulitic* limestones) were responsible for final marine Tethyan sedimentation before ocean closure at *ca.* 50 Ma (figure 7*d*, plate 1). Thick continental molasse overlies the Lower Eocene limestones and the whole sequence has been deformed by dominantly upright folding and N-directed backthrusting after deposition of the molasse (late Miocene–Pliocene, figure 3). Along the northern part of the ISZ the Indus molasse rests unconformably on the eroded Ladakh batholith (figures 5 and 6).

---

### DESCRIPTION OF PLATE 4

FIGURE 16. (*a*) Amphibolites, marbles and meta-pelites in a SW-verging recumbent nappe, the Donara nappe, looking N from Bobang Gali, eastern Kashmir, W of Panikar. (*b*) Upper intrusive contact of a Himalayan granite in zone 4, east of Kun. (*c*) Lower thrust contact of a sheet intrusive leucogranite in zone 3, Suru Valley near Shafat. Note ductile foliation bending into the thrust plane at base of the granite. (*d*) The Zanskar Shear Zone north of the Pensi-la showing condensed isograd sequence associated with normal faulting. (*e*) The anatectic zone 1 granite-leucogranite zone along the Chenab Valley near Dharwas. White bands are garnet–muscovite–tourmaline leucogranites, dark bands are sillimanite or kyanite-bearing gneisses. (*f*) Isoclinal folding and ductile shearing in a leucogranite mixing zone within sillimanite grade gneisses, Suru Valley.

## 5. ZANSKAR SHELF ZONE

### (a) Stratigraphy of the Tibetan-Tethys Zone

Early reconnaissance studies of the Zanskar Range were made by Lydekker (1883), Gupta & Kumar (1975), Nanda & Singh (1977), Fuchs (1977, 1979, 1982) and Sharma & Kumar (1978). The present stratigraphic framework is largely based on these works and also those of Gaetani *et al.* (1980, 1983, 1985), Kelemen & Sonenfeld (1983), Baud *et al.* (1984) and Garzanti *et al.* (1986).

A Lower Palaeozoic orogenic cycle has been identified by Garzanti *et al.* (1986) and is marked by passive margin, shallow water and tidal flat, terrigenous and dolomitic sequences (Phe and Karsha Formations) passing up into Upper Cambrian deep-water shales and turbidites (Kurgiak Formation) which are overlain by Ordovician continental molasse conglomerates and sandstones (Thaple Formation). Subsequent Palaeozoic sedimentation is represented by the Devonian aeolian sandstones of the Muth Formation, Early Carboniferous limestones and evaporites (Lipak Formation), and marine deltaic clastics (Po Formation). These units collectively comprise the Lahoul Supergroup.

The Zanskar Supergroup (figure 8, plate 2, and figure 9) represents sedimentation on the north Indian Plate margin from the inception of Neo-Tethyan rifting in the Permian to final collision of the Indian Plate with the Karakoram–Lhasa Blocks of the southern Eurasian Plate. Rift-related continental tholeiitic volcanics and volcaniclastics (Panjal Volcanic Group) rest unconformably on the Precambrian to Carboniferous sediments of the Lahoul Supergroup, although this contact has been locally modified by normal faulting associated with the Zanskar Shear Zone.

Following an early clastic phase (Kuling Formation), rapid deepening of the newly formed continental margin resulted in a shallowing-upward sedimentary sequence, from outer-shelf argillaceous limestones with minor carbonates (Tamba Kurkur and Hanse Formations) to shallow water carbonate environments (Zozar, Tsatsa and Kioto Formations) (figure 8*a*, plate 2).

Carbonate platform evolution ended in the middle Jurassic (? Bathonian) and, following a 10 Ma hiatus, a condensed sequence of ferruginous oolites, sandstones and shales accumulated (Laptal Formation; Jadoul *et al.* 1985), overlain by thin Oxfordian–Lower Cretaceous shales of the Spiti Formation, representing the local response to a regional Tethyan deepening event (figure 8*b*, plate 2). Coarser clastic input (Guimal Formation) up until the Albian marked the end of passive margin sedimentation patterns, and subsequent sedimentation is interpreted as a response to the late Cretaceous obduction of the Spontang ophiolite.

During the Cenomanian–Turonian, deep-water, nodular limestones (Chikkim Formation) were deposited in an interior continental margin environment, whereas along the platform margin, collapse of the continental margin caused an erosion surface to cut down to the Kioto Formation, above which deeper-water pelagic limestones (Shillakong Formation) were deposited. Intraformational unconformities and conglomerate horizons indicate increasingly unstable conditions along the continental margin. A rapid increase in sedimentation rates during the Campanian is recorded in the Kanji-la Formation (Gaetani *et al.* 1983), and is modelled as a response to foredeep development ahead of the Spontang ophiolite. The Kanji-la Formation is a diachronous, prograding sequence of hemipelagic marls and turbidites, which

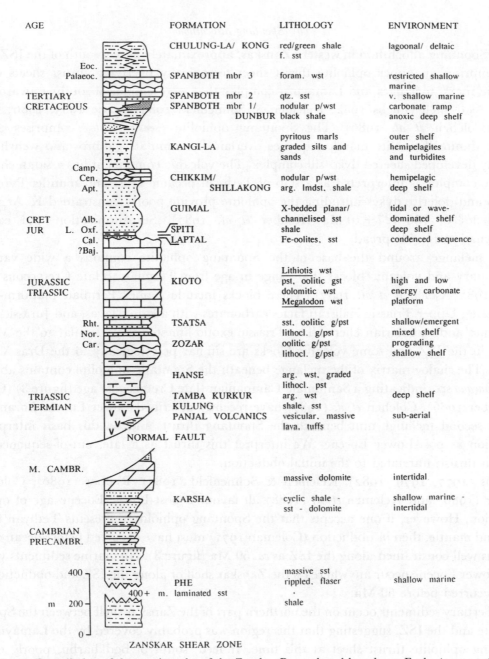

| AGE | | FORMATION | LITHOLOGY | ENVIRONMENT |
|---|---|---|---|---|
| | Eoc. | CHULUNG-LA/ KONG | red/green shale f. sst | lagoonal/ deltaic |
| | Palaeoc. | SPANBOTH mbr 3 | foram. wst | restricted shallow marine |
| TERTIARY | | SPANBOTH mbr 2 | qtz. sst | v. shallow marine |
| CRETACEOUS | | SPANBOTH mbr 1/ | nodular p/wst | carbonate ramp |
| | | DUNBUR | black shale | anoxic deep shelf |
| | | KANGI-LA | pyritous marls graded silts and sands | outer shelf hemipelagites and turbidites |
| | Camp. Cen. Apt. | CHIKKIM/ SHILLAKONG | nodular p/wst arg. lmdst. shale | hemipelagic deep shelf |
| CRET JUR | Alb. L. Oxf. Cal. ?Baj. | GUIMAL SPITI LAPTAL | X-bedded planar/ channelised sst shale Fe-oolites. sst | tidal or wave- dominated shelf deep shelf condenced sequence |
| JURASSIC TRIASSIC | | KIOTO | Lithiotis wst pst, oolitic gst dolomitic wst Megalodon wst | high and low energy carbonate platform |
| | Rht. Nor. Car. | TSATSA ZOZAR | sst. oolitic g/pst lithocl. g/pst oolitic g/pst lithocl. g/pst | shallow/emergent mixed shelf prograding shallow shelf |
| | | HANSE | arg. wst. graded lithocl. pst | |
| TRIASSIC PERMIAN | | TAMBA KURKUR KULUNG PANJAL VOLCANICS | arg. wst shale. sst vesicular. massive lava. tuffs | deep shelf sub-aerial |
| | | NORMAL FAULT | | |
| M. CAMBR. | | | massive dolomite | intertidal |
| CAMBRIAN | | KARSHA | cyclic shale - sst - dolomite | shallow marine - intertidal |
| PRECAMBR. | | | massive sst rippled. flaser | shallow marine |
| | | PHE | 400+ m. laminated sst shale | |
| | | ZANSKAR SHEAR ZONE | | |

FIGURE 9. Compilation of the stratigraphy of the Zanskar Range based largely on Fuchs (1979, 1982), Gaetani et al. (1983, 1985), Baud et al. (1984), and Garzanti et al. (1986).

is Campanian in the margin interior (SW) and Maastrichtian along the northern Zanskar shelf. A return to shallow water conditions is marked by the localized northwards progradation of a Maastrichtian carbonate ramp (Spanboth Formation, member 1) into an anoxic shale basin (Dunbur Formation). The Lower Palaeocene is missing in Zanskar (Gaetani et al. 1983) whereas above the unconformity the Spanboth (members 2 and 3)–Lingshet Formation is well dated by Foraminifera as Upper Palaeocene–Lower Eocene (Gaetani et al. 1983).

## (b) *Spontang allochthon*

The Spontang allochthon in western Zanskar, approximately 30 km south of the ISZ (figure 10) comprises an upper ophiolite thrust sheet tectonically overlying thrust sheets of Dras volcanic rocks, mélanges and Lamayuru complex that were emplaced southwestwards on to the Zanskar shelf (Fuchs 1982; Kelemen & Sonnenfeld 1983; Searle 1983, 1986; Reuber 1986; Colchen *et al.* 1986). The Spontang ophiolite (*sensu stricto*) comprises a harz-burgite–dunite–lherzolite mantle sequence overlain by cumulate gabbros and wehrlites and a poorly developed sheeted dyke–sill complex. The volcanic component has a MORB chemistry and the complex is interpreted as a slab of Tethyan oceanic crust and mantle. Pyroxenite, gabbro and dolerite dykes intruding the ophiolite provide poorly constrained K–Ar ages on amphiboles of $156-127\pm10$ Ma (Reuber *et al.* 1987). Serpentinization and calcium–metasomatism is widespread.

The mélange around the base of the Spontang ophiolite contains a wide variety of sedimentary and volcanic blocks that range in age from Permian to late Cretaceous (Fuchs 1979, 1982; Colchen *et al.* 1987). These blocks include Upper Permian platform-related carbonates, Upper Triassic Hallstatt facies carbonates with alkaline lavas, and Jurassic pelagic limestones and radiolarian cherts. The Triassic exotic limestones are similar to the Mulbeck exotics in the ISZ and some volcanic blocks are similar petrologically to the Dras Volcanic Group. The shaley matrix of the mélange beneath the Spontang ophiolite contains abundant *Globotruncana* sp., indicating a Santonian–Campanian (late Cretaceous) age (figure 3) (Colchen & Reuber 1986). Colchen *et al.* (1987) have recently described Lower Eocene Foraminifera from a second mélange unit beneath the Spontang thrust, and on this basis interpret the obduction as post-Lower Eocene. We interpret this thrust as a late, out-of-sequence, post-collision thrust, unrelated to the initial obduction.

Fuchs (1977, 1979, 1982), Kelemen & Sonnenfeld (1983), Reuber (1986), Colchen & Reuber (1986) and Kelemen *et al.* (1988) all favour a post-Lower Eocene age of ophiolite obduction. However, if one accepts that the Spontang ophiolite represents Tethyan oceanic crust and mantle, then its obduction (Coleman 1971) must have occurred before ocean closure, which is well constrained along the ISZ as *ca.* 50 Ma (figure 3). No marine sediments younger than Lower Eocene occur anywhere on the Zanskar shelf or along the ISZ and obduction must have occurred before 50 Ma.

No Tertiary sediments occur on the northern part of the Zanskar shelf between the Spontang ophiolite and the ISZ, suggesting that this region was probably covered by the Lamayuru and Spontang ophiolite thrust sheet at this time. Indeed, south of Bodkharbu, poorly exposed Lamayuru complex slates and limestones, with associated alkali basaltic sills, structurally overlie black slates and calcareous slates assigned to the Dunbur Formation (facies transitional to the Kanji-la Formation), implying probable Maastrichtian emplacement of the Spontang ophiolites.

Gaetani *et al.* (1983) have shown that extremely rapid sedimentation rates are recorded in the Kanji-la Formation, and Brookfield & Andrews-Speed (1984a) demonstrate the existence of a foredeep along the Zanskar continental margin during the late Cretaceous. We interpret this foredeep as the inboard response to loading of Tethyan thrust sheets on to the continental margin during a late Cretaceous obduction event. Obduction-related thrusting propagated from NE to SW and may have spanned 20 Ma (from 80 to 60 Ma; figure 3), comparable to the span of emplacing the Oman ophiolite and Bay of Islands, west Newfoundland ophiolite (Searle & Stevens 1984; Searle 1985, 1986, 1988).

### (c) Pre-collision thrusting

In the Zanskar Range, pre-collision (T1) thrusts are now only preserved within later, post-collision (T2, T3) thrust-bounded duplexes, and can only be demonstrated in the area around the Spontang ophiolite (figures 10 and 11). In this area at least three major phases of thrusting can be determined by simple cross-cutting relations. The recognition of out-of-sequence and breakback thrusts is crucial to the interpretation of tectonic events.

The Lamayuru complex rocks exposed in T1 thrust sheets around the base of the Spontang ophiolite are distal oceanic sediments (red radiolarian cherts and shales) and mélanges. The shaley sequences are hard to distinguish from the overlying Kanji-la and Dunbur Formations. However, the great thickness observed on the north side of the Spontang ophiolite when compared with the known stratigraphic thickness of the Dunbur Formation further to the west (ca. 150 m in Wakka Chu and Kangi Chu) suggests the presence of Lamayuru complex rocks below the ophiolite. A Jurassic ammonite from between the Spontang ophiolite and Photoksar thrust near the Spong River (Brookfield & Westermann 1982) further suggests these rocks belong to the Lamayuru complex. These shales extend westwards as narrow thrust-bounded slices towards Kanji (figures 12 and 13) where they are of more proximal oceanic facies (Goma Formation; figure 4). Thrust slices of Lamayuru rocks exposed at Singe-la (figure 10) tectonically overlie the complete Mesozoic shelf sequence along thrust contacts (T1) that pre-date the unconformably overlying Lingshet limestones (figure 11 a). Restoration of the folding in the Palaeocene–Lower Eocene Lingshet Formation shows that folds and thrusts in the Mesozoic sequence are not continuous upwards into the Lingshet limestones (figure 11 b). A major phase of shortening therefore occurred during the late Cretaceous, before deposition of the late Palaeocene–early Eocene shallow marine carbonates.

The same Lamayuru complex rocks exposed at Singe-la tectonically underlie the Spontang ophiolite and slices of Jurassic Lower Cretaceous Dras volcanics and mélanges around the margin of the Spontang Klippe. The pre-collision stacking order of thrust sheets around Spontang is directly comparable to that of the well-exposed and better-studied ophiolites such as the Oman and West Newfoundland examples, both of which were obducted on to passive continental margins. This tectonic stacking order during the T1 obduction phase is diagrammatically shown in the model presented in figure 14.

### (d) Post-collision thrusting

Following continental collision between India and the Karakoram–Lhasa Blocks and closing of the Indus Suture Zone at 50 Ma, thrusting of the complete Zanskar shelf and its overlying Tethyan thrust sheets caused widespread crustal shortening and thickening south of the ISZ. Initially, major thrusts propagated southwestwards across the Zanskar shelf, in a piggy-back sequence of thrusting. The youngest rocks affected by this deformation are the late Palaeocene–early Eocene Spanboth and Chulung-la Formations in the west (figure 15 a, b, plate 3), and the Lingshet and Kong Formations in the east.

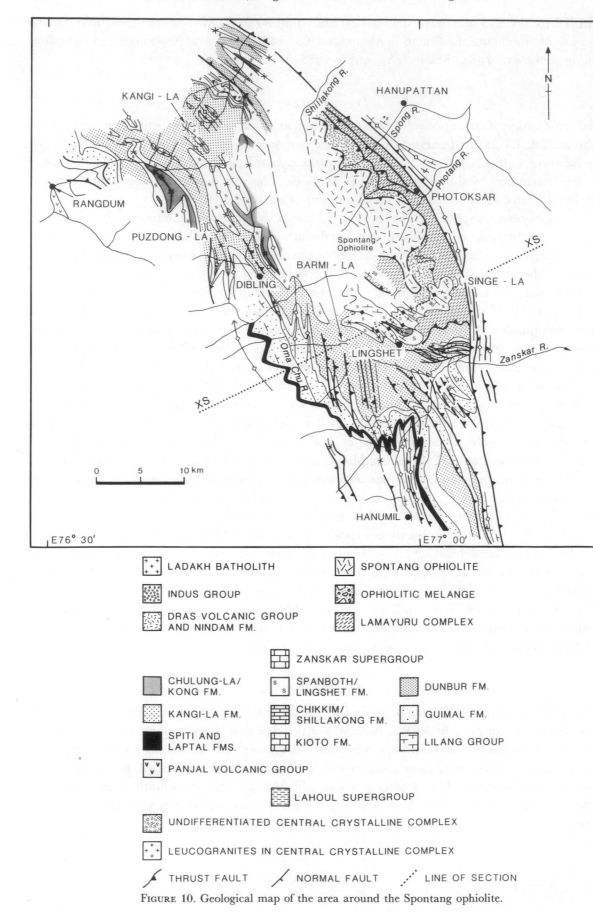

FIGURE 10. Geological map of the area around the Spontang ophiolite.

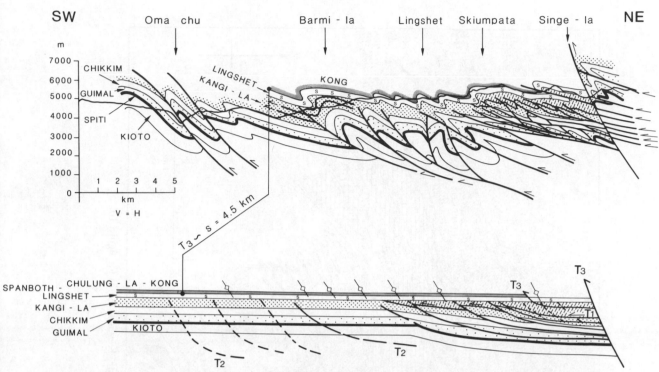

FIGURE 11. (a) Structural cross section and (b) partly restored section across the area shown in figure 10. The Photoksar thrust is the large-scale breakback thrust along the NE margin. T1, T2, T3 refer to thrust sequences described in text. Section (b) shows restoration of the post – lower Eocene folding and thrusting to show the obduction-related T1 and early collision-related T2 thrusts in the shelf sequence.

The structure of the Zanskar shelf sediments can be broadly divided into five zones (partly after Baud *et al.* 1984), each characterized by a differing structural style.

1. The Spontang Zone in western Zanskar is the only area where the pre-collision ophiolite obduction-related structures can be distinguished. These early thrusts (see previous section) are preserved within later SW-directed breakback thrusts such as the Photoksar thrust (figures 10 and 11), which truncates the earlier folds and thrusts along its footwall.

2. The Phugtal Zone along the SW margin is dominated by late NE-dipping normal faults associated with the Zanskar Shear Zone which downthrows Palaeozoic sediments on to sillimanite–kyanite grade gneisses of the High Himalaya (see following section). Earlier SW-facing folds (figure 15 *c*, plate) have in places been refolded by upright and NE-facing folds which are associated with gravitational collapse in response to the Miocene Himalayan uplift to the SW.

3. The Zangla Zone is the zone of maximum compression in eastern Zanskar. Tight, isoclinal, non-concentric, flexural slip and flexural shear folds and steep SW-directed thrusts are dominant (figure 15 *d*, plate 3). Out-of-sequence thrusts, placing younger rocks onto older, are inferred from a high degree of layer-parallel shortening and thickening.

4. The Northern Zanskar Zone along the north and northeastern margin is a 25–30 km wide 'pop-up' zone with steep SW-directed thrusts in the SW and steep NE-directed backthrusts along the NE. The SW-directed thrusts are related to the Photoksar breakback thrust (figure 11) and are late in the sequence. Backthrusting affects the northern part of the Zanskar shelf

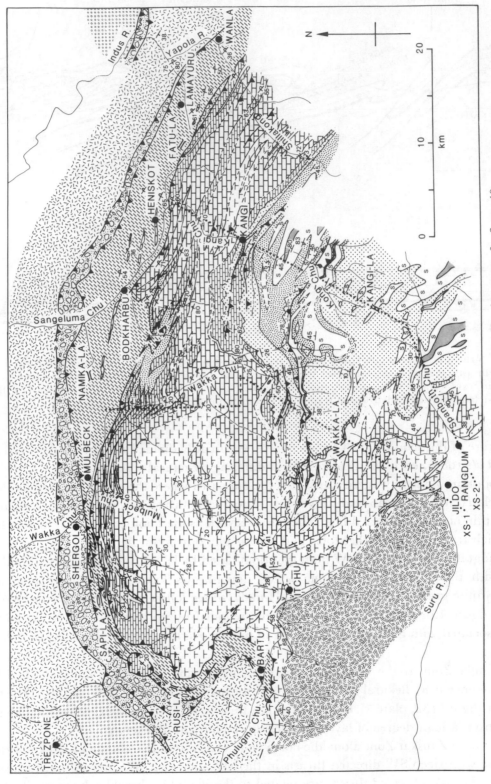

FIGURE 12. Geological map of western Zanskar. Key is same as for figure 10.

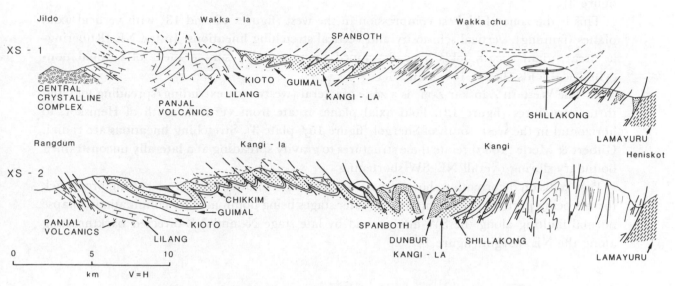

SW

NE

FIGURE 13. Cross section across western Zanskar, locations marked on figure 12.

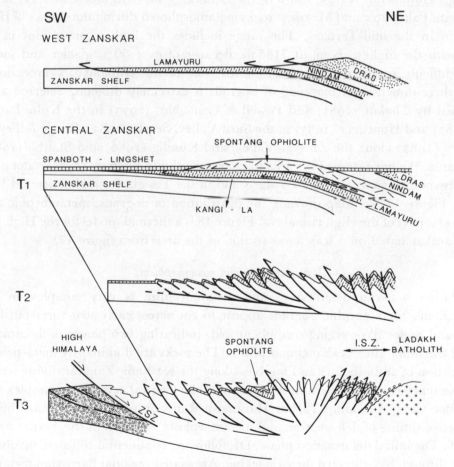

FIGURE 14. Schematic sections to explain the structural evolution of the Zanskar shelf structure.
See text for discussion.

and the Indus Suture Zone rocks and must therefore post-date the Indus molasse in age (see figure 3).

This is the zone of highest compression in the west (figures 12 and 13) with vertical axial planes (fanning), vertical schistosity and vertical stretching lineations. In the NE (Rubering-la area), giant sheath folds have developed by progressive simple shear of original non-cylindrical folds (figure 15e, plate 3).

5. The Western Zanskar Zone is a zone of lateral (westward) extruding/spreading fold and thrust structures (figure 12). Fold axial planes rotate from vertical, south of Heniskot, to horizontal in the west, south of Shergol (figure 15f, plate 3). Stretching lineations are radial. Gilbert & Merle (1987) relate these structures to gravity spreading at a laterally unconstrained boundary during overall NE–SW shortening.

Post-collision deformation shows a multi-stage thrust evolution with cross-cutting breakback thrusts being particularly important, the later stages being dominated by culmination-collapse normal faulting along the SW margin and by late stage N- and NE-directed backthrusting along the NE margin (figure 14).

## 6. High Himalayan Zone

The High Himalayan Range, south of the Zanskar Valley (figures 2 and 17) is a zone of Precambrian, Palaeozoic and Mesozoic rocks metamorphosed during the climax of Himalayan deformation in the mid-Tertiary. The range includes the highest mountains in Ladakh–Zanskar, with the highest, Nun, at 7135 m. Because of over 50% glacier and snow cover-age, and difficult access, it is the least-known zone of the Himalaya. Conventional map-ping and three-dimensional mapping of isograds is extremely difficult. Selected areas have been studied by Thakur (1981) and Powell & Conaghan (1973) in the Kulu–Lahoul area, Searle (1983) and Honegger (1983) in the Suru Valley, Searle (1983), Searle & Fryer (1986) and Herren (1987) along the Zanskar Valley, and Kundig (1989) and Staubli (1989) in the Kishtwar area. We have studied the area in eastern Kashmir–western Zanskar along the Suru, Doda, Kargyiak, Chenab and Kulu Valleys and in the Tos glacier region around White Sail (figure 2). Figure 17 is a map showing the distribution of isograds, metamorphic zones and granites in this part of the High Himalaya. Figure 18 is a thermal model for the High Himalaya south of Zanskar based on a scale cross section of the area from figure 17.

### (a) Structure and metamorphism

The relation between deformation and metamorphism is very complex in the High Himalaya Zone. Metamorphic isograds appear to cut across early structures but have been folded around major SW-verging recumbent folds indicating two phases of deformation, one before and the other after peak metamorphism. The rocks are dominantly meta-pelites with a high proportion of amphibolites and marbles along the Kashmir–Zanskar divide west of Nun-Kun. These represent the westernmost exposure of the Central crystalline complex until their reappearance along the Nanga Parbat–Haramosh Range in northern Pakistan (figure 2).

The relative timing of deformation and metamorphism is shown in the Tertiary time chart in figure 19. The initial deformation phase, D1, following continental collision, involved crustal thickening through SW-directed thrust stacking. Associated regional Barrovian metamorphism (M1) constitutes a prograde medium $P–T$ event reaching kyanite grade metamorphic

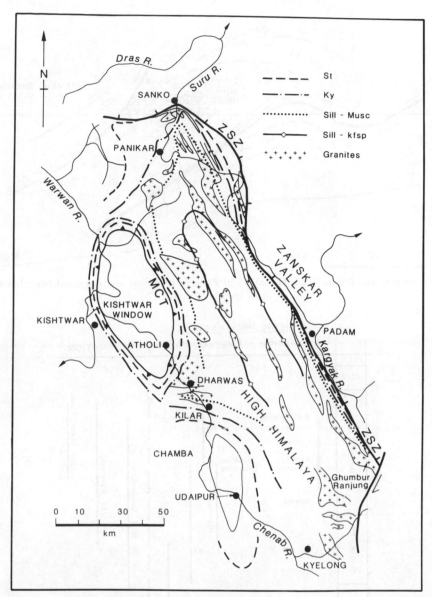

FIGURE 17. Metamorphic map of the Zanskar–Chamba–Kulu area of the High Himalaya showing approximate distribution of isograds from geologic mapping of Honegger (1983), Searle & Fryer (1986), Herren (1987), Staubli (1989), Kundig (1989), and this work. MCT, Main Central Thrust; ZSZ, Zanskar Shear Zone. St = staurolite, Ky = kyanite, Sill = Sillimanite, Musc = muscovite, kfsp = K-feldspar.

conditions. M1 assemblages are overprinted by a second stage of kyanite growth, fibrolite sillimanite and biotite (Staubli 1989; Kundig 1989). This M2 stage indicates an increase in temperature and/or a decrease in pressure. It is roughly contemporaneous with major SW-directed recumbent folds and thrusts along the Main Central Thrust (MCT) Zone.

The M2 event is contemporaneous with the onset of partial melting in the sillimanite zone and the generation of melt pods and leucogranitic sheets. Migmatites with *in situ* leucogranitic segregations are widespread in the deepest structural levels along the central part of the High Himalayan Range, around Dharwas in the Chenab Valley (figure 17) and the Manali–Tos glacier region in the SE. Continual NE–SW compression during the Oligocene–Miocene

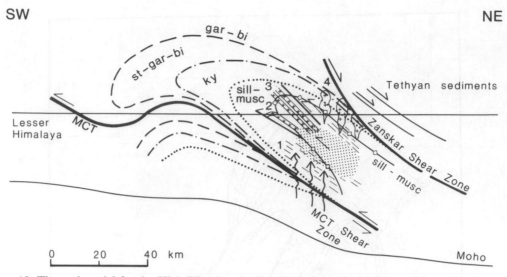

FIGURE 18. Thermal model for the High Himalaya in Zanskar proposed and discussed here, based on a scale cross section of figure 17.

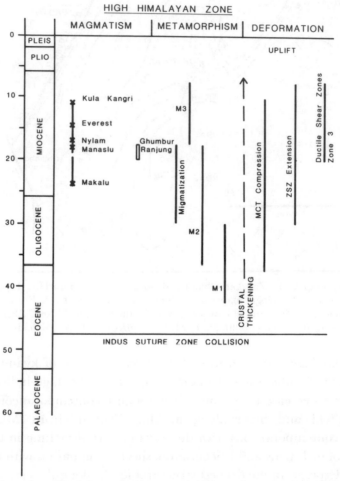

FIGURE 19. Tertiary time chart for deformation, metamorphism and magmatic events of the High Himalaya. Crosses represent U–Pb zircon, monazite ages. Ghumbur Ranjung is Rb–Sr age. See text for sources of data.

resulted in a cycle of NE-directed underplating, SW-directed overthrusting, metamorphism and partial melting at the highest $P-T$ conditions and subsequent exhumation on later SW-directed thrusts (D2). NE-dipping extensional faulting at upper crustal levels is also inherent to the thrust system. M3 is a retrograde metamorphism due to decreasing $P-T$ conditions during uplift or thrust culmination. Rapid uplift on SW-directed MCT-type thrusts preserved the inverted metamorphic isograds seen around the Kishtwar window (Staubli 1989); they are also described from the MCT Zone in Nepal (Le Fort 1975; Pêcher 1975; Caby *et al.* 1983) and Darjeeling (Sinha-Roy 1982).

The higher structural levels of the High Himalayan Zone can be seen to the west of Nun-Kun (figure 17). In eastern Kashmir–western Zanskar, three large SSW-verging recumbent folds or nappes are separated by ductile shear zones or brittle thrust faults. The isograds appear to be folded around these structures creating an inverted metamorphic profile in the lower limbs. The uppermost nappe (Sanko nappe) consists of garnet-biotite leucogranite in the core of a recumbent fold. The southernmost of these folds is the Donara nappe (figure 16a, plate 4) where interbanded pelites, marbles and garnet amphibolites are folded into a S-verging, SW-plunging nappe.

### (b) Himalayan crustally derived granites

Granites and leucogranites of the High Himalayan Zone south of the Zanskar Valley and along the Suru Valley are lithologically diverse and consist of K-feldspar + plagioclase + quartz + biotite ± muscovite ± garnet ± tourmaline. Most of these granites are considered to be Himalayan–Miocene in age. Older granites (possibly Cambrian) also occur in the High Himalaya and frequently form the cores of granite gneiss domes in the central part of the range (Le Fort *et al.* 1986; Kundig 1989). These older porphyritic granites are mantled by migmatites, probably of Himalayan age, within the sillimanite zone (figure 18).

Many compositional varieties of leucogranites occur along the Suru Valley and south of the Zanskar Valley, including garnet–biotite granite, K-feldspar megacrystic two-mica ± garnet leucogranite and tourmaline leucogranite. Tourmaline is rarer in these leucogranites than in other Himalayan leucogranites, such as Manaslu and Bhagirathi. South of the Zanskar and Kargyak Valleys, garnet–tourmaline–muscovite leucogranites occur, and near Dharwas in the Chenab Valley (figure 17) kyanite is also present in the 'lit-par-lit' injection granites (Searle & Fryer 1986).

Himalayan granites and leucogranites are spatially restricted to the sillimanite and kyanite zones but narrow dykes or sheets may extend up into the staurolite or garnet zones. We propose here that the granites and leucogranites within the High Himalayan Zone in the western Himalaya occur in four main tectonic settings. The four morphostructural granite types are schematically illustrated in the thermal model shown in figure 18. Zone 1 corresponds to the proposed location of partial melting within the sillimanite–muscovite and sillimanite–K-feldspar metamorphic terrane. Anatexis occurs at depths of *ca.* 15–30 km, at temperatures in the region of 600–750 °C and pressures of 4–5 kbar (400–500 MPa) (Searle & Fryer 1986; Pinet & Jaupart 1987). The composition and mineralogy of the *in situ* granites are dependent upon the composition of the source protolith and the ambient $P-T-X$ ($H_2O$) conditions. The amount of water available in the zone of partial melting is controlled by the localized metamorphic environment and by the abundance of externally derived fluid. The migration of volatile-enriched fluid derived from the cool, hydrous Lesser Himalayan slab, underlying the hotter High Himalayan slab, may be effective in lowering the 'wet' granite solidus. The migmatite terrane occurs along the highest peaks of the High Himalayan Range and deeper levels of migmatites and layered granitic melts are well exposed along the Chenab Valley near Dharwas (figure 16e, plate 4).

Zone 2 melts are incipient migmatitic mantles around the porphyritic granite gneiss domes, exposed for example around the Umasi-la, and N of the Kishtwar Window (Kundig 1989). The structural relationships indicate that these older (? Cambrian) porphyritic granites form a considerable component of the melting source for the Miocene leucogranites.

Zone 3 granites are the sheet intrusives, dominantly leucogranitic, that are well exposed along the Suru Valley. They have been thrust up to higher structural levels on fluid-rich ductile shear zones as a result of melt-enhanced deformation (Hollister & Crawford 1986) that juxtaposed rocks of different metamorphic grades and crustal levels. Along the Suru Valley these granites occur as NE-dipping sheet-like intrusions parallel to the regional schistosity in the surrounding sillimanite–garnet–biotite gneisses. Upper contacts are generally intrusive (figure 16b, plate 4) whereas basal contacts are brittle thrust faults (figure 16c, plate 4) or ductile shear zones often with migmatites and complex mixing zones in the hangingwalls (figure 16f, plate 4). Strain generally increases downwards within a leucogranite sheet and kinematic indicators (C–S fabrics, asymmetric pressure shadows around garnet or feldspars) in the gneisses indicate SW-directed thrusting, except along the top of the High Himalayan slab where brittle normal faults show displacement down to the NE. Some leucogranites show a high degree of internal ductile strain (figure 16f, plate 4), and are recumbently folded, indicating that intrusion and cooling were synchronous with the mid-Tertiary SW-verging folding and thrust culmination of the High Himalaya.

Zone 4 leucogranites are the higher-level migrated plutons exemplified in this area by the Ghumbur Ranjung pluton (figure 17) in SE Zanskar (Searle & Fryer 1986; Pognante et al. 1987). These plutons have actively migrated into higher structural levels of the High Himalayan crystalline slab and thus cross-cut the metamorphic isograds. Conditions of maximum thermal conductivity contrast exist along the footwall of the Zanskar Shear Zone (figure 18), which separates low thermally conductive Tethyan sediments from the high thermally conductive gneisses of the High Himalayan Zone (Jaupart & Provost 1985; Pinet & Jaupart 1987).

The Ghumbur Ranjung pluton is a garnet–muscovite–biotite–tourmaline leucogranite with a Rb–Sr (mica) isochron age of $20.7–18.4 \pm 0.6$ Ma and an initial $^{87}Sr/^{86}Sr$ ratio of 0.747–0.775 (Ferrara et al. 1986), which compares closely with many other Miocene Himalayan leucogranites (Le Fort et al. 1987). A preliminary Rb–Sr (whole rock–muscovite) age of 17.6 Ma was obtained on a tourmaline–muscovite coarse-grained leucogranite–adamellite from near Bardan gompa, 10 km SE of Padam (Searle & Fryer 1986). These Miocene dates (figure 19) are compatible with the well-dated leucogranite plutons of Makalu, 24 Ma (Schärer 1984); Manaslu, 25–18 Ma (Deniel 1985; Le Fort et al. 1987); Nyalam, 17 Ma; Everest, 14 Ma; and Kula Kangri, 11 Ma (Schärer et al. 1986).

It is important to dismiss the view that all granitic melts generated from intracrustal metasedimentary sources are of low-temperature, minimum-melt composition. All of the granitic–leucogranitic lithologies observed in the Suru Valley section, for example, may be derived from the melting of crustal protoliths. The amount of melting, the chemistry of the melts and the residue will depend not only on the composition of the protolith, but also upon the water budget of the source that is undergoing partial melting.

The lowest temperature, near minimum-melt leucogranites are produced under prograde metamorphism resulting from the breakdown of muscovite. However, higher-temperature melting involving biotite breakdown reactions in high-grade metasediments will also produce granites and leucogranites, but of variable, 'non-minimum' melt composition. The fluid content of the magma defines its mobility and emplacement level in the crust. Furthermore, the heat availability will have important consequences for melting with, for example, higher temperature melt fractions being produced from a source that has previously undergone a partial melting episode. These later melt fractions will be of a distinctly 'non-minimum' melt composition. As the thermal budget is of prime importance in the petrogenesis of High Himalayan melts, and localized thermal heterogeneity is strongly suspected (Pinet & Jaupart 1987; Hodges et al., this symposium), a variety of granite compositions are likely to be produced in this tectonic environment.

### (c) Main Central Thrust Zone

The existence of a major crustal-scale shear zone placing the high-grade rocks of the High Himalaya over the low-grade rocks and sediments of the Lesser Himalaya has long been known (Heim & Gansser 1939; Le Fort 1975). However, the exact location of the Main Central Thrust (MCT) with respect to the inverted metamorphic isograds is constantly disputed. Valdiya (1980) places it at the top of the Central Crystalline Complex (Vaikrita thrust) in Kumaon, Sinha-Roy (1982) places it at the base of the inverted metamorphic sequence in Darjeeling, and others place it along the kyanite isograd in Central Nepal (Le Fort 1975; Pêcher 1977). All these authors define the MCT on lithological or metamorphic conditions rather than strain. We prefer to define the MCT as a ductile shear zone ca. 5 km wide along which there are several zones of high strain with mylonites at higher levels. Brunel (1986) shows that the minimum translation on the MCT in the Everest–Makalu region is around 100 km.

The field relations in the High Himalaya of Zanskar, Kishtwar, Chamba and Kulu (figures 2 and 17) clearly indicate that the metamorphic isograds have been deformed by a later phase of SW-verging folding and thrusting. In eastern Kashmir–western Zanskar, the location of the MCT is contentious. It follows approximately the base of the Donara nappe around the Boktol Pass west of Panikar (figure 17) but further west has been rotated by subsequent thrust movements in the footwall to become vertical, and in places possibly even backthrusted, so that the Kashmir Palaeozoic–Mesozoic sediments have been re-thrust during D3 northwards over the metamorphic rocks of the Donara nappe.

Around the Kishtwar window, metamorphic isograds have been telescoped during SW-directed thrusting (Staubli 1989) and folded by subsequent lower thrust culminations of the Lesser Himalayan sheets. Searle & Fryer (1986) describe a decrease in metamorphic grade in eastern Zanskar–Lahoul southeastwards to the Chenab valley where Powell & Conaghan (1973) found Jurassic microfossils in low-grade marble at Tandi. The isograds have clearly been folded around the Chamba syncline and the Kishtwar anticline. Stretching lineations, in general, are aligned NE–SW, parallel to the displacement direction.

There are three main models to explain the inverted metamorphic isograds of the Himalaya. The first is the rapid thrusting of the hot High Himalayan slab on to a colder Lesser Himalayan slab (Le Fort 1975). The result of this would be downward decreasing $P$–$T$ conditions away from a major thrust plane in the footwall and upward decreasing retrogressive metamorphism in the hangingwall. Extremely rapid thrust rates and uplift–exhumation rates are implied in order to preserve the inverted metamorphism without thermal re-equilibration.

The second model involves two-stage metamorphism: an earlier emplacement of a 'hot'

High Himalayan hangingwall on to a 'cold' Lesser Himalayan footwall at intermediate pressures (kyanite grade) followed by higher $T$–lower $P$ (sillimanite grade) metamorphism related to granitic magmatism. We feel it is unlikely that any thermal imprint from partial melting would be so spatially extensive (up to 50 km away from the nearest granite source region). We also discount the effect of any shear heating involved outside the immediate vicinity of high-strain shear zones, mainly because of the extensive fluid and volatile fluxing within the High Himalaya.

The third model involves deformation of an earlier Barrovian metamorphic sequence with upward decreasing $P$–$T$ conditions by later crustal-scale folds and ductile shear zones. The result of this would be both downward (inverted limb of fold) and upward (upper limb) decreasing initial $P$–$T$ conditions. Unlike the Nepal and Darjeeling sectors of the Himalaya, the Zanskar area does show the upper limb with upward decreasing isograds (figure 18), and we therefore favour this third model to explain the inverted metamorphism.

### (d)  Zanskar Shear Zone

The contact between the High Himalayan crystalline rocks and the sediments of the Zanskar shelf sequence is a N-dipping normal fault zone (Searle 1986; Herren 1987). The fault extends along the 100 km long Zanskar Valley (figure 2) and separates Barrovian facies metamorphic rocks, migmatites and leucogranites in the south from Palaeozoic–Mesozoic sediments and Panjal volcanics in the north. Metamorphic isograds are structurally telescoped along the contact where a transition from upper amphibolite facies (sillimanite–K-feldspar gneisses) to lower greenschist facies occurs within 200 m. The normal fault reaches a maximum offset in the central part of the area around Padam (figure 17). Herren (1987) calculated a minimum horizontal extension of 16 km and a minimum vertical displacement of 19 km in this region. Towards the west along the Suru Valley (figure 16$d$, plate 4), isograds widen as the throw on the fault decreases. The axis of maximum $P$–$T$ conditions is parallel to the fault zone approximately 20 km south of the Zanskar Valley along the highest mountains of the Himalayan Range.

The Zanskar Shear Zone offsets all the rocks of the High Himalaya including the Miocene leucogranites. Garnet–muscovite–tourmaline leucogranites from the footwall of the Zanskar shear zone have K-Ar ages on micas (muscovite blocking temperature $ca.$ 350 °C, biotite $ca.$ 300 °C) of $30 \pm 2$ and $22 \pm 1$ Ma (D. C. Rex, personal communication). Kinematic indicators such as C-S fabrics and feldspar augen consistently indicate the shear sense as being northeast side down (Herren 1987). Stretching lineations are generally NE SW or N–S and plunge north. No indications of strike-slip motion have been found along the Zanskar Shear Zone.

The trailing edges of many thrust culminations are bounded by extensional normal faults although none have been described on such a large scale as the Zanskar Shear Zone. Searle (1986) interpreted this fault as a dorsal culmination collapse feature associated with, and synchronous with, rapid thrusting and uplift in the High Himalaya and MCT Zone to the south. The NE–SW extension is probably confined only to the upper crustal levels and does not imply extension in the lower crust (figure 18). Structural analysis of the High Himalaya and fault-plane solutions of earthquakes (Seeber et al. 1981; Molnar 1984) indicate only compressional tectonics in the High and Lesser Himalaya. This also implies that the Zanskar Shear Zone is listric and flattens out to the north into a mid-crustal detachment. This may also explain why there is no typical Indian lower crust (granulites) exposed in the Himalaya (see §7).

The same normal fault can be traced for 900 km eastwards across the Annapurna Range in Nepal (Le Fort 1975; Pêcher 1977; Caby *et al.* 1983) and across southern Tibet (Burg & Chen 1984; Burg *et al.* 1984). The low-angle, N-dipping normal fault along the northern flanks of Everest and at Nyalam on the Kathmandu–Lhasa road cuts Oligocene metamorphic rocks and leucogranite sheets dated radiometrically at 30–15 Ma (Allègre *et al.* 1984) and must therefore have been active during Miocene–Pliocene times (Burchfield & Royden 1985).

The thermal conductivity contrast across the ZSZ between sediments of low thermal conductivity on the hangingwall and gneisses and granites of high conductivity along the footwall (Jaupart & Provost 1985) is also important in maintaining a maximum heat flux at the top of the High Himalayan slab. The sediments of the Zanskar shelf would have acted as a thermal barrier to the upward transport of heat and could explain the concentration of Himalayan leucogranites at the top of the slab.

## 7. Discussion

### (a) Thrust sequences

It is clear that although the overall structural framework of the Himalayan collision is one of southward propagating thrusting from the ISZ in the north to the MCT Zone and MBT in the south, numerous complications arise. The ISZ has undergone compression since the mid-Eocene collision and is now dominated by steep N-directed backthrusts that post-date the Indus Molasse Group and must therefore be late Miocene to (?)Pliocene in age. Earlier SW-directed thrusts in the Lamayuru complex must have been passively rotated by underthrusting and continued compression so as to dip steeply to the south. Subsequent compression during the backthrusting phase has reactivated these faults as N-directed thrusts. The resulting internal folding and cleavage can therefore be extremely complex and difficult to restore fully. This rotation of earlier thrusts and folds also resulted in a tectonic inversion of the earlier stacking order and the widespread rethrusting of the Zanskar shelf rocks northwards over the ISZ rocks.

The structural evolution of the Zanskar shelf involves a continuum of crustal shortening and thickening by thrust stacking since collision at *ca.* 50 Ma. Within this continuum, cross-cutting thrust relations indicate that out-of-sequence and breakback thrusts are extremely important, particularly in the northern part of the Zanskar Range. Thrusts associated with the obduction of the Spontang ophiolite are the earliest in the sequence and are cut off along their trailing edges by late breakback thrusts that cut through the whole tectonic stack (Searle 1986). These late breakback thrusts are thought to be contemporaneous with the steep N-directed backthrusting creating a 25–30 km wide 'pop-up' zone along the northern margin of the Zanskar shelf.

Very tight to isoclinal folding within the Mesozic shelf carbonates suggests that major bedding-parallel detachments occur along the less competent horizons. For example, the Kioto Formation limestones in the Zangla zone are deformed mainly by intra-formational isoclinal folding with large-scale detachments along the underlying relatively incompetent beds of the Lilang Group below, and the Laptal and Spiti Formations above. This style of thrusting results in homogenous crustal thickening with thrusts placing younger rocks on to older.

The deformation in the High Himalayan zone southwest of the Zanskar Valley is characterized by ductile folding and numerous SW-directed ductile shear zones. The higher crustal levels around the Suru Valley and eastern Kashmir show SW-verging recumbent folds that are bounded by major shear zones involving Barrovian facies metamorphic rocks. The deeper structural levels in the core of the range show that crustal thickening occurred through flow folding and migmatization with remobilization of leucosome material. Syn- or post-metamorphic thrusting along the MCT Zone at the base of the slab and concomitant NE-directed culmination collapse normal faulting along the ZSZ have both resulted in telescoped metamorphic isograds. Rapid thrust rates in the High Himalaya during the late Oligocene–Miocene coupled with tectonic denudation on the NE dipping normal fault system along the ZSZ resulted in rapid uplift and preservation of the inverted metamorphism along the base of the slab and the high-grade metamorphic and anatectic core along the top of the slab. Post-metamorphic folding of isograds during SW-directed MCT-thrusting can be demonstrated in western Zanskar (figures 17 and 18).

### (b) Crustal shortening

The amount of crustal shortening across the ISZ and Zanskar shelf can only be estimated by the restoration of balanced cross sections. The section across the ISZ along the Zanskar River (figures 5 and 6) indicates 36 km of post-Indus molasse shortening, on restoration of the late-stage T3 thrusts and folds. Lack of sufficiently detailed stratigraphical and structural data at present makes it impossible to restore the pre-molasse sediments.

A balanced and restored cross-section across the Zanskar shelf and ISZ east of the Zanskar River shows that the present 98 km width from the Zanskar Valley normal fault to the Ladakh batholith restores to 250 km, implying shortening of 152 km (Searle 1986; Searle *et al.* 1987) of which 112 km is accommodated by shortening of the Zanskar shelf units. Shortening estimates from the western (Rangdum–Kanji-la section, *ca.* 56 km) and the eastern (Marang-la–Phugtal section, *ca.* 36 km) part of the Zanskar Range are considerably less (Searle & Cooper, in preparation). These balanced cross-section reconstructions have several con-straining factors, notably: (1) trailing edges of thrust sheets are rarely seen; (2) depth-to-basal detachments are not known; (3) folding is dominantly flexural slip or flexural flow, rarely parallel or concentric; (4) differing amounts of layer-parallel shortening are commonly observed between incompetent shaley beds that deform by internal thickening and competent limestones that deform by folding; and (5) strain rates increase towards the north, where ductile flow folding is observed.

No attempt has yet been made to determine the amount of shortening across the High Himalaya. Very complex ductile strain in these metamorphic rocks and anatectic granites makes it impossible to restore any cross section. The *P–T* estimates for these rocks in the High Himalaya also suggest a complex Tertiary thermal history with evidence of multi-stage thrusting and crustal thickening, accompanied by NE-dipping normal faults at upper crustal levels. Shortening estimates from the Lesser Himalaya in Kashmir, Chamba and Kulu are also unknown although preliminary mapping in eastern Kashmir suggests only minor shortening (34 km) in the section from the Vale of Kashmir to the Central crystalline complex (Searle & Cooper, unpublished data).

Despite major pre-collision palaeogeographic differences, the continuity of Himalayan thrust systems between India and Pakistan allows some comparisons to be made. Coward & Butler (1985) and Coward *et al.* (1987) constructed a balanced cross section from the MMT southwards to the Punjab foreland in an attempt to quantify the amount of crustal shortening in the Pakistan Himalaya. The rocks of the Indian Plate in this section restore to at least 720 km with an implied shortening of 470 km and they, and Butler (1986), argue for large-scale underthrusting of Indian lower crust beneath Kohistan and the Karakoram northwards to the Pamirs.

Our shortening estimates for the Zanskar shelf sequence must be matched by a similar amount of shortening in the lower and middle crust. In common with most present-day passive continental margins with relatively thick shelf sediments, the continental crust beneath the pre-collision Zanskar shelf was probably considerably thinner than the average 35 km thickness of the Indian foreland crust. This decreases the necessity for large-scale underthrusting of Indian crust beneath Tibet.

### (c) Crustal subduction

Balanced cross sections of the Ladakh–Zanskar Himalaya indicate that some crustal subduction must have occurred. The major question is how much, and how far northwards has it gone? The low density and great thickness of continental crust suggests that it is too buoyant to be subducted unless it was greatly attenuated (McKenzie 1969). However, Molnar & Gray (1979) suggested that if the upper and lower crust could be detached, the lithospheric mantle could be cold and dense enough to subduct the lower 10 km of continental crust. In the Himalaya, the MCT Zone could be considered as a mid-crustal plane along which such detachment may have occurred. Major whole crustal detachments as suggested by Molnar (1984) or crust–mantle detachments (Mattauer 1986) for the Himalaya can account for the shortening estimates (ca. 250 km) in the Zanskar–Ladakh Himalaya (figure 20) but could not account for the 470 km shortening across the Pakistan section estimated by Coward & Butler (1985) assuming standard 35 km thick crust. Either the 470 km shortening is an overestimate, or large volumes of Indian lower crust must have been subducted northwards, to achieve a crustal balance, or else the crust was greatly thinned prior to thrusting. Butler (1986) invoked a model of eclogite metamorphism in the subducting plate to facilitate this large-scale continental subduction, based on the radical difference in density across the metamorphic phase boundary between granulite–amphibolite and eclogite (Richardson & England 1979). In this model the subducting continental plate would be metamorphosed to dense eclogite and the geophysical Moho might only represent this phase boundary, making an area balance of the continental crust impossible.

Our model for the structural evolution of the Western Himalaya in India is based on the crustal-scale cross section shown in figure 20. This figure has been constructed with the deep crustal constraints of Lyon-Caen & Molnar (1983) and Molnar (1984), but also includes the Zanskar shelf and ISZ structure from Ladakh. The crustal thickness is known to be

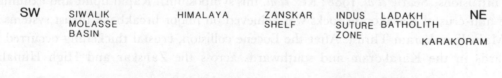

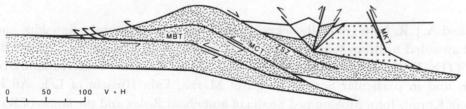

Figure 20. Crustal section of the western Himalaya based on Molnar (1984) but also showing the Zanskar shelf structure and ISZ from this work. See text for discussion. See Rex et al. (this symposium, figure 12) for a crustal section of the central Karakoram north of Ladakh.

approximately 35 km below the Indo–Gangetic Plains, and about 70 km below the Karakoram and southern Tibet (Molnar & Chen 1982). Fault-plane solutions and earthquake focal depths (Seeber *et al.* 1981) show that the Indian Plate underthrusts the Lesser Himalaya along the MBT as a coherent slab. Lyon-Caen & Molnar (1983) demonstrate from gravity anomalies in the Everest region that the dip of the Moho steepens to about 15° beneath the High Himalaya.

Molnar (1984) assumed a minimum 125 km underthrusting of Indian crust along the MCT system and a further 125 km along the MBT and Lesser Himalayan thrust systems, in central Nepal where both zones are narrower than in the western sector. In our model for the crustal structure of the West Himalaya (figure 20) we assume the MCT Zone descends to the base of the continental crust, and the minimum shortening in the High Himalayan Zone to be 200 km. Thrusting propagated southwards and the minimum amount of shortening along the MBT and Lesser Himalaya thrust sheets is likewise assumed to be 200 km. During this time, the MCT Zone was relatively inactive and being carried piggy-back by the MBT. The High Himalaya was undergoing rapid uplift by underplating beneath its southern margins, and rapid exhumation–erosion facilitated by gravitational collapse normal faulting. *P–T* estimates from metamorphic rocks of the High Himalaya southwest of the Zanskar Valley indicate that as much as 25–30 km of erosion has occurred during the last 20 Ma along the footwall of the ZSZ.

The 550 km of total shortening across the western Himalaya of India has, in our model, been taken up by crustal stacking in the Himalaya, and by underthrusting Indian crust northwards beneath the Ladakh Range north of the ISZ, and beneath the southern part of the Karakoram (bounded by the dextral Karakoram strike-slip fault that shows *ca.* 150 km of right lateral motion), but not beneath Tibet (see Rex *et al.*, this symposium, figure 12). Gravity data from the Karakoram suggest that the thickened crust is being pulled down by cold (Indian crust) material beneath it, and that the deep structure is different from that of Tibet (Molnar, this symposium).

Geological constraints from the Karakoram (Searle *et al.* 1988; Rex *et al.*, this symposium) and Tibet (Allègre *et al.* 1984) show that both have been emergent land masses since the mid-Cretaceous. England & Searle (1986) suggested that the Lhasa Block was an Andean-type margin, both in magma type and elevation (and therefore crustal thickness), before the Eocene collision. The main phase of crustal thickening and metamorphism in the Baltoro Karakoram occurred between 50–36 Ma, and resulted in intrusion of numerous syn- and post-collisional granite intrusions (Searle *et al.* 1988; Rex *et al.* this symposium). Rapid uplift and exhumation of these mid-crustal Karakoram rocks was achieved on major breakback thrust systems such as the Main Karakoram Thrust. After the Eocene collision, crustal thickening occurred both northwards in the Karakoram and southwards across the Zanskar and High Himalayan Ranges.

M. P. S. and A. J. R. both gratefully acknowledge NERC postdoctoral fellowships and grant GR3/4242 awarded to Brian Windley. D. J. W. C. gratefully acknowledges NERC fellowship and grant GTS/F/85/GS4. We thank all our Ladakhi, Zanskari and Kashmiri trekking companions and in particular Sonam Targis of Marka, Fida Hussein of Leh, Ali Hussein Shillikchey of Kargil, John Mohammed Shalla of houseboat Rolex and the lamas of Rangdum and Phugtal gompas. Sue Button is thanked for excellent cartographic work.

## References

Allègre, C. J. et al. 1984 Nature, Lond. 307, 17–22.

Andrieux, J., Arthaud, F., Brunel, M. & Sauniac, S. 1981 Bull. Soc. géol. Fr. 23, 651–661.

Bard, J. P. 1983 Earth planet. Sci. Lett. 65, 133–144.

Bassoullet, J.-P., Colchen, M., Guex, J., Lys, M., Marcoux, J. & Mascle, G. 1978 C.r. hebd. Séanc. Acad. Sci., Paris 287, 677–678.

Bassoullet, J.-P., Colchen, M., Marcoux, J. & Mascle, G. 1981 Riv. ital. Paleont. Stratigr. 86, 825–844.

Baud, A., Arn, B., Bugnon, P., Crisinel, A., Dolivo, E., Escher, A., Hammerschlag, J. G., Marthaler, M., Masson, H., Steck, A. & Tieche, J.-C. 1982 Bull. Soc. géol. Fr. 24, 341–361.

Baud, A., Gaetani, M., Garzanti, E., Fois, E., Nicora, A. & Tintori, A. 1984 Eclog. geol. Helv. 77, 171–197.

Brookfield, M. E. & Westermann, G. E. G. 1982 J. geol. Soc. India 23, 263–266.

Brookfield, M. E. & Andrews-Speed, C. P. 1984a Sediment. Geol. 40, 249–286.

Brookfield, M. E. & Andrews-Speed, C. P. 1984b Geol. Rdsch. 73, 175–193.

Brunel, M. 1986 Tectonics 5, 247–265.

Burchfield, C. & Royden, L. 1985 Geology 13, 679–682.

Burg, J.-P. & Chen, G. M. 1984 Nature, Lond. 331, 219–233.

Burg, J.-P., Brunel, M., Gapais, D., Chen, G. M. & Liu, G. H. 1984 J. struct. Geol. 6, 535–542.

Butler, R. W. H. 1986 J. geol. Soc. Lond. 143, 857–873.

Butler, R. W. H. 1987 J. geol. Soc. Lond. 144, 619–634.

Caby, R., Pêcher, A. & Le Fort, P. 1983 Revue Géogr. phys. Géol. dyn. 24, 89–100.

Colchen, M. & Reuber, I. 1986 C.r. Rebd. Séanc. Acad. Sci., Paris 303, 719–724.

Colchen, M., Mascle, G. & Van Haver, T. 1986 In Collision tectonics (ed. M. P. Coward & A. C. Ries), pp. 173–184. Geol. Soc. Lond. Spec. Publ. no. 19. London: Blackwell.

Colchen, M., Reuber, I., Bassoullet, J.-P., Bellier, J.-P., Blondeau, A., Lys, M. & De Wever, P. 1987 C.r. hebd. Séanc. Acad. Sci., Paris 305, 403–406.

Coleman, R. G. 1971 J. geophys. Res. 76, 1212–1222.

Coward, M. P. & Butler, R. H. W. 1985 Geology 13, 417–420.

Coward, M. P., Windley, B. F., Broughton, R., Luff, I. W., Petterson, M. G., Pudsey, C., Rex, D. & Khan, M. A. 1986 In Collision Tectonics (ed. M. P. Coward & A. C. Ries), pp. 203–219. Geol. Soc. Lond. Spec. Publ. no. 19. London: Blackwell.

Deniel, C. 1985 Ph.D. thesis, University of Clermont.

Dietrich, V. J., Frank, W. & Honegger, K. 1983 J. Volc. Geotherm. Res. 18, 405–433.

England, P. C. & Searle, M. P. 1986 Tectonics 5, 1–14.

Ferrara, G., Lombardo, B., Tonarini, S. & Turi, B. 1986 Abstract, Himalayan workshop, Nancy.

Frank, W., Gansser, A. & Trommsdorff, V. 1977 Schweiz. miner. petrogr. Mitt. 57, 89–113.

Fuchs, G. 1977 Jb. geol. B.-A 120, 219–229.

Fuchs, G. 1979 Jb. geol. B.-A 122, 513–540.

Fuchs, G. 1982 Jb. geol. B.-A 125, 1–50.

Gaetani, M., Nicora, A. & Premoli Silva, I. 1980 Riv. ital. Paleont. Stratigr. 86, 127–166.

Gaetani, M., Nicora, A., Premoli Silva, I., Fois, E., Garzanti, E. & Tintori, A. 1983 Riv. ital. Paleont. Stratigr. 89, 81–118.

Gaetani, M., Casnedi, R., Fois, E., Garzanti, E., Jadoul, F., Nicora, A. & Tintori, A. 1986 Riv. ital. Paleont. Stratigr. 91, 443–478.

Gansser, A. 1964 Geology of the Himalayas. 289 pages. London: J. Wiley.

Gansser, A. 1977 Sci. Terre Himalaya: C.N.R.S. 268, 147–166.

Garzanti, E., Casnedi, R. & Jadoul, F. 1986 Sediment. Geol. 42, 237–265.

Gilbert, E. & Merle, O. 1987 J. struct. Geol. 9, 481–490.

Gupta, V. J. & Kumar, S. 1975 Geol. Rdsch. 64, 540–563.

Herren, E. 1987 Geology 15, 409–413.

Heim, A. & Gansser, A. 1939 Denkschr. schweiz. naturf. Ges. 73, 1–245.

Hollister, L. F. & Crawford, M. C. 1986 Geology 14, 558–561.

Honegger, K. 1983 Ph.D. thesis, ETH Zürich.

Honegger, K., Dietrich, V., Frank, W., Gansser, A., Thoni, M. & Trommsdorff, V. 1982 Earth planet. Sci. Lett. 60, 253–292.

Honegger, K. & Raz, U. 1985 Abstract, Himalayan workshop, Leicester.

Honegger, K., Le Fort, P. & Mascle, G. 1985 Abstract, Himalayan workshop, Leicester.

Jadoul, F., Fois, E., Tintori, A. & Garzanti, E. 1985 Rc. Soc. geol. ital. 8, 9–13.

Jan, M. Q. 1985 Geol. Bull. Peshawar Univ. 18, 1–40.

Jaupart, C. & Provost, A. 1985 Earth planet. Sci. Lett. 73, 385–397.

Johnson, B. D., Powell, C. McA. & Veevers, J. J. 1976 Bull. geol. Soc. Am. 87, 1560–1566.

Kelemen, P. B. & Sonnenfeld, M. D. 1983 Schweiz. miner. petrogr. Mitt. 63, 267–287.

Kelemen, P. B., Reuber, I. & Fuchs, G. 1988 J. struct. Geol. 10, 129, 130.

Klootwijk, C. J. 1979 In *Structural geology of the Himalaya* (ed. P. S. Saklini), pp. 307–360. New Delhi: Today and Tomorrow Publishers.

Klootwijk, C. J. & Radhakrishnamurthy, C. 1981 In *Palaeo-reconstruction of the continents*. Geodynamic Series no. 2, pp. 93–105.

Kundig, R. 1989 *J. metamorph. Geol.* (In the press.)

Le Fort, P. 1975 *Am. J. Sci.* **275** A, 1–44.

Le Fort, P., Debon, F., Pêcher, A., Sonet, J. & Vidal, P. 1986 *Sci. Terre, Mem.* **47**, 191–209.

Le Fort, P., Cuney, M., Deniel, C., France-Lanord, C., Sheppard, S. M. F., Upreti, B. N. & Vidal, P. 1987 *Tectonophysics* **134**, 39–57.

Lydekker, R. 1883 *Mem. geol. Surv. India* **22**, 108–122.

Lyon-Caen, H. & Molnar, P. 1983 *J. geophys. Res.* **88**, 8171–8192.

Maluski, H. & Matte, P. 1984 *Tectonics* **3**, 1–18.

Mattauer, M. 1986 In *Collision tectonics* (ed. M. P. Coward & A. C. Ries), *Geol. Soc. Lond. Spec. Publ. no.* 19, pp. 37–50.

McKenzie, D. P. 1969 *Geophys. Jl R. astr. Soc.* **18**, 1–69.

Molnar, P. 1984 *A. Rev. Earth planet. Sci.* **12**, 489–518.

Molnar, P. & Tapponnier, P. 1975 *Science, Wash.* **189**, 419–426.

Molnar, P. & Gray, D. 1979 *Geology* **7**, 58–62.

Molnar, P. & Chen, W. P. 1982 In *Mountain building processes* (ed. U. Breigel & K. Hsu), pp. 41–57. New York: Academic Press.

Nanda, M. M. & Singh, M. P. 1977 *Himalayan Geol.* **6**, 364–388.

Patriat, P. & Achache, J. 1984 *Nature, Lond.* **311**, 615–621.

Pearce, J. A. & Cann, J. R. 1973 *Earth planet. Sci. Lett.* **19**, 290–300.

Pêcher, A. 1975 *Himalayan Geol.* **5**, 115–132.

Pêcher, A. 1977 *Colloques int. Cent. natn. Rech. scient.* **268**, 301–318.

Petterson, M. G. & Windley, B. F. 1985 *Earth planet. Sci. Lett.* **74**, 45–57.

Pickering, K. T., Searle, M. P. & Cooper, D. J. W. 1987 Abstract, Himalayan workshop, Nancy.

Pierce, W. J. 1978 *Geophys. Jl R. astr. Soc.* **52**, 277–311.

Pinet, C. & Jaupart, C. 1987 *Earth planet. Sci. Lett.* **84**, 87–99.

Pognante, U., Genovese, G., Lombardo, B. & Rossetti, P. 1987 *Rc. Soc. miner. petr. ital.* **42**, 95–102.

Powell, C. McA. & Conaghan, P. J. 1973 *J. Geol.* **81**, 127–143.

Pudsey, C. J. 1986 *Geol. Mag.* **123**, 405–423.

Radhakrishna, T., Rao, V. D. & Murali, A. V. 1984 *Tectonophysics* **108**, 135–153.

Reuber, I. 1986 *Nature, Lond.* **321**, 592–596.

Reuber, I., Montigny, R., Thuizat, R. & Heitz, A. 1987 Abstract, Himalayan workshop, London.

Richardson, S. W. & England, P. C. 1979 *Earth planet. Sci. Lett.* **42**, 183–190.

Schärer, U. & Allègre, C. J. 1982 *Nature, Lond.* **295**, 585–587.

Schärer, U., Hamet, J. & Allegre, C. J. 1984 *Earth planet. Sci. Lett.* **67**, 327–339.

Schärer, U. 1984 *Earth planet. Sci. Lett.* **67**, 191–204.

Schärer, U., Xu, R. H. & Allègre, C. J. 1986 *Earth planet. Sci. Lett.* **77**, 35–48.

Sclater, J. G. & Fischer, R. L. 1974 *Bull. geol. Soc. Am.* **85**, 683–702.

Searle, M. P. 1983 *Trans. R. Soc. Edinb.* **73**, 205–219.

Searle, M. P. 1985 *J. struct. Geol.* **7**, 129–144.

Searle, M. P. 1986 *J. struct. Geol.* **8**, 923–936.

Searle, M. P. 1988 *J. struct. Geol.* **10**, 130–132.

Searle, M. P. & Fryer, B. J. 1986 In *Collision tectonics* (ed. M. P. Coward & A. C. Ries), *Geol. Soc. Lond. Spec. Publ. no.* 19, pp. 185–201. London: Blackwell.

Searle, M. P., Rex, A. J., Tirrul, R., Rex, D. C. & Barnicoat, A. 1988 *Bull. geol. Soc. Am.* (In the press.)

Searle, M. P. & Stevens, R. K. 1984 In *Ophiolites and oceanic lithosphere* (ed. I. G. Gass, S. J. Lippard & A. W. Shelton), *Geol. Soc. Lond. Spec. Publ. no.* 13, pp. 185–201. London: Blackwell.

Searle, M. P., Windley, B. F., Coward, M. P., Cooper, D. J. W., Rex, A. J., Li Tingdong, Xiao Xuchang, Jan, M. Q., Thakur, V. C. & Kumar, S. 1987 *Bull. geol. Soc. Am.* **98**, 687–701.

Seeber, L., Armbruster, J. & Quittmeyer, R. C. 1981 In *Zagros, Hindu Kush, Himalaya, geodynamic evolution* (ed. H. K. Gupta & F. M. Delany), Geodynamics Series no. 3, pp. 215–242. Washington, D.C.: American Geophysics Union.

Shah, S. K., Sharma, M. L., Gergan, J. T. & Tara, C. S. 1976 *Himalayan Geol.* **6**, 534–556.

Shams, F. A. 1980 *Geol. Bull. Peshawar Univ.* **13**, 67–70.

Shams, F. A., Jones, G. C. & Kempe, R. D. C. 1980 *Mineralog. Mag.* **43**, 941–942.

Sharma, K. K. & Kumar, S. 1978 *Himalayan Geol.* **8**, 252–287.

Sinha-Roy, S. 1982 *Tectonophysics* **84**, 197–224.

Srimal, N., Basu, A. R. & Kyser, R. K. 1987 *Tectonics* **6**, 261–274.

Staubli, A. 1989 *J. metamorph. Geol.* (In the press.)

Tahirkheli, R. A. K. & Jan, M. Q. 1979 *Geol. Bull. Peshawar Univ.* Spec. Issue II, pp. 1–30.

Tapponnier, P. *et al.* 1981 *Nature, Lond.* **294**, 405–410.

Tewari, B. S. & Sharma, S. P.  1972  *Bull. Indian geol. Ass.* **5**, 52–62.
Thakur, V. C.  1981  *Trans. R. Soc. Edinb.* **72**, 890–897.
Thakur, V. C.  1987  *Tectonophysics* **134**, 91–102.
Van Haver, T.  1984  Ph.D. thesis. Grenoble.
Wadia, D. N.  1937  *Rec. geol. Surv. India* **72**, 151–161.
Virdi, N. S., Thakur, V. C. & Kumar, S.  1977  *Himalayan Geol.* **7**, 479–482.
Valdiya, K.  1980  *Tectonophysics* **66**, 323–348.

## Discussion

Eveline Herren (ETH-Zentrum, Zürich, Switzerland). Field investigations in the Zanskar region made over the past 5 years indicate that there are no large-scale overthrusts present as indicated on the cross section drawn and presented by Dr Searle. For example, between the anticlinal harmonic and polyharmonic folds located east of Zosar (between Tongde and Zangla in the Zanskar region) developed in the Lilang group and the overlying Kioto limestones, no evidence has been found that indicated that thrust planes cut the folds (as shown by Dr Searle in his spoken presentation and illustrated by him in his 1986 paper). Local thrusts occur only at the base of the Permian Panjal Trap and appear to be best explained as the effect of different deformation styles dependant of the competence contrasts of the different lithologies (Herren 1987).

Careful and detailed mapping and field investigations of more than one third of the High Himalayan crystallines shown on Dr Searle's map indicate that volumetrically there are fewer Himalayan age leucogranites present than he suggests (Honegger *et al.* 1982; Honegger 1983; Herren 1987; Kündig 1989; Stäubli 1989). The amount of leucogranites in the Zanskar–Suru–Kashmir–Kishtwar area is very small compared with the great granite bodies of the eastern part of the Himalayas and these leucogranites are limited to the Haptal Tokpo Valley (Herren 1987) and to the Gumburanjon leucogranite in the Tsarap Lingti Chu Valley (Srikantia *et al.* 1978; Pognante *et al.* 1987).

*References*

Herren, E.  1987  Ph.D. thesis, E.T.H. Zürich.
Srikantia, S. V., Ganesan, T. M., Rao, P. N., Sinha, P. K. & Tirkey, B.  1978  *Himalayan Geol.* **8**, 1009–1033.

M. P. Searle and A. J. Rex. Less than one tenth of the High Himalayan area in figure 17 has been mapped in detail. There are no accurate radiometric dates on any granites from this region yet, only preliminary Miocene Rb–Sr dates by Searle & Fryer (1986) and Pognante *et al.* (1987). It is therefore not possible, without extensive geochronological studies, to state what proportion of granites are of Himalayan age. However, within the high-grade metamorphic terrane SW of the Zanskar Valley, there is a high density of two-mica, garnet and tourmaline-bearing granite, leucogranite and adamellite, with some intrusions, such as the White Sail–Papsura peaks in east Kulu, being of similar proportions to the better known Manaslu or Gangotri plutons. The fact that many of the Zanskar granites are restricted to the sillimanite and kyanite zones of Himalayan metamorphism, are spatially related to migmatites of Himalayan age, and contain mineralogy, geochemistry, O and Sr isotopic compositions compatible with the Himalayan leucogranites to the east, suggest to us that their age is also Himalayan. Heterogeneities in the composition, fluid content, and heat-production potential of the source can all produce a variety of crustally derived granites, not necessarily solely of leucogranitic variety. We do not, however, state that all the granites marked on our map are Himalayan; indeed we do not specify age simply because reliable radiometric data have yet to be obtained.

M. COLCHEN (*UFR Sciences, Poitiers, France*).

1. Dr Searle reports an unconformity below the Lower Eocene sediments, which in his opinion belongs to the T1 thrusts, and in his model accompanies a thrust sheet of Lamayuru sediments. In fact, the slates he reports as the Lamayuru Formation resemble those of Trias-Jurassic age in the north of the Shillakong Range, but are of Cretaceous age here. Furthermore, they are concordantly overlying the Upper Cretaceous Fatu La Formation of the Shillakong Range, and thus are not overthrusted (Colchen *et al.* 1986, 1987). This fact set right, the Eocene limestone is conformably overlying those Upper Cretaceous slates; this does not exclude numerous sedimentary gaps, but in no place does an angular unconformity give evidence of a tectonic phase.

Accordingly, no deformation on the North Indian Plate is observed before collision. Obduction itself must have occurred after the Lower Eocene, as the Spongtang ophiolite is overlying mélange series containing Limestone layers and radiolarian chert of Lower Eocene age overlying serpentinized harzburgites (Colchen *et al.* 1987; Reuber *et al.* 1988). The whole is thrusted over the nummulitic limestone discussed above. We think that this is enough evidence for obduction being post Lower Eocene and coeval to collision, a process taking a long time, between 50 Ma (initial collision) and 36 Ma, when India resumes a stable Northward drift (Achache & Patriat 1984, numerous communications this meeting).

2. My second remark concerns the deformation of the Zanskar series: I do not agree with three phases of deformation. As demonstrated above, there is no evidence of a T1 deformation before collision on the North Indian Plate.

There is a syncollisional south-vergent emplacement of ophiolite and ophiolitic mélange that, however, is not accompanied by any penetrative deformation in the substratum. The penetrative schistosity is concordant in the mélange and in the underlying cretaceous slates (Reuber *et al.* 1988). The only deformation we observe is one continuous schistose deformation of the Zanskar series resulting in folds of varying vergence from one area to the adjacent one. South-vergent folds SE of Photaksar become vertical between Yapola and south of Bodkharbu and turnover towards the north and northwest in Wakka Chu and Mulbeck Chu (Colchen & Reuber 1988). This deformation post-dates the collisional period.

A corollary is that the vergence of folds is not significant of a deformation phase and neither is cylindrical folding along an orogen.

M. P. SEARLE AND D. J. W. COOPER. The controversy over the timing of obduction of the Spontang ophiolite has recently been discussed (see Kelemen *et al.* 1988; Searle 1988) and we do not wish to repeat these points here. We disagree with Dr Colchen that the present Spontang ophiolite and mélange sole thrust is a primary emplacement feature, as the amount of shortening in the Lower Eocene limestones is considerably less than the shortening in the Mesozoic shelf sequence. Fold axial planes and cleavage in the Kangi-la Formation slates are truncated at the unconformity and are not conformable around Lingshet although further inboard (SW) they are.

We do not base our deformation phases on differing fold facing (not vergence) directions as implied by Dr Colchen. We base them on cross-cutting thrust and fold relations and sequential restoration of balanced cross sections. The Photoksar thrust is the largest and most impressive such feature and clearly truncates earlier folds and thrusts along its footwall. Although we compartmentalize fold and thrust events, we fully recognize that the post-collision orogenic process represents a continuum.

Phil. Trans. R. Soc. Lond. A **326**, 151–175 (1988)     [ 151 ]

Printed in Great Britain

# Tectonics and evolution of the central sector of the Himalaya

## By K. S. VALDIYA

### Kumaun University, Nainital-263 001, U.P., India

[Plates 1–4]

Following the India–Asia collision, intracrustal movements along the Main Central Thrust (MCT) and Main Boundary Thrust (MBT) in a piggy-back-style, thrust duplexes developed that uplifted the Vaikrita (Central) crystallines of the basement to more than 8000 m elevation. Blocking of subduction on the suture and slowing down of movement on the MCT led to the formation of the Trans-Himadri (Malari) Thrust between the Vaikrita basement and the Tethyan cover sediments, and to gravity-induced backfolds and backthrusts in the latter. The Vaikrita crystallines underwent upper amphibolite to lower granulite facies metamorphism at 600–650 °C and more than 5 kbar (1 kbar = $10^8$ Pa) and migmatistation associated with 28–20 Ma old S-type granites that formed at 15–30 km depth during the culmination of metamorphism and thrust deformation. Delimited by the MCT and MBT, the Lesser Himalaya is made of Proterozoic sediments beneath the Almora nappe constituted of low- to medium-grade metamorphics and $1900 \pm 100$ Ma old granitic gneisses and $560 \pm 20$ Ma old granites. The Lesser Himilaya underwent considerable neotectonic rejuvenation during differential movements along the MBT. The frontal Siwalik molasse below the MBT was severely thrusted and folded in the late Holocene, and continued underthrusting of the Indian Shield beneath the Himalaya is manifest in the development and activation of the deep Himalayan Front Fault (HFF), which separates the Siwalik from the subRecent–Recent alluvial plain of the Ganga Basin.

## 1. INTRODUCTION

For 300 km between Nepal and Himachal Pradesh, the Kumaun Himalaya in the centre of the Himalayan arc (figure 1) is stratigraphically the most representative and structurally a very revealing segment of the mountain chain. It is also the most keenly explored sector since the days of geologists such as G. D. Herbert (1842), who carried out the first-ever mineralogical survey in the Himalaya and Richard Strachey (1851) who crossed the formidable mountain barrier to probe the mysteries of the 'roof of the world'.

A product of repeated deformation of the great pile of sediments at the northern continental margin of the Indian Shield, which collided with the Asian continent some 50 Ma ago, the Himalaya is divisible into five lithotectonically and physiographically distinct domains or subprovinces (figure 2). The dividing surfaces are thrusts of regional dimensions and varying activity. On the extreme south lie the sprawling swampy plains of Tarai, perennially wetted by the springs that emerge from the edge of the gently sloping piedmont gravel fans fringing the hills. The latter, called the 'Bhabhar', is separated from the abruptly rising Siwalik Hills (900–1500 m) by the Himalayan Front Fault (HFF). The ruggedly youthful Siwalik domain, made of late Tertiary to Pleistocene molasse is characterized by steep slopes, swift-flowing consequent streams and deep valleys of antecedent rivers. Synclinal valleys, wherever filled

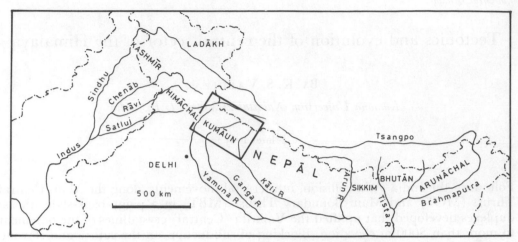

FIGURE 1. Location of Kumaun Himalaya in the central segment of Himalayan chain between Nepal in the east and Himachal Pradesh in the west. Rivers Kali and Tons define the natural boundaries.

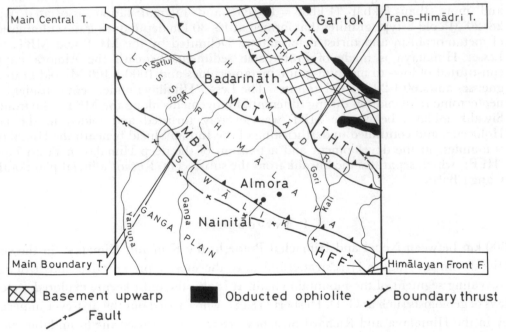

FIGURE 2. Division of the Kumaun Himalaya into five geomorphically and lithotectonically distinctive subprovinces, separated by intracrustal faults and thrusts.

with subrecent–recent gravels eroded from rapidly wasting hills, form flatter tracts called 'duns'.

The Siwalik in turn is separated from the Lesser Himalaya by the Main Boundary Thrust (MBT). The thrust zone is marked by very wide valleys, characterized by fans and cones of landslide debris and by recent–subrecent fault-fashioned triangular facets on spurs. The Lesser Himalaya is made up of Precambrian and partly Palaeozoic sedimentaries overthrust by vast thick sheets of metamorphics and their injected Precambrian–early Palaeozoic granites. Rising to an elevation of 1500–2500 m, the Lesser Himalayan subprovince shows a mild and mature topography with gentle hill slopes. Major transverse rivers, however, flow in deeply dissected valleys still being strongly eroded as a consequence of subrecent tectonic resurgence.

North of the Lesser Himalaya, the Vaikrita or Main Central Thrust (MCT) demarcates the southern (lower) boundary of the Great Himalayan (Himadri) complex of high-grade metamorphics that are extensively injected and migmatized by mid-Tertiary granite. Rising to a height over 6500–7000 m, the Himadri is characterized by extremely rugged and youthful topography, precipitous scarps, sharp peaks, deep gorges with vertical to convex walls and very steep gradients.

Further north, the Malari or Trans-Himadri Thrust (T-HT) marks the boundary between the Himadri basement complex and its thick sedimentary cover of the Tethys domain. Ranging in age from late Precambrian to late Cretaceous, the sediments represent the distal continental margin of the Indian Shield. The synclinorial Tethys subprovince has an extremely rugged topography particularly adjacent to the Himadri. Desolate and virtually bare of vegetation, this domain of frigid climate is a cold desert.

Finally, the Indus–Tsangpo Suture (ITS), passing through Darchen in the Mansarovar area, marks the junction of the Indian and Asian Plates. It is a very deep fault zone characterized by vertically disposed, greatly sheared and shattered seafloor material (ophiolites) and deep-sea sediments.

In this paper attention is focused on five geodynamic belts: (i) the ITS Zone where the leading edge of the northward drifting Indian Plate exhibits pronounced buoyant resistance to slip beneath the Asian Plate as testified by large-scale domal upwarp of the crystalline basement; (ii) the basement-cover contact that broke out into a regional lag thrust (T-HT) following blocking of movements on the suture and on the intracrustal boundary thrust at the

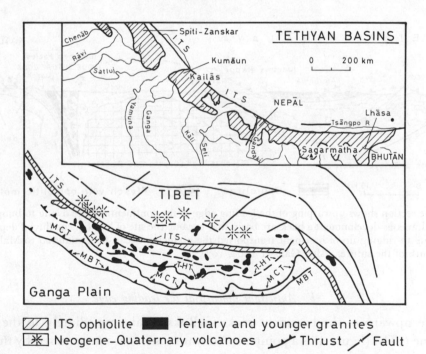

FIGURE 3. Convergence of the Indian and Asian Plates is reflected in widespread acid volcanism (50–48 Ma) at the Andean-type magmatic arc (110–45 Ma) in southern Tibet. (After Zhou *et al.* 1981.) The inset shows discontinuous Tethyan basins with practically uniform pattern of sedimentation and stratigraphy, and with their northern part slightly truncated. (After Valdiya 1988*b*.)

base of the basement; (iii) the MCT zone where repeated slicing with crustal stacking in piggy-back manner has uplifted a deep-seated basement to make the loftiest mountain Himadri; (iv) southern rampart of the Lesser Himalaya intimately related to the MBT where repeated intraplate movements (including neotectonic) resulting from the under-thrusting of the Indian Shield under the Himalaya and attendant strike-slip and vertical movements on tear faults in the thrust sheet have caused spectacular block uplift and subsidence; and (v) the emerging new deep fault (HFF) at the truncated edge of the Siwalik prism against the alluvial plain of the Ganga Basin.

This paper is not intended to be a general review of the extensive work carried out by other workers, but summarizes my observations and deductions in recent years.

## 2. COLLISION AND BUOYANCY RESISTANCE

### (a) Obduction and ophiolite nappes

As the Indian and Asian Plates converged, an oceanic trench developed in front of the Asian continent. This deepening is reflected in the accumulation of flysch with radiolarian chert of the Giumal and Sangcha Malla formations in Malla Johar. The continental collision (*ca.* 50 Ma) initiated widespread acid volcanism genetically related to the Andean-type magmatic arc of the Kailas–Ladakh Ranges constituted of commonly 80–48 Ma old hornblende-bearing granodiorite–diorite–tonalite association (Sr isotopic ratio 0.7033–0.7036). This mag-matic–volcanic arc (figure 3) frames the collision zone, the Indus–Tsangpo Suture (ITS), represented by intensely deformed, and vertically disposed trench sediments tectonically intruded with ocean-floor ophiolitic material (figure 4).

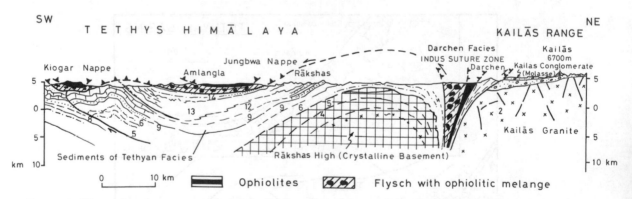

FIGURE 4. The section shows upwarping of the leading edge of the continental plate owing to buoyancy resistance and simultaneous development a back-arc basin that became the site of late Eocene fluvial deposition. Gravity sliding due to ridging up of basement transported the obducted ocean-floor material to Malla Johar, about 80 km south of the suture. (After Gansser 1964, 1974.)

### (b) Anticlinal upwarp at the leading edge

The large upwarp of the basement rocks in Rakshastal, immediately to the south of the collision zone (ITS) (figure 4) is a pointer to buoyant resistance encountered by the continental plate to slide any further under Asia (Valdiya 1984*a*). This domal structure, constituted of gneisses, schists and phyllites of the Great Himalayan Vaikrita Group and intruded by older gneissose porphyritic granites as well as younger leucoadamellites, extends southeast from

Nimaling and Tso Morari in Ladakh to Gurla Mandhata in extreme northwestern corner of Nepal. In Nepal it is represented by the 'axial rise of the basement crystallines between Manangobhot and Phijor (Hagen 1956) and the 900 km long Lhagoi Kangri Range embracing 16 gneissic domes (figure 3) just 70–80 km south of the ITS. These gneissic domes in Nepal consist of Lower Ordovician porphyritic granite intruded by Upper Cainozoic (6–15 Ma) leucogranites of Kangmar (Pham *et al.* 1986; Le Fort *et al.* 1986). Both these suites are remarkably similar in mineral–chemical composition (including strontium isotopic ratio) to those of the Tibetan Slab (Vaikrita). The anticlinal ridge of gneissic domes at the leading edge of the northward moving Indian Plate must have evolved as a consequence of the resistance to further subduction (Valdiya 1984 *a*, 1987, 1988 *b*). An analysis of gravity anomaly data by Lyon-Caen & Molnar (1983) suggests that the Indian Plate was weakened and bent to 10–15° at a position 50 km south of the ITS, presumably as a result of collision that detached the crust from the mantle, as the cold mantle part of Indian lithosphere slid beneath the Himalaya. The detached crust must have been thus folded at its front.

The ridging up of the basement at the leading edge was accompanied by the formation of a backarc basin in the Sindhu–Tsangpo valleys, now represented by the 2000–4000 m thick fluvial accumulations of the Kailas Conglomerate of late Eocene age (figures 4 and 5).

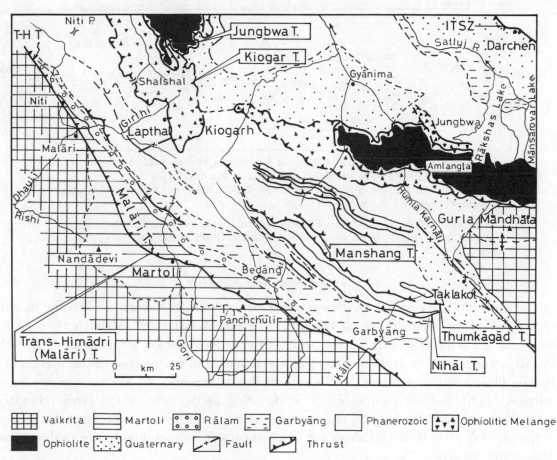

FIGURE 5. Tectonic map of the synclinorial Tethys Zone showing imbrication of thrust sheets in the northeastern area adjacent to the wrench fault of the Humla Karnali valley, and decoupling of the basement and cover along the Malari Thrust (Trans-Himdari Thrust). (After Heim & Gansser 1939; Valdiya 1979, 1987, 1988 *b*.)

## 3. Deformation of Tethys synclinorium and basement-cover decoupling

### (a) Thrust-stacking on the northern margin

The 10–15 km thick sedimentary pile of the Tethys Zone to the south of the basement upwarp is split up in the eastern part into a multiplicity of imbricate thrust sheets (figures 5 and 6a), each characterized by isoclinal or overturned to recumbent folds with attendant disharmonic deformation near planes of dislocation (Heim & Gansser 1939; Valdiya & Gupta 1972). To the east is a major NW–SE trending wrench fault, which has not only sinistrally offset the MCT, but also formed this schuppen zone (Valdiya 1979, 1981). It is surmounted in the western part (Malla Johar) by a succession of two nappes constituted of ophiolitic melanges and ophiolites transported 80 km south from their root zone in the ITS. The central part of the synclinorium comprises upright folds, and domes and basins (figure 6b).

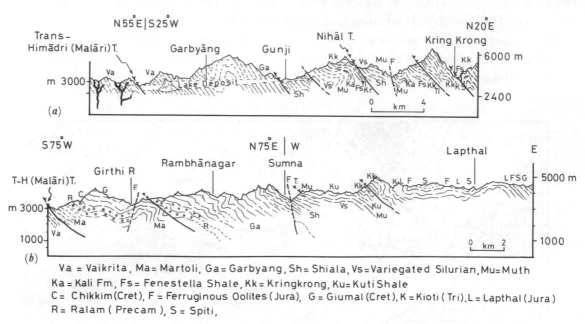

FIGURE 6. Cross section of the Tethyan Zone along the rivers Kali in the east and
Girthi (Dhauli) in the west. (After Valdiya & Gupta 1972; Valdiya 1979.)

### (b) Basement-cover decoupling

The southern margin of the sedimentary domain is sharply defined against the basement crystallines of the Himadri (Great Himalaya) by the Malari Fault (Valdiya 1979), a local name for the Trans-Himadri Thrust, which extends for 1600 km from Kashmir to Sikkim (Valdiya 1987, 1988b). It originated as a detachment fault on the continental margin of the Indian Plate as a consequence of blocking of the movements (45 Ma ago) in the zone of collision (ITS) and the slowing down of thrusting at the base of the Great Himalayan crystalline complex. The buoyant resistance appears to have been so strong that the back part of the moving plate broke up along the basement-cover contact, giving rise to this plane of decoupling (figures 5–7).

Genetically related to the gravity-induced backfolding and backthrusting of the sedimentary cover, the Trans-Himadri Thrust has not only caused tremendous shearing and mylonitization of the Vaikrita basement metamorphics, but also attenuated and even eliminated lithostratigraphic units of both the basement and the cover (e.g. Budhi Schist in Dhauli Valley, Martoli Flysch and Ralam Conglomerate in Darma and Kali Valleys). The termination of the movement on the Main Central Thrust at the base of the 10–20 km thick Vaikrita Slab must have reactivated the T-HT as evident from the shearing of the Miocene intrusive leucogranite in the Dhauli Valley (Valdiya 1979, 1987). This reactivation implies upthrusting of the basement to a great height, the sedimentary cover lagging behind and its terminal part sliding down under gravity to give rise to north-vergent backfolds and minor backthrusts (figure 6b). Significantly, the NNE-plunging early mesoscopic folds of axial foliation in the basement metamorphics (Vaikrita), which are folded on the NW–SE axis, are absent in the Tethyan sediments (Thakur & Chaudhury 1983).

## 4. STRUCTURAL EVOLUTION OF HIMADRI

### (a) Tectonic design

Below the Tethyan sedimentary cover is a 10–20 km thick pile of high-grade metamorphics of the Vaikrita Group, intruded extensively by Miocene leucogranites. Bounded by the T-HT at the top and the Vaikrita Thrust (MCT) at the base, the Vaikrita lithotectonic domain is a huge homocline (figures 7 and 13) (Valdiya 1979, 1981). It has been described in Nepal

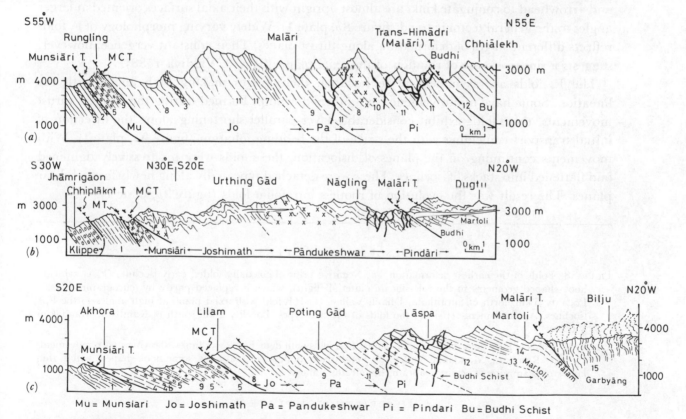

FIGURE 7. Cross sections across the Himadri (Great Himalaya) revealing its structural architecture and the nature of the delimiting thrusts. Numbers: 1, Deoban carbonates; 2, Berinag quartzites; 3, amphibolite; 4, Munsiari sericite–chlorite–schist; 5, Munsiari augen gneiss; 6, Munsiari micaceous quartzite; 7, Vaikrita two-mica–garnet–kyanite–schist; 8, streaky garnet–kyanite gneiss (Vaikrita); 9, biotite quartzite; 10a, migmatite; 10b, adamellite–aplogranite; 10c, pegmatite; 11, calc-silicatefels; 12, biotite-porphyroblast calc-schist; 13, sericite–staurolite schist; 14, Martoli flysch; 15, carbonaceous graphitic phyllite.

by French workers as the Tibetan Slab. The Vaikrita is singularly devoid of large-scale folds, save for macroscopic reclined folds (Roy & Valdiya 1988) recognized on the southern slope of the Himadri (Great Himalaya) and for the fault-bounded granite dome of the Badarinath area (Heim & Gansser 1939; Gansser 1964). Transverse to the general tectonic trend, the large-sized reclined folds are represented by the NNE/NE-striking vertical or nearly vertical S-surfaces in a zone where the general strike is WNW/NW–ESE/SE and the dip rather shallow.

The penetrative synmetamorphic deformation pattern of the Vaikrita is the result of polyphase folding and deep-level ductile shearing involving repeated transposition of foliation planes. The bulk strain was non-coaxial and attributable to variation in the ease of slip on the shear planes during thrust movements (Roy & Valdiya 1988).

### (b) Early deformation

The superposed folds of the Vaikrita are represented by coaxially folded hooks with subparallel to diversely oriented axial planes and complex folds that show strong curvature of hinges and axial surfaces of the earlier folds (figure 8, plate 1) (Roy & Valdiya 1988).

The earliest deformation is represented by down-dip plunging appressed isoclines with their thickened hinges parallel to mineral lineation and attenuated limbs ($F_1A$) (figure 8a, plate 1). The second category of early folds ($F_1B$) varying in morphology from sharp or round hinged and arrowhead to conjugate kinks are almost upright with their axial surfaces oriented at larger angles to the general tectonic trend (figure 8b, plate 1). Widely varying morphology of $F_1$ folds reflects different stages of deformation along thrust planes. Their constant vergence, however, suggests a sinistral sense of rotation of the axial planes (Roy & Valdiya 1988).

The $F_2$ folds are asymmetrical to isoclinal, their hinges making high angles with the lineation. Some have updip vergence and others downdip. Formed during the sluggish thrust movements, these folds exhibit considerable layer-parallel shortening along the direction of initial transport coinciding with the strike of the bedding foliation (figure 9a, plate 1). With movements continuing on the planes of dislocation, these folds were progressively tightened and flattened into stacks of isoclines. These were detached repeatedly along newly formed shear planes. The result was the evolution of rootless intrafolial folds (figure 9b, plate 1).

---

### DESCRIPTION OF PLATE 1

FIGURE 8. Folds of the earliest deformation. (a) Negative print of coaxially folded early isoclines ($F_1A$) evincing hook-shaped geometry in the calc-silicate material (light), which is replaced partly by leucogranite (dark). Locality 5 km north of Suraithota, Dhauli Valley. (b) $F_1B$ fold with axial plane at high angles to the $F_1A$ isoclines. A decollement separates the folds of the two types. Locality 1 km south of Rambara, Mandakini Valley.

FIGURE 9. Folds of later deformation. (a) Isoclinal $F_2A$ folds with detached lower limbs, showing updip movement. Locality Rambara, Mandakini Valley. (b) Small-scale rootless isoclinal folds, some hook-shaped, within thin layers of psammites. Locality Dabrani, Bhagirathi Valley.

### DESCRIPTION OF PLATE 2

FIGURE 11. Garnet crystals in the Vaikrita and Munsiari metamorphics showing mutual compositional and structural differences. (a) and (b) Smooth and rounded edges of ellipsoidal Vaikrita garnet crystals wrapped around by mica flakes. The core and rim show different kinds of inclusions. Locality: Dabrani, Bhagirathi Valley. (c) and (d) The Munsiari garnet, in contrast, is rotated and characterized by sigmoidal inclusion trails. Locality: Kyarkikhal, Bhilangana Valley. (Photos by S. S. Bhakuni.)

*Phil. Trans. R. Soc. Lond. A, volume 326*

*Valdiya, plate* 1

FIGURE 8. For description see opposite.

FIGURE 9. For description see opposite.

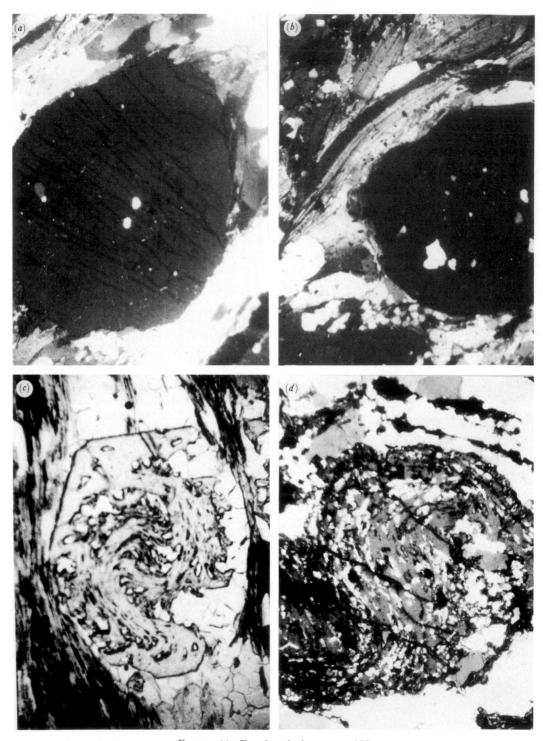

FIGURE 11. For description see p. 158.

*Phil. Trans. R. Soc. Lond. A, volume* 326

*Valdiya, plate* 3

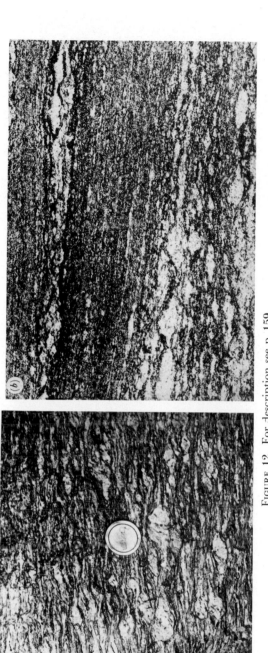

FIGURE 12. For description see p. 159.

FIGURE 22. For description see p. 159.

FIGURE 24. The northward tilted (2–6°) uppermost Pleistocene to early Holocene gravel deposits in the Dabka Valley.

## (c) Formation of Main Central (Vaikrita) Thrust

It is evident that later deformation, obviously caused by the revival of convergence of the continental plates in the Middle to Upper Miocene, is related to movements along thrust planes that developed particularly in the lower part of the Vaikrita Slab. It was most pronounced on the Vaikrita Thrust dipping northwards at an angle of 30–45°. This thrust, representing the MCT in Kumaun (Valdiya 1980a, b, 1981), evolved as the northern part of the Indian crust was sheared following the arrest of movement on the ITS. According to Lyon-Caen & Molnar (1983), the MCT began as a listric fault, detaching the upper crust along a subhorizontal zone in the crust and registering underthrusting of the order of 125 km. The MCT is a wide (ca. 5 km) zone of severe shearing and mylonitization on a multiplicity of dislocation planes, giving rise to a conspicuous schuppen zone (Valdiya 1980a, b). Continued movements along these planes have not only brought the Great Himalayan high-grade metamorphics on to the metasediments and Precambrian sedimentary rocks of the Lesser Himalaya but also lifted the Vaikrita from a zone of ductile deformation to that of brittle deformation (figure 13).

The upward thrusting along the 30–45° northward-dipping Vaikrita Thrust is manifest in the progressive increase in height of the Himadri. Estimated rate of uplift varies from 0.7–0.8 mm $a^{-1}$ (Mehta 1980) to 1.1 mm $a^{-1}$ (Saini 1982). Possibly, the diapiric rise of granitic bodies from deeper levels also contributed to some extent the uplift and increased height of the Himadri peaks. Seismic slip rates of 0.05 mm $a^{-1}$ along the MCT and 0.02 mm $a^{-1}$ along the T–HT deduced by Ye Hong et al. (1981) testify to the continuing uplift of the Great Himalaya. However, the majority of epicentres lie just south of the surface trace of the MCT. Excepting a few, almost all fault-plane solutions indicate thrust movements along a 30° north-dipping shallow (10–20 km) plane, quite different from the MCT, but possibly delineating the interface between the underthrusting Indian Plate and the overlying Himalayan mass (Molnar & Chen 1982; Ni & Barazangi 1984).

## 5. METAMORPHIC DEVELOPMENT AND THERMODYNAMIC CONDITIONS

### (a) Synkinematic metamorphism

The uniformity of mineralogical and chemical compositions of the Vaikrita Group is remarkable. Mineral assemblages developed during the early phase of metamorphism are intimately related to folds, foliations and mineral lineations. This early metamorphism probably occurred in the period 70–50 Ma as indicated by the concentration of isotopic mineral dates (K–Ar and Rb–Sr) and fission-track studies (Mehta 1980). The sillimanite-kyanite–garnet–biotite–quartz–felspar (±muscovite) assemblage in the metapelites and

---

## DESCRIPTION OF PLATE 3

FIGURE 12. (a) Intense mylonitization of the Joshimath (Vaikrita) gneiss in an intraformational shear zone. The streaky gneiss is formed of a row of small isoclinal hinges. Locality: Near the MCT, Ransi, northeast of Ukhimath, Madhyamaheshwar Valley. (b) Mylonitized porphyritic granodiorite (augen mylonite) of the Munsiari Fm. Locality: Kalamuni, south of Munsiari, Gori Valley.

FIGURE 22. Neotectonic evidence discernible in the Kosi Valley (Valdiya 1987b).

metapsammites, and the mineral associations of calcite–hornblende–labradorite $(An_{50-65})$–grossularite and hornblende–diopside–andesine–quartz in the calc-gneisses and calcsilicatefels suggest (figure 10) metamorphism of the Vaikrita Group in the upper amphibolite facies condition within $P$–$T$ range of 5–5.7 kbar† and 600–650/670 °C (Valdiya & Goel 1983). The intimately associated anatectic granite and migmatites in the upper part of the Vaikrita

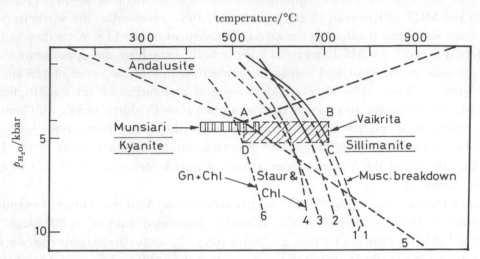

FIGURE 10. The thermodynamic conditions under which the Vaikrita and Munsiari rocks evolved. Curves refer to (1) breakdown of muscovite + quartz, (2) breakdown of staurolite, (3) breakdown of paragonite with quartz, (4) reaction staurolite + chlorite = muscovite + sillimanite + biotite + quartz, (5) Al-silicates and (6) reaction garnet + chlorite + muscovite = sillimanite + biotite. (See Valdiya & Goel 1983.)

succession bear testimony to this thermodynamic deduction. The situation is comparable with that prevailing in the Tibetan Slab (Vaikrita) in central Nepal where the Barrovian-type metamorphism occurred at 650–700 °C and 8 kbar, reflecting a normal geothermal gradient compatible with the total thickness of 20–27 km (Pecher & Le Fort 1986). In the east, the Darjeeling Gneiss (lower Vaikrita), recorded the metamorphic conditions of 6.5–7.0 ($\pm$0.5) kbar and 570 $\pm$ 10 °C to 680 $\pm$ 20 °C (Lal *et al.* 1981). The cooling of the Vaikrita metamorphics presumably occurred *ca.* 38 Ma ago as indicated by isotopic studies ([40]Ar–[39]Ar) on hornblende from Suraithota in thc Dhauli Valley of the study area (D. S. Silverberg, personal communication).

Garnet provides an insight into the mode of deformation during late-kinematic phase. Rounded to elliptical garnets (figure 11 *a*, *b*, plate 2) have their elongation coinciding with the mineral lineation, suggesting plastic deformation during flattening normal to foliation. The garnets show two-stage growth manifest in different orientation of inclusions in the core and the rim. Commonly, there is clustering of small-sized inclusions of rounded to subrounded quartz near the core, the margin remaining virtually free (Roy & Valdiya 1988). Significantly, a similar feature was noticed in the garnets of the Himalayan Gneiss (Vaikrita in Nepal) (Arita 1983). In the Tibetan Slab, garnets also show inverse type of zoning: pyrope-rich cores (up to 40% mole) and almandine-rich (up to 70% mole) rim (Pecher & Le Fort 1986).

Muscovite and biotite, occurring in book form and grown discordant to the foliation and lineation, recrystallized during the late-kinematic phase of deformation (Roy & Valdiya 1988).

† 1 kbar = $10^8$ Pa.

## (b) Late-tectonic retrograde metamorphism

As already suggested, the high-grade progressive metamorphism in the Vaikrita Slab, particularly at its base, is overprinted by retrograde metamorphism related to rapid movements on the thrust planes of the MCT zone. The fracturing of porphyroblasts, enmeshing of the porphyroclasts of kyanite, garnet and micas in the phyllonitic matrix, and the stretching and streaking out of constituent minerals (figure 12) including quartz-producing ribbon structure in mylonites are seen in the narrow horizon of the Vaikrita Thrust (MCT) as clearly discernible within a 2 km wide zone between Joti and Gangnani in the Bhagirathi Valley. A similar zone of intense mylonitization (figure 12, plate 3) and pervasive retrogressive metamorphism 0.5–1.0 km in width, occurs further up about 6 km north from the Vaikrita Thrust (MCT), suggesting another shear zone between the Joshimath and Pandukeshwar units.

## (c) Anatexis and high-temperature metamorphism

In the upper part of the Vaikrita succession, early progressive metamorphism is overprinted by high-temperature low pressure metamorphism. This part is intruded extensively by discordant bodies of 28–18 Ma old leucogranite–adamellite and anastomosing dykes and veins of aplite and pegmatite, all intimately associated with widespread migmatization of metapelites. The occurrence of cordierite in the Badarinath area (Gupta 1978, 1980) and of diopside and elongate garnet, kyanite, and sillimanite and orthoclase/microperthite throughout the belt indicate polymetamorphic recrystallization of the Vaikrita rocks. Goel & Bhakuni (1988) mention skarns and anthophyllite–tremolite rocks adjacent to the granite bodies in the Saraswati Valley. There is thus evidence for high-temperature metamorphism overprinting the amphibolite facies assemblages.

The granite is characterized by cordierite, sillimanite, garnet, kyanite, tourmaline, and by a high value of initial strontium ratio (0.743–0.789). They were derived by partial melting of the Vaikrita metamorphics (Powar 1972; Gupta 1978; Valdiya & Goel 1983; Roy & Valdiya 1988) at a depth of 15–30 km (figure 10). The wide range of strontium isotope ratio indicates that a wide variety of rock types of an evolved continent were involved in their genesis. Melting of the crustal material is related to crustal shortening by stacking of thrust nappes in the MCT zone (Windley 1983; Mattauer 1986). It may also be attributed to thrusting of a hot Tibetan Slab over the sedimentary rocks of the Lesser Himalaya that provided fluid necessary to induce partial melting in the overheated slab (Le Fort 1981, 1986). A third possibility is that the accumulation of heat in certain horizons at the top of the Tibetan Slab due to refraction and diversion of heat flux resulting from low thermal conductivity of the Tethyan sediment cover (Jaupart & Provost 1985) caused this heating and resultant differential melting.

The appreciably high concentration of mineral dates (K–Ar, Rb–Sr, fission track) in the periods 40–25 and 20–10 Ma (Mehta 1980) is suggestive of the thermal events leading to anataxis, genesis of granites and attendant late-kinematic thermal metamorphism. That the granitic activity continued even after the main orogenic event, is evident from undeformed intrusives betraying no sign of strains, emplacement of granite in neck zones of boudins in calc-silicate rocks, very young age of the granite of the Kangmar area (6–15 Ma) north of Sagarmatha, and young biotite dates (8.5–3.7 Ma) from the granites in northeastern Nepal.

## 6. Duplex zone below the MCT

### (a) Structure

Below the MCT (Vaikrita Thrust), lies a thick succession of imbricate thrust sheets evolved in a piggy-back style (Valdiya 1978, 1979, 1980b). The schuppen zone (figure 13) is particularly conspicuous and wide in the belt to the west of the Ganga Valley in northern Garhwal (figure 14). Also involved in the tectonics of repeated imbrication are the Lesser Himalayan lithological units of the sedimentary zone, the epimetamorphics associated with basement porphyroids, and the mesometamorphics of the Munsiari Formation at the top. The Munsiari rocks exhibit extreme shearing, mylonitization and related post-crystallization cataclasis. The flattening of mesoscopic folds increases progressively towards the Munsiari Thrust where it is highest (over 60% in the sedimentaries and over 90% in crystalline rocks) (Bhattacharya & Siawal 1985).

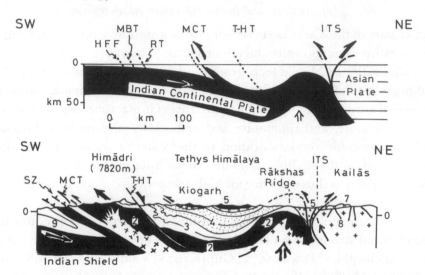

FIGURE 13. The origin of the Main Central Thrust (MCT) and the Trans-Himadri (Malari) Thrust (T-HT) as a result of blocking of movement on the Indus–Tsangpo Suture (ITS) and buoyancy resistance of the Indian Plate to slide under Asia. Numbers: 1, Mid-Tertiary granite; 2, Precambrian Vaikrita crystallines; 3, late Precambrian sedimentary succession; 4, Phanerozoic Tethyan sediments; 5, nappes of ophiolites and ophiolitic mélanges; 6, oceanic trench sediments and seafloor material of the subduction zone; 7, Kailas conglomerate of the back-arc basin; 8, Andean-type magmatic arc (80–45 Ma); 9, Lesser Himalayan Precambrian sediments.

In the duplex system of the area between the Mandakini and Bhagirathi rivers, the crustal shortening resulting from the stacking of nappes is of the order of 55.7% (9.4 km) in the Bhagirathi, 49% (10.9 km) in the Balganga, 82% (34.1 km) in the Bhilangana and 105% (27.9 km) in the Mandakini Valley (Saklani & Bahuguna 1986).

### (b) Metamorphism

The Munsiari consists of mesograde metamorphics, predominant augen gneisses of granodiorite–tonalite parentage and phyllonites. Four samples from the Alaknanda and Bhagirathi Valleys conform to the linear array corresponding to an age of 1950 Ma and reveal small enrichment of radiogenic strontium and low initial strontium ratio of 0.7006 (K. Gopalan, personal communication). Rb–Sr dating by Singh *et al.* (1986) has shown that the gneissic granites of the Munsiari are $1900 \pm 100$ Ma old and the strontium isotope ratio varying from 0.703 to 0.725, being usually on the lower side.

The sericite–chlorite–quartz and muscovite–chlorite–chloritoid–garnet–quartz assemblages in metapelites and epidote–actinolite–oligoclase $(An_{20})$–quartz and epidote–hornblende–andesine $(An_{29}) \pm$ quartz in the metabasites suggest greenschist metamorphism (figure 10) taking place at 450–500 °C and 4 kbar (Valdiya & Goel 1983).

There is thus a demonstrable difference in the physical conditions of metamorphism across the MCT (VT) that is also reflected in the morphology and composition of garnets. The Munsiari schist is characterized by synkinematic growth of garnet that shows rotational fabric (figure 11$c$, $d$). Moreover, in sharp contrast to the pyrope–almandine zoning in Vaikrita garnet, the Munsiari garnet shows normal spessartine core and almandine rim (Pecher & Le Fort 1986).

There is no doubt that there is a pronounced metamorphic discordance across the Vaikrita Thrust (MCT).

## 7. FAULTING AND THRUSTING IN LESSER HIMALAYA

### (a) Dislocation of autochthonous sedimentary succession

The larger part of the Lesser Himalayan terrane embodies Lower Riphean to Vendian (late Precambrian) sediments divisible into two groups: the lower flysch and quartzarenite with basic volcanics (Damtha–Jaunsar Groups) and the upper carbonates–shales assemblages (Tejam–Mussoorie Groups) (Valdiya 1964, 1980a, 1988a). These two groups bear considerable lithological similarity and seem to be homotaxial with the Proterozoic sediments (Martoli–Ralam–Garbyang) of the Tethys Basin. The basal flysch (Rautgara Fm) either rests upon the basement of porphyritic granite ($1900 \pm 100$ Ma old), or perhaps this granite is intruded in the lower part of the sedimentary succession.

The autochthonous subprovince (figures 14 and 15) shows open upright to overturned folds

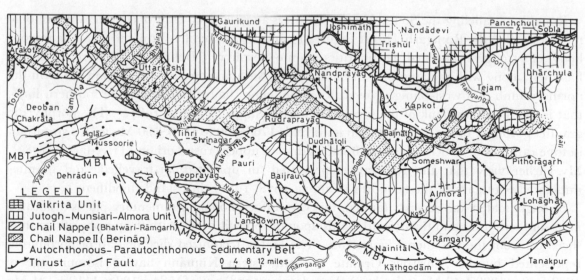

FIGURE 14. Simplified tectonic map of the Lesser Kumaun Himalaya, delimited by the MCT and the MBT. (Revised and modified after Valdiya 1980a, 1981.)

that are locally tight or even isoclinal in the proximity of thrust planes. Immediately to the north of the North Almora Thrust bounding the synclinal nappe of the crystallines, the folds are fan-shaped and extremely tight, where the flattening being as high as 90% (Yedekar & Powar 1986), the sedimentary belt extending from the Indo–Nepal border to the Ganga Valley (Valdiya 1980a) has been uplifted to a ruggedly high range over the nappe to the south.

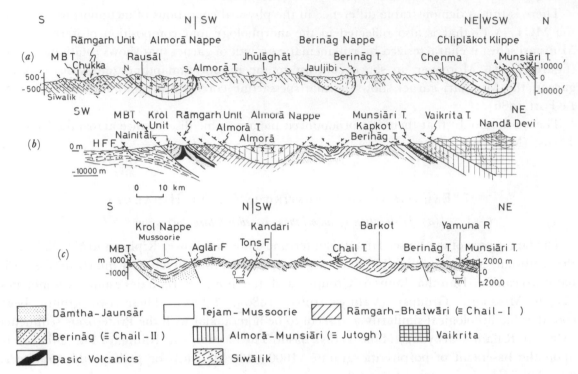

FIGURE 15. Profile of the Lesser Kumaun Himalaya showing in a simplified form the structural architecture of the approximately 20 km thick lithological succession. (Modified after Valdiya 1980b.)

Roughly paralleling the South Almora Thrust is another deep fault of considerable tectonic moment called the Ramgarh Thrust (figures 14 and 15), which has brought up a voluminous body of strongly deformed porphyritic granite–porphyry (Debguru Porphyroid) of the basement along with metamorphosed basal flysch (Nathuakhan Fm ≡ Rautgara) to form the northern limb of a west-plunging overturned anticline (figure 16), above the southern limb, that is, the northern flank of the synclinorial Krol Nappe (Valdiya 1987b). The porphyritic granite–porphyry of the Ramgarh unit is also old: $1765 \pm 60$ to $1875 \pm 90$ Ma with a strontium isotope ratio equal to $0.7335 \pm 0.0046$, indicating an upper crustal origin (Trivedi et al. 1984). My earlier view that the Ramgarh represented an uprooted far-travelled, and granite-implanted Rautgara of the far north (Valdiya 1978, 1981) thus stands modified.

Strongly folded and severely faulted, this vast autochthonous sedimentary pile is dislocated at its southern front and thrust southwards upon the Siwalik along with the porphyritic granite sliced off from the basement (figure 15a). The highly tectonized slice or slab of porphyroid represented by the Amritpur Granite in southeastern Kumaun having isotopic age of $1880 \pm 40$ to $1330 \pm 40$ Ma (mica ages) (Varadarajan 1978) and of $1584 \pm 192$ to $1110 \pm 131$ Ma with a strontium isotope ratio of 0.748–0.741 (Singh et al. 1986) is caught in the duplex

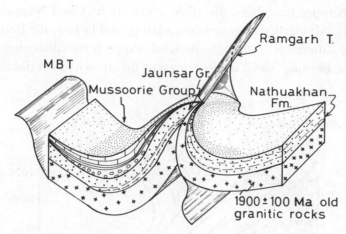

FIGURE 16. Conceptual diagram of a possible shape and structure of the Ramgarh Thrust zone, and faulting up of the basement porphyroids.

structure of the MBT. The dislocated (overthrust) sedimentary succession, the Krol Nappe of Auden (1934, 1937), rides 4–23 km over the Siwaliks in Kumaun. Shah & Merh (1978) and Powar (1980) are of the opinion that the porphyroids form an intrusive body within quartzites and that a part of the Ramgarh unit is really a limb of the Krol succession. However, I regard them to be part of the basement element, which might have remained at a middle crustal level for hundreds of million years, and brought to the surface by thrust faulting.

The intracrustal fault that separates the sub-Himalayan Cainozic subprovince (including the Siwaliks) from the Lesser Himalayan Precambrian rocks has been recognized as the Main Boundary Thrust (Valdiya 1980b). In the regional perspective of the Himalaya, the MBT is a series of disparate thrusts which constitutes the tectonic boundary between the two subprovinces. The MBT dips steeply (over 50°) near the surface but flattens to less than 20° at depth. It is characterized by development of 2–7 km wide schuppen zone or duplex made up of imbricate sheets of the Lesser Himalayan formations involved in the severe folding and splitting (figure 17). In some places, wedges and fragments of the granitic basement are caught between the tectonic slabs (Valdiya 1980b). The MBT appears to be active seismotectonically and is considered to mark the plane along which the Indian Plate slips beneath the Himalaya.

### (b) Uprooted basement and basal units

The northern part of the sedimentary belt is involved in the duplex tectonics of the MCT. The sedimentary succession below the Munsiari Thrust is severely and repeatedly sliced, giving rise to imbricate stacks of tectonic slabs (Valdiya 1978, 1980b, 1981). These slabs are constituted of sedimentary, epimetamorphic and basement gneisses dated $1900 \pm 100$ Ma; ($2120 \pm 60$ Ma; Raju et al. 1982). One of the sheets, the Bhatwari Nappe comprising metamorphosed Rautgara intricately associated with porphyroid, extends westward and joins up with the Chail Nappe (Valdiya 1978) in Himachal Pradesh. The other sheet made predominantly of quartzarenite and intimately associated with penecontemporaneous basic volcanics, covers a vast stretch of the inner Lesser Himalaya in the form of the Berinag Nappe (Valdiya 1979, 1981). These two sheets represent the uprooted and metamorphosed overthrust parts of the lower argillo-arenaceous Damtha Group of the sedimentary succession. The

Bhatwari–Berinag Nappes have been described correctly as Chail Nappes by Fuchs (1980; Fuchs & Sinta 1978). My earlier suggestion correlating and linking the Berinag with the Krol Nappe is apparently untenable. Although the Krol Nappe is the dislocated frontal part of the sedimentary pile, the Berinag–Chail sheets represent the uprooted overfolded back part of the same pile.

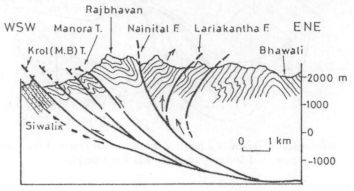

FIGURE 17. The movements along the MBT are transmitted on the thrusts and faults of the schuppen zone of the mountain front (Valdiya 1981).

According to Johnson (1986), the Ramgarh thrust sheet is discontinuous and the separated lenticular masses are horses attached to the base of the Almora–Munsiari sheet at the top of the succession. During the active slip on the basal thrust (or slide of the Munsiari sheet) a lenticular horse 100 km long and less than 9 km thick was cut off from the Ramgarh Formation in the process of 'ramp erosion' on the Munsiari Thrust. This Ramgarh horse has travelled forward for a great distance. Applying the piggy-back model of thrust tectonics, Johnson conceived the evolution of the Lesser Himalayan thrusts in a hinterland-to-foreland sequence in which the Vaikrita Thrust (MCT) developed first, followed in time by the Krol Thrust, the Berinag Thrust and the MBT. According to him, the Lesser Himalaya is underlain by what he called 'Himalayan Frontal Thrust' and which has ramped up towards the surface and has functioned as the sole fault along which the Lesser Himalaya and the Siwaliks have advanced over the foreland.

### (c) Far-travelled sheets and tectonic rejuvenation

The thick pile of mesograde metamorphics of the Almora Nappe (figures 14 and 15) comprises carbonaceous/graphitic schist, garnetiferous mica schists, and amphibolites with concordant gneissose granite bodies ($1820 \pm 130$ Ma and having a Sr isotope ratio of $0.7144 \pm 0.0118$) at the base and syntectonic tonalite–granodiorite–granite plutons towards the upper part. These later granites are $560 \pm 20$ Ma old with Sr isotope ratio of $0.7109 \pm 0.0013$ (Trivedi et al. 1984). The Baijnath–Dharamghar–Askot klippen (figures 14 and 15) are made predominantly of the basal augen gneisses ($1810 \pm 20$ Ma old with Sr isotope ratio of $0.7092 \pm 0.0015$) like the root zone unit the Munsiari (figures 14 and 15), which is constituted dominantly of augen gneisses dated $1830 \pm 200$ Ma (Sr isotope ratio 0.725), $1890 \pm 155$ and $1950 \pm 200$ Ma (Bhanot et al. 1977). The Munsiari is thus the much deformed (and compressed) root of the Almora Nappe and its klippen. So far the lower Ordovician granite has not been located in the klippen and the root. The Munsiari extends westward and joins with the Jutogh Nappe of Himachal Pradesh (figure 14).

Significantly, the wide zone of the Almora Nappe exhibits very mature topography recalling those of the Vindhya and Aravali provinces of the Peninsular India. The gentle slopes have thick soil profiles indicating a prolonged period of weathering and tectonic stability until very recent times. The sluggish meandering streams in their wide mature valleys abruptly assume the form of deep gorges, and then descend in rapids or waterfalls as they approach the South Almora Thrust. These features imply resurgent tectonics in subrecent times, and slight northward tilt of the terrane of the Almora crystallines.

## 8. Evolution of the southern front of Lesser Himalaya

### (a) The underthrusting Indian Plate

The Indian Plate bearing a large prism of the Siwalik molasse at its forefront continues to slide beneath the Lesser Himalaya. The repeated movements over a long period (20 Ma) has caused extremely severe brittle deformation leading to flattening of folds, their splitting along a multiplicity of thrust planes, and stacking of lithotectonic slabs and wedges giving rise to a highly elevated mountain front. Geomorphic features including stream gradient and fluvial and colluvial landforms, eloquently demonstrate continued, albeit episodal, neotectonic activity on the MBT and related faults and thrusts (Valdiya 1981, 1986; Valdiya et al. 1984). Uplift (ca. 30 m) along the MBT of the Siwalik block (figure 18) in the Nainital area, for example, has not only forced some streams to abandon their old channels but also caused the subrecent fluvial terraces and recent colluvial cones to be cut and differentially uplifted (45–85 m) on the Siwalik side (Valdiya et al. 1984; Valdiya 1984b, 1986). In some segments the Siwalik has risen up relative to the Lesser Himalaya. Elsewhere it is the Lesser Himalaya that has advanced southward (figure 19) over the deposits as young as the subrecent scree and fluvial gravels (Jalote 1966; Valdiya 1981, 1986, 1987b). On the southern slope of the Mussoorie Hills, the subrecent Dun fan has been uplifted (290 ± 76 m) on the Lesser Himalayan side (Nossin 1971). Obviously, the listric thrust has been reactivated as a normal fault in the Nainital area, presumably because of the geometric similarity near the surface between the steeply dipping MBT and a normal fault. The squeezing in of the tectonic slab of the Siwalik adjacent to the MBT has caused rotational movement and uplift of the Siwalik block.

### (b) Tear faulting

The Lesser Himalayan terrane particularly its southern front, is dissected by a large number of tear faults, forming conjugate pairs oriented transverse to the Himalayan trend (Valdiya 1976, 1981, 1988a). Following the 'locking' of movement on the MBT, and/or possibly because of differential loading caused by unequal thickness of rock formations, the whole pile of the thrust sheets started deforming by tear faults on fractures oriented transverse to the orogen. Incidently, these fractures and faults are in the line of the basement ridges or promontories of the Peninsular shield extending NNE under the Ganga alluvium and prodding the mountain chain. The tear faults are thus indent-linked strike-slip faults developed particularly in the MBT zone of shortening and attendant uplift.

Many of the transverse tear faults are linked with or become strike faults, as commonly seen in the Nainital Hills (figure 20). In the interior, they are linked with the reactivated weak zone of phyllonites and mylonites (figure 14). For example, in the zone of North Almora Thrust in

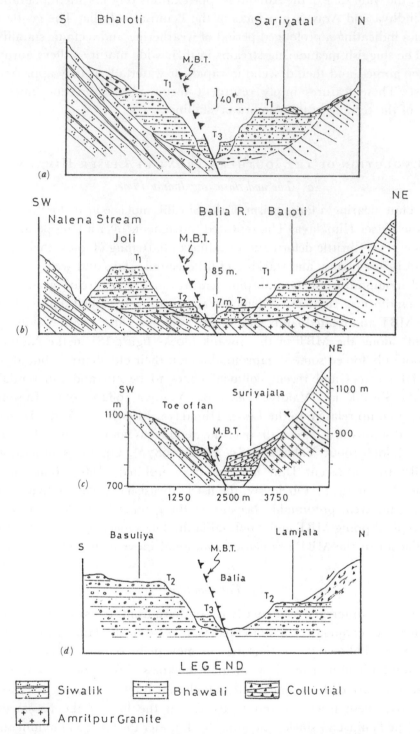

FIGURE 18. A variety of geomorphic features indicate that the MBT is an active fault in the segment SE of Nainital. Where the MBT has been reactivated as a normal fault, the Siwalik has risen up. (After Valdiya *et al.* 1984; Valdiya 1986, 1988*a*.)

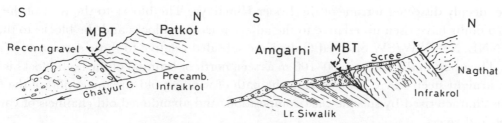

FIGURE 19. In the western segment of the MBT zone in the Nainital Hills, the late
Precambrian sediments have advanced over recent fluvial gravels.

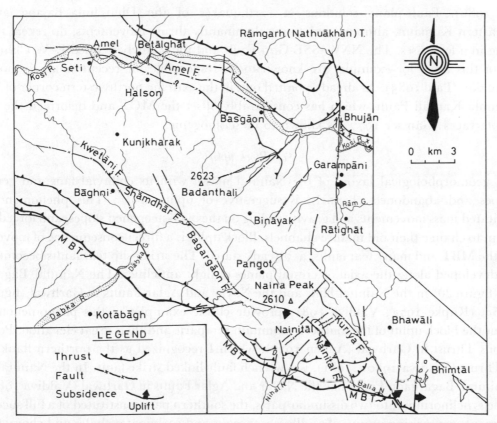

FIGURE 20. The faulted Nainital massif. The tear faults, offsetting even the active MBT, become strike faults along
their northwesterly extension. Northeast of the Nainital–Bagargaon Fault, the folds are vergent northwards,
whereas to the south they exhibit the normal southwards vergence. (From Valdiya 1988a.)

the Saryu Valley in the east and Dwarahat–Chaukhutia belt in the west (Valdiya 1976) where
the resurgence of tectonic movements took place episodically along the transition of brittle and
ductile rocks. This explains the extreme deformation and elevation of the underlying
autochthonous rocks higher than the overthrust sheets of the Almora crystallines.

The tear faults have offset even the very young and still active MBT (late Pliocene to middle
Pleistocene) as discernible in the Nainital Hills (figure 20), and in the Dehradun–Haridwar
area by distances varying from less than 1 km to more than 12 km. The NNW/N–SSE/S
trending faults are predominantly dextral whereas the NNE/NE–SSW/SW oriented ones are
sinistral. Neotectonic movements on these wrench faults have caused significant uplift and
subsidence of the faulted blocks, giving rise to remarkably elevated transverse spurs in the old

mature, deeply dissected terrane of the Lesser Himalaya. The blocks to the west of the NW/N–SE/S faults have risen up relative to the adjacent eastern blocks and the blocks to the west of the NNE/NE–SSW/SW oriented faults have subsided relative to those eastwards (Valdiya 1986). The uplift to the extent of 90–100 m as seen north of Nainital of the fault blocks is borne out by straight steep scarps, waterfalls in the path of winding old streams, uplift of the fluvial deposits characterized by specific clast composition, and abandoned old channels of vanished streams and rivers.

Although the tear faults of the outer Lesser Himalaya, particularly those that are linked with the MBT, have been active, the wrench faults of northern part seem to be seismogenic (Valdiya 1981, 1986). Fault-plane solutions of earthquakes of the Dharchula–Bajang area in northeastern Kumaun, although indicating dominantly thrust movements, do reveal normal faulting in a few cases. The NNW–SSE Gori Fault and the vaguely defined E–W Chhiplakot Fault of this area, for example, are known to be responsible for recent small to moderate earthquakes (Paul 1985). As already pointed out, in the extreme northwestern corner of Nepal, the Humla Karnali Fault, which has considerably offset the MCT and deformed the recent fluvial terraces (Gansser 1977), is known to be seismogenic.

### (c) Block uplift

The geomorphological layout of the Nainital massif with its colluvial fans and cones on hillslopes and abandoned channels, is suggestive of uplift *en bloc*. This phenomenon has precipitated mass movements and gravity sliding on the over-steepened slopes, and forced many a stream to change their old mature channels. Block uplift is a direct consequence of movements along the MBT and many tear faults, as already stated. The strike-slip tear faults become strike faults developed along the axial or crestal planes of tight anticlines. The Nainital–Bagargaon Fault (figure 20) in the Nainital Hills and the Nayar and Aglar Faults in Garhwal (figures 14 and 15c) (Rupke 1974; Valdiya 1981) provide classic examples of this phenomenon. This explains the block uplift of the southern mountain ramparts and their great elevation. Possibly, the Tons Thrust in Garhwal (Auden 1937), which I recognized as the northern flank of the Krol Thrust (Valdiya 1980a, 1981), is a wrench fault-linked strike fault. In the Nainital Hills, the Nainital–Bagargaon Fault, like the Nayar and Aglar Faults in Garhwal (Valdiya 1981) has split the synclinorium into two dissimilar parts, the southern part constituted of a full succession of sediments comprising groups of argillo-arenaceous and carbonates-shales and characterized by normal south-vergent folds, and the northern part made up of only the lower argillo-arenaceous group of sediments exhibiting pronounced north-vergent folds, resulting presumably from gravity sliding (figure 21).

The northern limit of the Nainital massif is defined by a deep fault of regional extent (Amel Fault) associated with a number of parallel to subparallel strike faults. These faults register differential vertical movements, giving rise to a minigraben and a minihorst in the extraordinarily wide valley of the west-flowing Kosi (figure 20). The Amel Fault continues to be active as evidenced by multiple levels of fluvial terraces, sharply truncated fluvial fans of recent origin (figure 22, plate 3), triangular fault facets devoid of vegetation and lined with colluvial cones at the foot, uplifted (by more than 250 m) channel fills of a river that has vanished or changed course (Valdiya 1987b), and debris avalanches of gigantic proportion generated by landslides that repeatedly ravage the terrain.

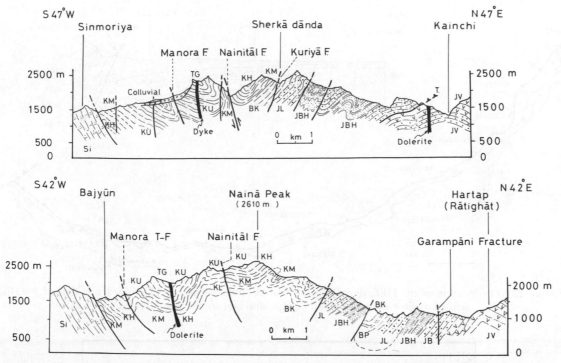

FIGURE 21. The tear fault-linked strike fault, Nainital Fault, has split the Krol synclinorium into two dissimilar parts. The northern part constituted of only the lower group of the sediments is characterized by north-vergent folds resulting from gravity sliding. The southern part exhibits south-vergent folds. (From Valdiya 1988a.)

## 9. A DEVELOPING BOUNDARY FAULT

### (a) Himalayan Front Fault

The multicyclic geomorphology of the strongly folded and faulted Siwalik subprovince contrasts sharply with the monocyclic geomorphology of the undeformed Ganga alluvium of Quaternary to Recent age, implying existence of a tectonic break between the two domains (figure 23). The Himalayan Front Fault (HFF) has obliquely truncated the folds and intrabasinal thrusts and eliminated at least two major structural belts of the Siwalik (Valdiya 1984a, 1986). The abrupt rise (60 m) in front of the Ganga Plain of the gravel deposits of uppermost Pleistocene to early Holocene age, and their northward tilting (2–6°) as seen in the Dabka and Kosi Valleys (figure 24, plate 4) clearly demonstrate the active nature of the HFF (Nakata 1972; Valdiya 1986, 1988a).

Significantly, the Siwalik between the MBT and the HFF is being considerably compressed and thus uplifted at the rate of 0.8 mm a$^{-1}$ in the Dehradun Valley (Chugh 1974). The continuing deformation is evident from the active tear faults connected with the perceptibly active intrabasinal Dhikala–Sarpaduli Thrust in the Ramnagar–Kotabagh area (figure 25). Differential vertical movements along the fault by as much as 20–30 m have given rise to depressions and rises within the basin in the Dehradun area (Nakata 1972; Rao 1977). The depressions have been rapidly filled up by great volumes of gravel deposits of Holocene to Recent age.

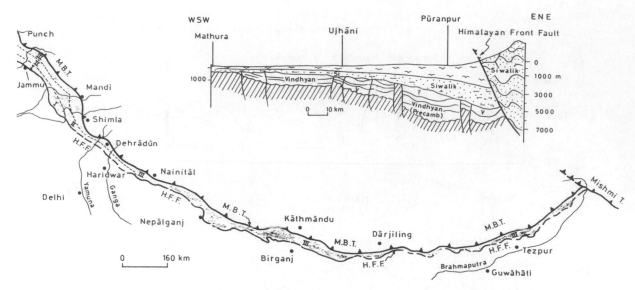

FIGURE 23. The location of the HFF, delimiting the Siwalik from the plains. The folded and faulted Siwalik is defined against the undeformed Ganga alluvium by the Himalayan Front Fault.

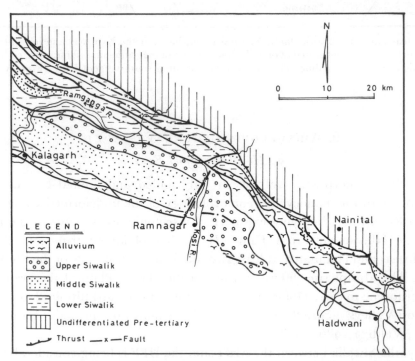

FIGURE 25. The continuing neotectonic activity in the Kotabagh–Ramnagar belt is evident from, among other things, advance of the Siwaliks over recent alluvial cover and the formation of intrabasinal depressions and rises due to normal faults. The depressions have been filled up by subrecent to recent gravel deposits and given rise to flat 'duns'.

### (b) Shifting belt of deformation

Looking back into the history of evolution of the Himalaya, it is realized that along with the basin of sedimentation, the belt of deformation has progressively shifted southwards through the time. As the foreland basin of the early Tertiary (Subathu, Dagshai–Kasauli) was deformed in the middle Miocene times, another foreland basin (Siwalik) evolved south of the MBT.

Possibly the flexing down of the Indian lithospheric plate under the load of Himalaya created the foredeep basin. The middle Pleistocene saw deformation and attendant crustal shortening of the Siwalik and formation of the third foreland basin to its south: the Ganga Basin. The logical culmination of the continuing northward push of the Indian Shield would be the deformation of the Ganga Basin sediments and southward advance of the Siwaliks along the delimiting HFF, which will eventually assume the attitude of a listric thrust.

Generous financial assistance from the Department of Science and Technology (for work in the Great Himalaya) and the University Grants Commission under COSIST Programme (for studies in outer Lesser Himalaya) is gratefully acknowledged. I am indebted to Professor V. K. Gaur and Dr R. S. Sharma for critically going through the manuscript and making valuable suggestions.

## REFERENCES

Arita, K. 1983 *Tectonophysics* **95**, 43–60.

Auden, J. B. 1934 *Rec. geol. Surv. Ind.* **67**, 357–454.

Auden, J. B. 1937 *Rec. geol. Surv. Ind.* **71**, 407–435.

Bhanot, V. B., Singh, V. P., Kansal, A. K. & Thakur, V. C. 1977 *J. geol. Soc. India* **18**, 90, 91.

Bhattacharya, A. R. & Siawal, A. 1985 *Geologie Mijnb.* **63**, 159–165.

Chugh, R. S. 1974 Presented at Int. Symp. Recent Crustal Movements. Zurich. (Unpublished.)

Fuchs, G. 1980 In *Stratigraphy and correlations of the Lesser Himalayan Formations* (ed. K. S. Valdiya & S. B. Bhatia), pp. 163–173. Delhi: Hindustan Publishing Corp.

Fuchs, G. R. & Sinha, A. K. 1978 *Jb. geol. Bundesanst, Wein* **121**, 219–241.

Gansser, A. 1964 *Geology of the Himalayas.* (289 pages.) London: Interscience Publishers.

Gansser, A. 1974 *Eclog. geol. Helv.* **67**, 479–507.

Gansser, A. 1977 In *Himalaya, Mem. Sciences de la Terre* no. 268, pp. 181–192. Paris: CNRS.

Goel, O. P. & Bhakuni, S. S. 1988 *Contr. Miner. Petr.* (In the press.)

Gupta, L. N. 1978 *Himalayan Geol.* **8** (II), 717–727.

Gupta, L. N. 1980 In *Structural geology of the Himalaya* (ed. P. S. Saklani), pp. 59–74. New Delhi: Today & Tomorrow Publishers.

Hagen, T. 1956 *Geographica helv.* **5**, 217–219.

Heim, A. & Gansser, A. 1939 *Central Himalaya, Geological Observations of the Swiss Expedition, 1936.* In *Mem. Soc. Helv. Sci. Nat.* **73** (1), 1–245.

Herbert, G. D. 1942 *J. Asiat. Soc. Beng.*, p. 11.

Jalote, P. M. 1966 In *Proc. Third. Symp. Earthquake Engineering*, pp. 455–458. Roorkee: University of Roorkee.

Jaupart, C. & Provost, A. 1985 *Earth planet. Sci. Lett.* **73**, 385–397.

Johnson, M. R. W. 1986 In *Himalayan thrusts and associated rocks* (ed. P. S. Saklani), pp. 27–39. New Delhi: Today & Tomorrow Publishers.

Lal, R. K., Mukherji, S. & Ackermand, D. 1981 In *Metamorphic tectonites of the Himalaya* (ed. P. S. Saklani), pp. 231–278. New Delhi: Today & Tomorrow Publishers.

Le Fort, P. 1981 *J. geophys. Res.* **86**, 10545–10568.

Le Fort, P. 1986 In *Collision tectonics* (ed. M. P. Coward & A. C. Ries), pp. 159–172. London: Geological Society.

Le Fort, P., Debon, F., Pecher, A., Sonet, J. & Vidal, P. 1986 In *Orogenic evolution of Southern Asia* (Mem. Sci. Terre), pp. 191–209. Paris: CNRS.

Lyon-Caen, H. & Molnar, P. 1983 *J. geophys. Res.* **88**, 8181–8191.

Mattauer, M. 1986 In *Collision tectonics* (ed. M. P. Coward & A. C. Ries), pp. 37–50. London: Geological Society.

Mehta, P. K. 1980 *Tectonophysics* **62**, 205–217.

Molnar, P. & Chen, W. P. 1982 In *Mountain building processes* (ed. U. Briegel & K. J. Hsu), pp. 41–57. New York: Academic Press.

Nakata, T. 1972 *Geomorphic history and crustal movements of the foothills of Himalaya*, pp. 39–177. Tohoku University: Institute of Geography.

Ni, J. & Barazangi, M. 1984 *J. geophys. Res.* **89**, 1147–1163.

Nossin, J. J. 1971 *Z. Geomorph.* **12**, 18–50.

Paul, S. K. 1985 Ph.D. thesis: Kumaun University, Nainital.

Pecher, A. & Le Fort, P. 1986 In *Orogenic evolution of southern Asia* (ed. P. Le Fort, M. Colchen & C. Montenat), Mem. Sci. Terre no. 47, pp. 285–309. Paris: CNRS.

Pham, V. N., Boyer, D., Therme, P., Yuan, X. C., Li, L. & Jin, G. Y. 1986 *Nature, Lond.* **319**, 310–314.

Powar, K. B. 1972 *Himalayan Geol.* **2**, 34–46.

Powar, K. B. 1980 In *Stratigraphy and correlations of the Lesser Himalayan Formations* (ed. K. S. Valdiya & S. B. Bhatia), pp. 49–58. Delhi: Hindustan Pub. Corp.

Raju, B. N. V., Bhabria, T., Prasad, R. N., Mehadevan, T. M. & Bhalla, N. S. 1982 *Himalayan Geol.* **12**, 196–205.

Rao, D. P. 1977 *Photonirvachak* **5**, 5–40.

Roy, A. B. & Valdiya, K. S. 1988 *J. geol. Soc. India* **31**. (In the press.)

Rupke, J. 1974 *Sedim. Geol.* **11**, 81–265.

Saini, V. S. 1982 *Proc. Indian Acad. Sci.* **148**, 539–540.

Saklani, P. S. & Bahuguna, V. K. 1986 In *Himalayan thrusts and associated rocks* (ed. P. S. Saklani), pp. 1–25. New Delhi: Today & Tomorrow Publishers.

Shah, O. K. & Merh, S. S. 1978 *J. geol. Soc. Ind.* **19**, 91–105.

Singh, R. P., Singh, V. P., Bhanot, V. B. & Mehta, P. K. 1986 *Indian J. Earth Sci.* **13**, 197–208.

Singh, V. P., Singh, R. P. & Bhanot, V. B. 1986 *Indian J. Earth Sci.* **13**, 189–196.

Strachey, R. 1851 *Q. Jl geol. Soc. Lond.* **7**, 292–310.

Thakur, V. C. & Chaudhury, B. K. 1983 In *Himalayan shears* (ed. P. S. Saklani), pp. 45–57. New Delhi: Himalayan Books.

Trivedi, J. R., Gopalan, K. & Valdiya, K. S. 1984 *J. geol. Soc. India* **25**, 641–654.

Valdiya, K. S. 1964 In *Report 22nd Int. Geol. Congr.*, vol. 11, pp. 15–36. Calcutta: Geological Survey of India.

Valdiya, K. S. 1976 *Tectonophysics* **32**, 353–386.

Valdiya, K. S. 1978 *Indian J. Earth Sci.* **5**, 1–19.

Valdiya, K. S. 1979 *J. geol. Soc. India* **20**, 145–157.

Valdiya, K. S. 1980a *Geology of Kumaun Lesser Himalaya*. (291 pages.) Dehradun: Wadia Institute of Himilayan Geology.

Valdiya, K. S. 1980b *Tectonophysics* **66**, 323–348.

Valdiya, K. S. 1981 In *Zagros–Hindukush–Himalaya: geodynamic evolution* (ed. H. K. Gupta & F. M. Delaney), pp. 87–110. Washington, D.C.: American Geophysics Union.

Valdiya, K. S. 1984a *Tectonophysics* **105**, 229–248.

Valdiya, K. S. 1984b *Aspects of tectonics: focus on South-Central Asia*. (319 pages.) New Delhi: Tata–McGraw-Hill.

Valdiya, K. S. 1984c In *Proc. Colloquium Tectonics of Asia, 27th Inter. Geol. Congr.*, Moscow, pp. 110–137.

Valdiya, K. S. 1986 In *Proceedings Intern. Symp. Neotectonics in South Asia*, pp. 241–267. Dehradun: Survey of India.

Valdiya, K. S. 1987 *Curr. Sci.* **56**, 200–209.

Valdiya, K. S. 1988a *Geology and natural environment of Nainital Hills*. Nainital: Gyanodaya Prakashan.

Valdiya, K. S. 1988b *Geol. Soc. Am. Spec. Publ.* (In the press.)

Valdiya, K. S. & Goel, O. P. 1983 *Proc. Indian Acad. Sci.* **92**, 141–163.

Valdiya, K. S. & Gupta, V. J. 1972 *Himalayan Geol.* **2**, 1–34.

Valdiya, K. S., Joshi, D. D., Sanwal, R. & Tandon, S. K. 1984 *J. geol. Soc. India* **25**, 761–774.

Varadarajan, S. 1978 *J. geol. Soc. India* **19**, 380–381.

Windley, B. F. 1983 *J. geol. Soc. Lond.* **140**, 849–865.

Ye Hong, Zhang Wen-Yu, Zhi-Shui & Xia Qin 1981 In *Geological and ecological studies of Qinghai-Xizang Plateau*, vol. 1, pp. 65–80. Beijing: Science Press.

Yedekar, D. B. & Powar, K. B. 1986 In *Himalayan thrusts and associated rocks* (ed. P. S. Saklani), pp. 41–69. New Delhi: Today & Tomorrow Publishers.

Zhou-Yun-Sheng, Zhang Qi, Jim Cheng-Wei & Deng Wan-Ming 1981 In *Proc. Symp. Quinghai-Xizang Plateau*, vol. 1, pp. 363–378. Beijing: Science Press.

*Note added in proof* (25 *April* 1988). The 'Trans-Himadri Thrust' is a normal detachment fault in a large part of its extent in the northwestern and central sectors but is a thrust in the eastern part.

## Discussion

V. S. Cronin (*Department of Geology, and Center for Tectonophysics, Texas A&M University, Texas, U.S.A.*). Professor Valdiya noted a right-lateral component of displacement on the Main Boundary Thrust and on other faults with a northwest trend in northern India (Valdiya 1976, 1981). The oblique convergence of India with the western Himalaya suggests the possibility of a right-lateral sense of shear within the western part of the Himalayan thrust prism. What is his current thinking related to strike-slip displacements in the Himalaya?

The Karakoram Fault is a right-lateral strike-slip fault that is currently active and separates the Karakoram–Himalaya Mountains from the Kun Lun in western Tibet. Total displacement along the Karakoram Fault is thought to be *ca.* 200–250 km (Srimal 1983). The Karakoram Fault is commonly depicted on regional maps as merging with the Indus–Tsangpo Suture Zone. Does Professor Valdiya have any field data that might indicate how displacement along the Karakoram Fault is accommodated by other structures along its southeastern end? Is there evidence for right-lateral faulting along the Indus–Tsangpo Suture Zone?

*Reference*

Srimal, N. 1983 *Geol. Soc. Am., Abstr. Progr.* **15**, 694.

K. S. VALDIYA. I have already expressed my views in my papers (Valdiya 1976, 1981, 1984*b*, 1986).

*Phil. Trans. R. Soc. Lond.* A **326**, 177–188 (1988)     [ 177 ]

*Printed in Great Britain*

# Palaeomagnetic constraints on Himalayan–Tibetan tectonic evolution

BY J. LIN[1] AND D. R. WATTS[2]

[1]*Institute of Geology, Academia Sinica, Beijing, People's Republic of China*
[2]*Department of Geology, University of Glasgow, Glasgow G12 8QQ, U.K.*

Palaeomagnetic data from the Lhasa, Qiangtang and Kunlun Terranes of the Tibetan Plateau are used with data from stable Eurasia, eastern China and Indochina, to test different models of crustal thickening in the Tibetan Plateau, to attempt a Carboniferous palaeogeographic reconstruction, and to calculate the relative motion between the South China Block and the Indochina Block. The data suggest that since the onset of the India–Eurasia collision, the Lhasa Terrane has moved $2000 \pm 800$ km north with respect to stable Eurasia. This indicates that strong internal defomation must have taken place in southern Eurasia since the collision, and thus challenges the model of large-scale underthrusting of the Indian subcontinent beneath the Tibetan Plateau as the mechanism for crustal thickening in Tibet. Palaeomagnetic results from the Kunlun Terrane show that it was at $22°$ south latitude during the Carboniferous. A Carboniferous reconstruction is presented in which the Kunlun and Qiangtang Terranes, several Indochina terranes, and the North and South China Blocks are grouped together. These units of continental crust all share the specific tropical and subtropical Cathaysian flora, and the group is therefore called the Cathaysian composite continent. To test the model of propagating extrusion tectonics, we have used newly available palaeomagnetic results from South China and Indochina to calculate probable displacements. This exercise suggests a rotation of about $8°$ of Indochina with respect to the South China Block that is smaller than the predicted rotation of $40°$. A large eastward translation of the South China Block relative to the Indochina Block of about 1500 km is consistent with the palaeomagnetic data.

## INTRODUCTION

Before the 1985 Tibetan Geotraverse, palaeomagnetic data from the Himalayan–Tibetan area were confined to the Himalayan belt, in the areas of Nepal, Kashmir and the Western Himalaya Syntaxis (Klootwijk *et al.* 1985, 1986*a*, *b*) and southern Tibet (Achache *et al.* 1984; Besse *et al.* 1984). Klootwijk *et al.* (1985, 1986*a*, *b*) demonstrated that the palaeomagnetic results show a consistent pattern of rotations of the Himalayan Arc relative to the Indian Shield, varying from $45°$ clockwise in the northwestern Himalayas to slightly counterclockwise in the Lhasa region, indicating that the Himalayan Arc formed through oroclinal bending. Based on their study of the mid-Cretaceous Takena Formation and the Lower Tertiary Linzizong volcanics of the Lhasa Terrane, Achache *et al.* (1984) determined crustal shortening of $1900 \pm 850$ km north of the Lhasa Terrane and $1400 \pm 1200$ km south of this area since the onset of the India–Eurasia collision. Further palaeomagnetic data collected during the 1985 Tibetan Geotraverse (Lin & Watts 1988) enable us to refine calculations of crustal shortening and extend this calculation to the central Tibetan Plateau. All palaeomagnetic data used in this paper from the Tibetan Plateau are summarized in table 1.

TABLE 1. TIBETAN PALAEOMAGNETIC DATA

| | age | unit | B(N) | lat./deg | long./deg | $A_{95}$/deg | dp/deg | dm/deg |
|---|---|---|---|---|---|---|---|---|
| | | | | *Lhasa Terrane* | | | | |
| 1 | Pal–E | Linzizong | 8 (46) | 72 | 299 | 10.5 | — | — |
| 2.1 | Ku | Takena (S) | 8 (51) | 68 | 279 | — | 3.5 | 6.9 |
| 2.2 | Ku | Takena (S) | 8 (61) | 71 | 288 | 7.9 | — | — |
| 2.3 | Ku | Takena (N) | 6 (49) | 64 | 326 | 8.9 | — | — |
| 3 | Ku | Nagqu vol | 9 (33) | 78 | 283 | — | 4.0 | 6.9 |
| 4 | Ku | Qelico vol | 4 (20) | 74 | 318 | — | 11.1 | 19.1 |
| | | | | *Qiangtang Terrane* | | | | |
| 5 | E | Fenghuoshan | 15 (84) | 61 | 253 | — | 3.8 | 7.4 |
| 6 | E | Fenghuoshan | 12 (57) | 55 | 201 | — | 3.2 | 5.6 |
| | | | | *Kunlun Terrane* | | | | |
| 7 | Tru | Kunlun dykes | 15 (69) | 76 | 237 | — | 3.7 | 5.9 |
| 8 | C | Dagangou | 4 (17) | −12 | 146 | — | 19.8 | 33.2 |
| 9 | C | Dagangou | 7 (40) | −6 | 40 | — | 16.0 | 26.4 |

$B$ is the number of sites and $N$ is the number of samples from which the palaeomagnetic data are derived; lat. and long. are the latitude and longitude of the pole position; $A_{95}$ is the apical half angle of the cone of 95 % confidence; dp and dm are the semiaxes of the oval of 95 % confidence. Poles 1, 2.2 and 2.3 are from Achache *et al.* (1984) with 2.2 and 2.3 representing poles from the south Lhasa Terrane and the north Lhasa Terrane respectively. All of the other poles are from Lin & Watts (1988). Poles 5 and 6 are respectively from the Fenghuoshan and the Erdaoguo sections of the Fenghuoshan Formation. Poles 8 and 9 are from two separate sections (P39–40, P41 respectively) from the Dagangou Formation as discussed in Lin & Watts (1988).

Palaeomagnetic data from Mesozoic and Upper Palaeozoic units from the Tibetan Plateau and new data from China allow us to address the problem of pre-Cretaceous reconstructions of the Tethys area with greater confidence. The palaeomagnetic data constrain the crustal blocks in latitude and azimuth. Palaeobotanical data can be used to constrain the positions in longitude. Palaeozoic and Mesozoic reconstructions of the Tethys area are important in understanding the tectonic evolution of Southeast Asia, as it is evident that the Himalayan mountain belt is only a relatively recent stage of a long history of orogeny in this part of the world.

Molnar & Tapponnier (1975) suggested that the major Cainozoic tectonic features in Asia may be understood as a consequence of the India–Eurasia collision. The idea is supported by a well-known planar indentation experiment on thin, confined blocks of plasticine that has resulted in the model of propagating extrusion (Tapponnier *et al.* 1982, 1986). According to this model, several large left-lateral strike-slip faults were activated successively as India drove northward into weaker Asian crust. During the first 20–30 Ma of the collision, the penetration of India into Asia activated the Red River Fault, along which the Indochina Block rotated about 25° clockwise and translated (extruded) 800 km southeastward. This led to the opening of the South China Sea before late Miocene time. After the late Miocene, the Altyn Tagh fault was activated and became the second major left-lateral strike-slip fault, along which the South China Block has moved eastward. The model predicts a total 40° clockwise rotation of the Indochina Block with respect to the South China Block. Other events explained by this model include the opening of the Mergui Basin and Andaman Sea, and the generation of rifts and grabens in North China and Yunnan.

As the model predicts up to 40° of clockwise rotation of the Indochina Block with respect to the South China Block, it can be tested palaeomagnetically (Achache *et al.* 1983; Achache

& Courtillot 1985; Maranate & Vella 1986). Further inquiry into this problem is possible with recently available data from the South China Block (Lin *et al.* 1985; Lin 1987 *a, b*) and a re-examination of the data from the Indochina Block.

## CRUSTAL SHORTENING NORTH OF THE HIMALAYA ARC

The Cretaceous palaeomagnetic data from the Lhasa Block include data from Achache *et al.* (1984) and Lin & Watts (1988) as summarized in table 1. To calculate the crustal shortening north of these sites, we calculate Cretaceous palaeomagnetic poles from the Lhasa Block. Accounting for the possibility of rotation between the south Lhasa Terrane and the north Lhasa Terrane (Achache *et al.* 1984) we calculate poles representative of these respective parts. For the north Lhasa Terrane we calculate a mean Cretaceous pole from the combined data from the Takena Formation in the vicinity of Amdo (Achache *et al.* 1984) and Cretaceous volcanics from Qelico and Nagqu (Lin & Watts 1988). For the south Lhasa Terrane we calculate a pole position combining the data from Achache *et al.* (1984) and Lin & Watts (1988) collected from the vicinity of Lhasa. The poles are given in table 2.

### TABLE 2. MEAN CRETACEOUS PALAEOMAGNETIC DIRECTIONS AND POLES FROM THE NORTH AND SOUTH LHASA TERRANE

| | $N$ | dec/deg | inc/deg | $k$ | $\alpha_{95}$/deg | pole lat./deg | pole long./deg | dp/deg | dm/deg |
|---|---|---|---|---|---|---|---|---|---|
| N Lhasa | 19 | 349 | 32 | 36.6 | 5.6 | 73 | 308 | 3.6 | 6.3 |
| S Lhasa | 16 | 355 | 19 | 52.0 | 5.2 | 69 | 283 | 2.8 | 5.4 |

$N$ is the number of site-mean directions used to calculate the resultant direction; dec is the mean declination; inc is the mean inclination; $k$ is the estimate of Fisher's precision parameter; $\alpha_{95}$ is the apical half angle of the cone of 95% confidence. Pole lat. and pole long. are the latitude and longitude of the corresponding pole position; dp and dm are the semiaxes of the oval of 95% confidence about the pole positions; dp may also be considered the error in the palaeolatitude determination.

In figure 1, the Cretaceous poles from the Lhasa Terrane are compared with the apparent polar wander path of Eurasia (Irving & Irving 1982). The figure clearly shows the poles are distinguishable from the apparent polar wander path of Eurasia, implying a considerable relative motion.

A computation of the amount of relative translation of the Lhasa Terrane with respect to stable Eurasia requires a reference palaeomagnetic pole for stable Eurasia. From the data compilation of Irving & Irving (1982) we find that nine poles (no. 72–no. 80) fall in the time interval from 60 to 109 Ma that spans the ages of the units used to calculate the Cretaceous poles for the north and south Lhasa Terrane. Among these, no. 72 is from Spitsbergen, no. 78 and no. 80 are from the Crimea. Spitsbergen and the Crimea have both undergone deformation since the Cretaceous and may not have palaeomagnetic poles representative of stable Eurasia. Using the other six reference poles we arrive at a mean Cretaceous pole position for stable Eurasia at 69.5° N, 167.4° E, K = 61, $A_{95}$ = 8.7°. The mean age is 84 Ma, which is near the estimated ages of the Lhasa Terrane units from which the palaeomagnetic data are derived.

The translation of the Lhasa Terrane with respect to stable Eurasia was calculated by using

FIGURE 1. Mean Cretaceous palaeomagnetic poles from the north and south Lhasa Terrane compared with the apparent polar wander path of Eurasia (Irving & Irving 1982). The Lhasa Terrane poles are with the ovals of 95 % confidence.

the procedures given by Beck (1980) and Demarest (1983). Recognizing that the result depends considerably on the data from Eurasia, we also carry out the calculation using the mean pole derived above and using the single palaeomagnetic pole from the Munster Basin of Germany (Heller & Channell 1979), which is of the appropriate age and may be the most reliably determined pole in terms of the laboratory treatment of the samples. Furthermore we calculate the translation of the Lhasa Block with respect to the South China Block using the Cretaceous pole of Kent et al. (1986). The results of these exercises are summarized in table 3.

The various choices in the reference pole for Eurasia all give the same answer, within the errors of the determination. There seems to be an average 2000 ± 800 km of translation of the Lhasa Terrane with respect to stable Eurasia and the South China Block since the Cretaceous. Calculations of translation since the early Tertiary using data from the Linzizong volcanics (Achache et al. 1984) yield a similar amount of translation for the Lhasa Terrane. Palaeomagnetic data from the early Tertiary Fenghuoshan Formation from the Qiangtang Terrane are also consistent with this magnitude of northward translation (Lin & Watts 1988). The Tibetan terranes thus appear to remain at a southerly position in the Eurasian frame of reference until at least the onset of collision with India.

Geological evidence indicates the Banggong Suture that separates the Lhasa Terrane from the Qiangtang Terrane closed by the latest Jurassic (Chang et al. 1986) and sutures to the north closed at even earlier times. As there is no evidence for the existence of ocean crust north of the Lhasa Terrane since the Jurassic, the northward translation must involve the internal deformation of continental crust. This is consistent with the deformation of the early Tertiary Fenghuoshan Formation in the vicinity of Erdaogou that may account for up to 40 % crustal shortening (Chang et al. 1986). This is inconsistent with a model of long-distance

TABLE 3. TRANSLATION OF THE LHASA TERRANE RELATIVE TO EURASIA SINCE
THE LATE CRETACEOUS

Eurasian reference pole at 69.5° N, 167.4° E, $A_{95}$ = 8.7°, $N$ = 6

|  | obs. lat./deg | pred. lat./deg | Δlat./deg | Δ/km |
|---|---|---|---|---|
| N Lhasa | 17.4±2.8 | 34.3±6.8 | 16.9±7.3 | 1860±810 |
| S Lhasa | 9.8±2.2 | 32.7±6.8 | 22.9±7.1 | 2520±790 |

Eurasian reference pole at 75.8° N, 181.1° E, $A_{95}$ = 3.8°, $N$ = 11
(Heller & Channell 1979)

|  | obs. lat./deg | pred. lat./deg | Δlat./deg | Δ/km |
|---|---|---|---|---|
| N Lhasa | 17.4±2.8 | 30.6±3.0 | 13.3±4.1 | 1460±450 |
| S Lhasa | 9.8±2.2 | 28.9±3.0 | 19.2±3.7 | 2110±410 |

South China reference pole at 76.3° N, 172.6° E, $A_{95}$ = 10.3°, $N$ = 11
(Kent et al. 1986)

|  | obs. lat./deg | pred. lat./deg | Δlat./deg | Δ/km |
|---|---|---|---|---|
| N Lhasa | 17.4±2.8 | 32.7±8.1 | 15.4±8.6 | 1690±950 |
| S Lhasa | 9.8±2.2 | 31.0±8.3 | 21.2±8.4 | 2340±930 |

The obs. lat. and pred. lat. are the latitude of the terrane inferred from palaeomagnetic data and the latitude predicted from the reference pole, respectively; Δ lat. is the difference between the observed and predicted latitude; Δ is the minimum crustal shortening (in kilometres) with respect to the Eurasian or South China reference frame. The reference site is chosen at 31.6° N, 91.5° E for the north Lhasa Terrane, and at 29.9° N, 91.25° E for the south Lhasa Terrane.

underthrusting of the Indian continental crust beneath a passive Tibet (Holmes 1965; Barazangi & Ni 1982), as there is no prediction of relative motion of the observed magnitude between southern Tibet and Eurasia.

## A CARBONIFEROUS RECONSTRUCTION

Many Asian terranes could not be accurately positioned in previously published Carboniferous reconstructions (Ziegler et al. 1979; Smith et al. 1981) because of the lack of palaeomagnetic data. Carboniferous palaeomagnetic data have only recently become available from the North and South China Blocks (Lin et al. 1985), the Tarim Block (Bai et al. 1987) and the Tibetan Kunlun Terrane (Lin & Watts 1988). The definition of individual terranes will remain a considerable problem until further systematic geological data are available from this part of the world. A sketch map of the individual tectonic terranes that is based on the analysis and interpretation of Chinese literature (Huang et al. 1980; Zhang et al. 1983; Zhang et al. 1984; Li et al. 1982) is shown in figure 2. The detailed justification of the boundaries of these terranes will be given elsewhere (Lin 1989).

For the Tibetan Kunlun Terrane we combine the inclination data from the two localities in the Dagangou Formation (Lin & Watts 1988) with results from the Carboniferous Dagangou volcanics (Lin & Watts, unpublished data) to calculate an average palaeolatitude for the Kunlun Terrane from only the palaeomagnetic inclinations (McFadden & Reid 1982). The combined data give an inclination of 40±7° corresponding to a palaeolatitude of 22±5°. The mean declinations from the three localities are scattered, probably because of the effect of plunging folds in Dagangou Valley, but the results are consistent with a reversed, Southern Hemisphere magnetization (Lin & Watts 1988). Because of the uncertainty in the declination, the palaeoazimuth of the Kunlun Terrane in our reconstruction is uncertain and we select that

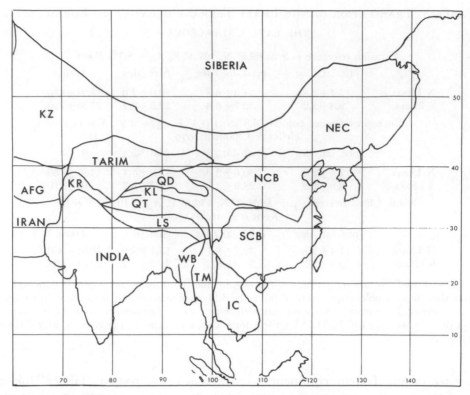

FIGURE 2. Tectonic units in China and its environs. The abbreviations of the terranes are: AFG, Afghanistan
Terrane; IC, Indochina Terrane; KL, Kunlun Terrane; KR, Karakoram Terrane; KZ, Kazakhstan Terrane;
LS, Lhasa Terrane; NBC, North China Block; NEC, Northeast China Terrane; QD, Qaidam Terrane; QT,
Qiangtang Terrane; SCB, South China Block; TM, Thai–Malay Terrane; WB, West Burma Terrane. A
Mercator projection is used.

indicated by the data from the Dagangou Formation, locality P39–40, only because it is based
on the most samples.

A Carboniferous reconstruction similar to that of Lin & Watts (1988) is shown in figure 3.
This is based on a combination of the available palaeomagnetic data and inference from
geological and palaeontological data. For the position of Gondwana with respect to Laurasia
we adopt the configuration of Pangaea A2 (Van der Voo & French 1974), as this
reconstruction is favoured by Hallam (1983) and Livermore et al. (1986). The palaeolatitude
and palaeoazimuth of the North and South China Blocks, Tarim Block, Turkey, Iran and the
Kunlun Terrane are defined by palaeomagnetic data. The positions of the Qiangtang
Terrane, Lhasa Terrane and the Southeast Asia terranes are suggested purely on geological
and palaeontological data summarized below.

Carboniferous tillites, similar to those that occur in the Indian Peninsula are found in the
vicinity of Urulung and Damxung in the Lhasa Terrane (Chang et al. 1986). Therefore the
Lhasa Terrane is placed along the northern margin of Gondwana connected to the Indian
Peninsula.

The Qiangtang Terrane of Tibet has a very similar flora to the South China Block (Li et al.
1985) and is therefore placed in the vicinity of this unit. The Jinsha Suture between the Kunlun
and Qiangtang Terranes was closed by the late Triassic, but it is not known if there was a
considerable separation between these areas.

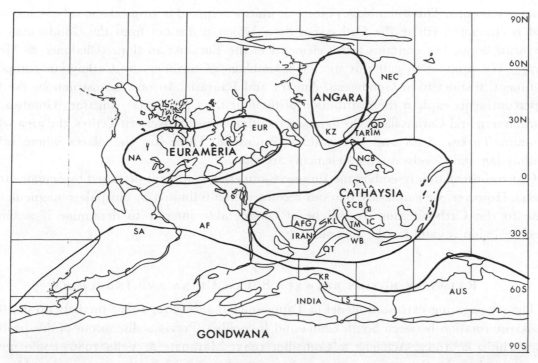

FIGURE 3. A global reconstruction for the Carboniferous. The abbreviations, in addition to those of figure 2 are: AF, Africa; AUS, Australia; EUR, Europe; NA, North America; SA, South America; T, Tarim. The approximate boundaries of the Angaran, Euramerian, Cathaysian and Gondwana floras are delineated. A cylindrical equidistant projection is used.

Our division of the Southeast Asian area follows that of Hamilton (1979) and includes the Indochina Block, the Thai–Malay Terrane and the West Burma Terranes. The former two contain a typical Cathaysian flora and were therefore likely connected with the South China Block in the late Palaeozoic. A preliminary palaeomagnetic result from the Baoshan area (25.0 °N, 99.2 °E) of western Yunnan yields a late Carboniferous palaeolatitude of 34.1 °S (Zhang & Zhang 1986). The Baoshan area and the Shan State Plateau of Burma belong geologically to the Thai–Malay Terrane as shown in figures 2 and 3. This new result is therefore consistent with our reconstruction. However, the provenance of the West Burma Terrane is uncertain. It may have been connected with the South China Block together with the Indochina Block and the Thai–Malay Terrane as shown in figure 3, or it may have been associated with northwestern Australia until the mid-Jurassic. The mid-Jurassic magnetic anomalies (M22–M25) offshore of northwestern Australia (Norton & Sclater 1979) imply that the continental crust moved northwest away from Australia at this time.

The late Carboniferous palaeomagnetic data from the Tarim, Kazakhstan and Siberian Blocks suggest that this assemblage may have come together by this time, as we show in figure 3. However, Zonenshain et al. (1984) indicate that Kazakhstan had first collided with the East European Block by the late Carboniferous, and that it was in the late Permian that the Siberian Block joined the combined East European–Kazakhstan Block.

The reconstruction in figure 3 differs from previous reconstructions in that what was previously a large triangular void between the Indian–Australian northern margin of Gondwana and the southern Asian margin of Laurasia contains a considerable amount of continental crust. This assemblage of tectonic units was spread in tropical and subtropical areas and is characterized by the Cathaysian flora, which is distinct from the Gondwanan and Angaran floras, but contains a few elements of the Euramerian flora (Chaloner & Meyen 1973). We propose to call this specific assemblage of terranes the Cathaysian composite continent, distinct from Gondwana, Angara and Laurasia. In our reconstruction, the four supercontinents explain the distribution of the four flora provinces: Angaran, Gondwanan, Euramerian and Cathaysian. The four supercontinents met in western Tethys, the area which contains Turkey, Iran and the Arabian Peninsula. These are the places where mixed Cathaysian and Gondwana flora elements are known.

Our reconstruction resembles the Jurassic Cimmerian continent proposed by Sengör (1979, 1984). However, we emphasize that our reconstruction is limited by our palaeomagnetic data base to the Carboniferous. It will be of considerable interest to determine if a similar reconstruction is valid for the Jurassic.

## Relative motion between South China and Indochina

The propagating extrusion model of Tapponnier *et al.* (1982, 1986) predicts up to 40° of clockwise rotation between South China and Indochina. Previous discussions of this problem (Achache *et al.* 1983; Achache & Courtillot 1985; Maranate & Vella 1986) could not be conclusive because the appropriate data base from South China was not available. Here we consider the propagating extrusion model in the light of the data from China now available (Lin *et al.* 1985; Lin 1987 *a*, *b*) and the data from Indochina (Bunopas 1981; Achache & Courtillot 1985; Maranate & Vella 1986).

The entire data set from Indochina is derived from the Khorat Group of the Khorat Plateau, Thailand. The Khorat Group consists of lacustrine and fluvial sediments over 5000 m thick ranging in age from middle to late Triassic up to Cretaceous. We employ the late Triassic pole determined by Achache & Courtillot (1985) and we calculate mean poles for the latest Triassic–earliest Jurassic and the mid-Jurassic from the data of Bunopas (1981) and Maranate & Vella (1986). Table 4 shows the resultant Indochina poles as well as the relevant South China Block poles. The data for the Cretaceous have a considerable dispersion and are not used. The poles from the South China Block and the Indochina Terrane are plotted in figure 4 with the ovals of 95 % confidence.

It is clear from figure 4 that the poles from the respective blocks fall on comparatively smooth tracks that are significantly different, indicating a relative displacement since the mid-Jurassic. The amounts of translation and rotation of the Indochina Block with respect to the South China Block are given in table 5. These are calculated according to the method of Beck (1980) and Demarest (1983). They indicate that the difference in pole positions can be explained by an 8° clockwise rotation of the Indochina Block and a 1500 km translation relative to the South China Block. The rotation is demonstrably smaller than the 40° predicted by the propagating extrusion model but it is in the correct sense.

Another way of looking at the problem is to determine the Euler poles and the amount of rotation about them which are necessary to bring the two polar wander tracks in figure 4 into coincidence. Fold trends and major faults (Gansu Fault, Kunlun Fault and Kang Ding Fault) appear to form small circles about a pole (28° N, 95° E) near the Eastern (Assam) Syntaxis of the Himalayan ranges. This pole is chosen as an Euler pole to rotate the palaeopoles of

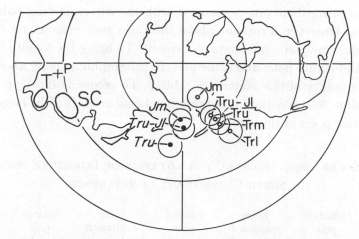

FIGURE 4. Palaeomagnetic poles from the South China Block (shown as open circles) and from the Indochina Terrane (shown as closed circles). The cross labelled P shows the first Euler pole near the Assam Syntaxis. Also shown are the positions of the Khorat Plateau of Thailand (T) and the South China Block (SC). The ages of the palaeomagnetic poles are abbreviated as follows: Jm, mid-Jurassic; Tru–Jl, latest Triassic to earliest Jurassic; Tru, late Triassic; Trm, mid-Triassic; Trl, early Triassic. The approximate apical half angles of 95% confidence are shown for each of the pole positions.

TABLE 4. PALAEOMAGNETIC POLES FOR SOUTH CHINA AND INDOCHINA

*Indochina*

| age | N(B) | lat./deg | long./deg | K | $A_{95}$/deg | reference |
|---|---|---|---|---|---|---|
| Jm | 3 (44) | 62.1 | 178.5 | 260 | 7.7 | 3, 4, e |
| Tru–Jl | 4 (48) | 57.4 | 181.6 | 334 | 5.0 | 5, 6, f, g |
| Tru | 1 (39) | 49.0 | 172.0 | 106 | 7.0 | h |

*South China*

| age | N(B) | lat./deg | long./deg | K | $A_{95}$/deg | reference |
|---|---|---|---|---|---|---|
| Jm | 6 | 71.5 | 201.0 | 135 | 5.8 | Lin (1987a) |
| Tru–Jl | (5) | 60.3 | 207.2 | | 6.0 | interpolated |
| Tru | (5) | 57.5 | 208.8 | | 6.0 | interpolated |
| Trm | 1 (2) | 54.6 | 209.7 | 95 | 5.7 | Lin (1987a) |
| Trl | 3 (22) | 48.8 | 214.4 | 198 | 8.8 | Lin (1987b) |

Age is the likely age of the palaeomagnetic pole. $N$ is the number of independent studies from which the mean is determined. $B$ is the number of sites; lat. and long. are the latitude and longitude of the palaeomagnetic poles. $K$ is the estimate of Fisher's precision parameter. $A_{95}$ is the apical half angle of the cone of confidence around the pole position. For the Indochina poles, the numbers and letters in the reference column give the labels of the poles in table 1 and table 2 of Maranate & Vella (1986), which include results from Bunopas (1981). The Tru–Jl and Tru South China poles are interpolated between the Jm and Trm poles.

TABLE 5. TRANSLATION AND ROTATION OF INDOCHINA WITH RESPECT TO SOUTH CHINA

| age | obs. lat. deg | pred. lat. deg | Δlat. deg | Δ/km | obs. dec. deg | pred. dec. deg | rot./deg |
|---|---|---|---|---|---|---|---|
| Jm | 21±4.0 | 13±4.5 | 8±6.0 | 900±670 | 29±7.3 | 19±4.6 | 10±8.7 |
| Tru–Jl | 20±2.5 | 7±4.6 | 12±5.3 | 1370±580 | 34±4.7 | 29±4.7 | 5±6.6 |
| Tru | 26±4.4 | 5±4.6 | 20±6.4 | 2240±700 | 43±7.8 | 31±4.6 | 12±9.1 |

Age is the age of the units from which the palaeomagnetic results are derived. The observed latitude is determined from the palaeomagnetic data from Indochina. The predicted latitude is what would be expected if Indochina had remained stationary in the South China reference frame since the Triassic. Δlat. is the difference between them and is the amount of relative translation between the two blocks since the age of the palaeomagnetic poles. Δ is the amount of translation in kilometres. The obs. dec. is the observed palaeomagnetic declination of Indochina, and pred. dec. is the predicted declination if Indochina was stationary in the South China reference frame. The difference between these is the amount of relative rotation which is given in the rot. column. These rotations are in a clockwise sense. The reference site for the calculation is 16.5° N, 102.5° E in Thailand.

Indochina so as to bring them into agreement with the corresponding palaeopoles from the South China Block. However, a rotation about the Euler pole near Assam can only bring the latitudes of the palaeomagnetic poles into agreement. To bring the longitudes into agreement a rotation of the Indochina poles about the present geographic pole by about 20° is required, as shown in the summary of this exercise in table 6. To reconcile the Upper Triassic poles a rather larger rotation (36°) is required, but this may be an artefact of the larger error associated with these particular poles of this age.

TABLE 6. EULER POLES (DEGREES) FOR ROTATION OF INDOCHINA POLES INTO THE SOUTH CHINA FRAME OF REFERENCE

| age | Indochina pole lat. | Indochina pole long. | Euler rotation 1 lat. | Euler rotation 1 long. | $\theta$ | rotated pole lat. | rotated pole long. | Euler rotation 2 lat. | Euler rotation 2 long. | $\theta$ | rotated pole lat. | rotated pole long. | S China pole lat. | S China pole long. |
|---|---|---|---|---|---|---|---|---|---|---|---|---|---|---|
| Jm | 62 | 179 | 28 | 95 | 8 | 69 | 181 | 90 | 0 | 20 | 69 | 201 | 72 | 201 |
| Tru–Jl | 57 | 182 | 28 | 95 | 8 | 65 | 185 | 90 | 0 | 20 | 64 | 205 | 60 | 207 |
| Tru | 49 | 172 | 28 | 95 | 8 | 56 | 173 | 90 | 0 | 36 | 56 | 210 | 58 | 209 |

The age and position of the Indochina poles are given in the first three columns. The third to fifth columns (Euler rotation 1) give the first (Assam Syntaxis) Euler pole and amount of rotation $\theta$ (positive for a counterclockwise rotation) for rotating the Indochina poles into the South China frame of reference with the resulting pole given in the sixth and seventh columns (rotated pole). The second Euler pole is the present geographic pole and $\theta$ is the amount of this second rotation required to bring the Indochina and South China poles into agreement. The final result of the rotation of the Indochina poles can be compared with the South China poles of equivalent age.

We emphasize that the Euler poles selected are not unique because the two segments of the apparent polar wander paths are too short. But these rotations seem straightforward in view of the fact that relative movement between these two blocks must have involved movements on the major faults in the region. Also, as reliable, unscattered Cretaceous and Tertiary palaeomagnetic data are still not available for the Indochina Block, the timing of the relative movement is not well constrained.

The palaeomagnetic data are consistent with a smaller rotation of the Indochina Block and a larger translation of the South China Block than predicted by the plasticine experiments. However, we note that the placticine experiments also seemingly fail to account for the apparent present dextral sense of movement on the Red River Fault, as discussed by Tapponnier et al. (1982), or apparent reversal of the sense of movement along this fault (Tapponnier et al. 1986). Various combinations, in time, of the 8° rotation of Indochina and the translation of the South China Block could produce a dextral movement on the Red River Fault, or even a reversal of the sense of movement across this structure. It has been pointed out that the published experiment considerably simplified the collision process between India and Eurasia (Achache et al. 1983, 1984). For example, the Indian Plate was rotating counterclockwise as it moved northward, colliding with Eurasia. This may have resulted in competing senses of motion for the extruded blocks: counterclockwise rotation in sympathy with India against southeastward extrusion and clockwise rotation. Achache et al. (1983) have suggested that a rotating indenter be used in future experiments.

## Conclusions

The Mesozoic and Cainozoic palaeomagnetic results from the Lhasa and Qiangtang Terranes in the Tibetan Plateau are discordant with those of stable Eurasia and the South China Block. The data are consistent with $2000 \pm 800$ km of northward translation of the Lhasa Terrane with respect to stable Eurasia since the India–Eurasian collision began. The palaeomagnetic data support the internal deformation within the Eurasian continental crust as a principal mechanism of crustal thickening. The model of long-distance underthrusting of the Indian subcontinent beneath a passive Tibetan Plateau does not predict the 2000 km of northward convergence of continental crust north of the Himalaya Arc as indicated by the palaeomagnetic data. This conclusion is also consistent with a result obtained from numerical analysis of a thin viscous sheet (England & Houseman 1986).

Palaeomagnetic data indicate that the North China Block, South China Block, Kunlun Terrane and Thai–Malay Terrane were in equatorial or low latitudes during the Carboniferous. Combining palaeomagnetic results with geological and palaeontological data, we propose a reconstruction showing that the traditional wedge-shaped Tethys was divided into two halves by a Cathaysian composite continent consisting of the above terranes as well as the Indochina Block. These terranes all contain elements of the Cathaysian flora. The southern supercontinent Gondwana and the northern supercontinents Laurasia and Angara may have been loosely connected via the Cathaysian composite continent. If the Tibetan Terranes could be restored to their original sizes, taking into account crustal shortening, both the northern and southern halves of Tethys would approach epicontinental seas in size. This would resolve the Tethys paradox, i.e. why all attempts to find relics of the vast Palaeotethys proposed to have existed during the late Palaeozoic have failed (Carey 1976; Stocklin 1984). In our reconstruction, such an ocean never existed.

To interpret the difference between the Mesozoic palaeomagnetic poles from South China and from Indochina, we propose that following the India–Eurasia collision, the Indochina Block rotated 8° clockwise about a pole within the Assam Syntaxis, and the South China Block extruded over 1500 km eastward. If the rotation dominated the earlier phases of movement, the sense of movement on the Red River Fault would have been left lateral. If the extrusion of the South China Block dominated the relative movement at some point, the sense of movement on the Red River Fault would be right-lateral. Hence the observed apparent reversal of sense of movement on this fault could be explained by the dominance of one or other components of motion. This retains the salient features of the extrusion model of Tapponnier *et al.* (1982) but would suggest that initial conditions of the collision other than that originally used would be appropriate, e.g. a rotating indenter as proposed by Achache *et al.* (1983).

We thank all participants in the Academia Sinica–Royal Society Geotraverse for many discussions. Professor J. C. Briden and Dr E. McClellan kindly provided laboratory facilities at the University of Leeds. Dr A. G. Smith of Cambridge University allowed us use of his ATLAS program and assisted with the mechanics of palaeogeographic reconstruction. J. L. thanks Professor R. Shackleton, F.R.S., Professor J. Dewey, F.R.S., and Professor B. Windley for the invitation to attend the Royal Society Discussion on the evolution of the Himalaya and Tibet. D. R. W. acknowledges support from the Natural Environment Research Council. Professor K. Hsu commented on an early draft of this paper.

188 J. LIN AND D. R. WATTS

REFERENCES

Achache, J. & Courtillot, V. 1985 *Earth planet. Sci. Lett.* **73**, 147–157.
Achache, J., Courtillot, V. & Besse, J. 1983 *Earth planet. Sci. Lett.* **63**, 123–136.
Achache, J., Courtillot, V. & Zou, Y. X. 1984 *J. geophys. Res.* **89**, 10311–10339.
Bai, Y., Cheng, G., Sun, Q., Sun, Y., Li, Y., Dong, Y. & Sun, D. 1987 *Tectonophysics* **139**, 145–153.
Barazangi, M. & Ni, J. 1982 *Geology* **10**, 179–185.
Beck, M. E. Jr 1980 *J. geophys. Res.* **85**, 7115–7131.
Besse, J., Courtillot, V., Pozzi, J. P., Westphal, M. & Zhou, Y. X. 1984 *Nature, Lond.* **311**, 621–626.
Bunopas, S. 1981 Ph.D. thesis, Victoria University of Wellington, New Zealand.
Carey, W. 1976 *The expanding Earth.* Amsterdam: Elsevier.
Chaloner, W. G. & Meyen, S. V. 1973 In *Atlas of paleogeography* (ed. A. Hallam). Amsterdam: Elsevier.
Chang Chengfa *et al.* 1986 *Nature, Lond.* **323**, 501–507.
Demarest, H. H. Jr 1983 *J. geophys. Res.* **88**, 4321–4328.
England, P. C. & Houseman, G. 1986 *J. geophys. Res.* **91**, 3664–3676.
Hallam, A. 1983 *Nature, Lond.* **301**, 499–502.
Hamilton, W. 1979 *Tectonics of the Indonesian region.* U.S. Geol. Surv. Prof. Paper no. 1078.
Heller, F. & Channell, J. E. T. 1979 *J. Geophys.* **46**, 413–427.
Holmes, A. 1965 *Principles of physical geology.* 2nd edn. London: Thomas Nelson Ltd.
Huang, J., Ren, J., Jiang, Ch, Zhang, Zh. & Qin, D. 1980 *The geotectonic evolution of China.* Beijing: Science Press.
Irving, E. & Irving, G. A. 1982 *Geophys. Surv.* **5**, 141–188.
Kent, D. V., Xu, G., Huang, K., Zhang, W. Y. & Opdyke, N. D. 1986 *Earth planet. Sci. Lett.* **79**, 179–184.
Klootwijk, C. T., Conaghan, P. J. & Powell, C. McA. 1985 *Earth planet. Sci. Lett.* **75**, 167–183.
Klootwijk, C. T., Nazirullah, R. & de Jong, A. 1986*a* *Earth planet. Sci. Lett.* **80**, 394–414.
Klootwijk, C. T., Sharma, M. L., Gergan, J., Shah, S. K. & Gupta, B. K. 1986*b* *Earth planet. Sci. Lett.* **80**, 375–393.
Li, Ch., Wang, Q., Liu, X. & Tang, Y. 1982 *Tectonic map of Asia* (scale 1:8,000,000). Beijing: Cartographic Publishing House.
Li, X. X. *et al.* 1985 *Acta paleont. sin.* **24**, 150–170.
Lin, J. 1987*a* *Scientia geol. sin.* **2**, 182–187.
Lin, J. L. 1987*b* *Scientia geol. sin.* **4**, 295–305.
Lin, J. L. 1989 (In preparation.)
Lin, J. L., Fuller, M. & Zhang, W. Y. 1985 *Nature, Lond.* **313**, 444–449.
Lin, J. L. & Watts, D. R. 1988 *Phil. Trans. R. Soc. Lond.* A. (In the press.)
Livermore, R. A., Smith, A. G. & Vine, F. J. 1986 *Nature, Lond.* **322**, 162–165.
Maranate, S. & Vella, P. 1986 *J. southeast Asian Earth Sci.* **1**, 23–31.
McFadden, P. L. & Reid, A. B. 1982 *Geophys. Jl R. astr. Soc.* **69**, 307–319.
Molnar, P. & Tapponnier, P. 1975 *Science, Wash.* **189**, 419–426.
Norton, I. O. & Sclater, J. G. 1979 *J. geophys. Res.* **84**, 6803–6830.
Sengör, A. M. C. 1979 *Nature, Lond.* **279**, 590–593.
Sengör, A. M. C. 1984 *The Cimmeride orogenic system and the tectonics of Eurasia.* Geol. Soc. Am. Spec. Paper no. 195.
Smith, A. G., Hurley, A. M. & Briden, J. C. 1981 *Phanerozoic paleocontinental world maps.* Cambridge University Press.
Stocklin, J. 1984 In *Plate reconstruction from Paleozoic paleomagnetism* (ed. R. Van der Voo, C. R. Scotese and N. Bonhommet), Geodynamics Series vol. 12, pp. 27–28. Washington, D.C.: American Geophysical Union.
Tapponnier, P., Peltzer, G., Le Dain, A. Y., Armijo, R. & Cobbold, P. 1982 *Geology* **10**, 611–616.
Tapponnier, P., Peltzer, G. & Armijo, R. 1986 In *Collision tectonics* (ed. M. P. Coward & A. C. Ries), pp. 115–157. Geological Society Publication no. 19. London: Geological Society.
Van der Voo, R. & French, R. B. 1974 *Earth Sci. Rev.* **10**, 99–119.
Zhang, W. Y. *et al.* 1983 *The marine and continental tectonic map of China and its environs* (scale 1:5,000,000). Beijing: Science Press.
Zhang, Zh. M., Liou, J. G. & Coleman, R. G. 1984 *Bull. geol. Soc. Am.* **95**, 925–312.
Zhang, Zhengkun & Zhang, Jingxin 1986 *Bull. Inst. Geol. Chin. Acad. geol. Sci.* **15**, 183–189.
Ziegler, A. M., Scotese, C. R., McKerrow, W. S., Johnson, M. E. & Bambach, R. K. 1979 *A. Rev. Earth planet. Sci.* **7**, 473–502.
Zonenshain, L. P., Korinevsky, V. G., Kazmin, V. G., Pechersky, D. M., Khain, V. V. & Matveenkov, V. V. 1984 *Tectonophysics* **109**, 95–135.

*Phil. Trans. R. Soc. Lond.* A **326**, 189–227 (1988)    [ 189 ]

*Printed in Great Britain*

# Late Palaeozoic biogeography of East Asia and palaeontological constraints on plate tectonic reconstructions

By A. B. Smith

*Department of Palaeontology, British Museum (Natural History), Cromwell Road,
London SW7 5BD, U.K.*

Biogeographical patterns of late Palaeozoic rugose coral genera are analysed for the Lower Carboniferous (Visean), early Lower Permian (Asselian/Sakmarian), late Lower Permian (Qixian) and early Upper Permian (Maokoan) of East Asia. Boundaries to the biotic regions are defined to coincide with tectonically significant suture zones to test rival hypotheses about the plate tectonic reconstruction of that region. Three numerical techniques are employed to cluster areas on the basis of shared endemic taxa; parsimony analysis of endemism, principal coordinates analysis and single linkage cluster analysis. Geographical variation in overall diversity is also considered. These results are compared with empirically derived patterns based on other groups of organisms.

Major conclusions from this work are as follows. (i) During the Carboniferous and early Permian, the Cathaysian region (North and South China Blocks, Tarim Terrane, Kunlun Terrane, Qiangtang Terrane) formed one cohesive biotic region lying tropically or subtropically; it did not start to fragment until the Upper Permian. (ii) This region was biotically isolated from Central Asia at least during the Carboniferous and Lower Permian. (iii) The southern boundary to the Cathaysian region does not coincide with a single suture zone through time, nor is it sharply defined. Instead there appears to be a gradual faunal impoverishment southwards across the Tibetan Plateau. This implies that faunal ranges are controlled only by the prevailing global climatic regime and not by a geographical barrier. (iv) The Lhasa and Himalaya Terranes shared a similar fauna until the mid-Permian, when a marked faunal disjunction developed coincident with the Zangbo Suture. (v) For terrestrial floras, the barrier to biotic exchange between the North China Block and Angaraland started to break down in the late Permian.

It follows that no major oceanic break ('Palaeotethys') can be recognized within the Cathaysian region during the late Palaeozoic on palaeontological evidence. This region then formed an integral part of the Gondwanaland craton, extending up into broadly tropical latitudes, and did not become separated from it until the late Lower Permian. The Tienshan–Yinshan Suture is the most likely site of 'Palaeotethys', which appears to have occupied a broadly equatorial latitude. Combined with evidence on the ages of the various Asian sutures, this raises significant problems for those who demand a large ocean in their Carboniferous to early Permian palaeogeographical reconstructions of this region.

## 1. Introduction

Palaeontology is an important source of data for reconstructing palaeogeographies. It provides three lines of evidence that can be used to identify changing patterns of relationship between geographical regions through geologic time. Firstly, palaeontological data still provide the best and most refined timescale by which strata can be dated and disparate geological phenomena set in a temporal framework. It is through palaeontology that the timing of ophiolite obduction, mélange formation, and a whole host of other tectonically significant events are actually dated and related, although this is rarely acknowledged.

Information concerning the habitat preferences and modes of life of fossil organisms (their autecology) is a second source of data, providing palaeoenvironmental parameters about the sediments in which they are discovered. Such palaeoecological data are usually combined with sedimentological data to create an overall view of the palaeoenvironment of sedimentary deposition. The analysis of sedimentary environments on a geographical scale and through time constitutes 'facies analysis' and can be used to identify, for example, the position of land masses (terrestrial or near-shore environments) or oceanic tracts (open oceanic or continental rise environments) in the geological past. It is this combination of sedimentological and palaeontological data that is most commonly used to reconstruct palaeogeographies and establish relative positions of plates.

A third source of information comes from the analysis of taxonomic distribution data in a pure form, without palaeoenvironmental interpretation. In this approach faunal or floral lists of taxa are compiled for predetermined areas of interest and the degree of biotic similarity between sample areas calculated. A relatively low degree of biotic similarity indicates that there was some significant barrier to biotic dispersal between regions, whereas a high degree of similarity suggests the converse. Facies analysis evidence can then be used to infer the nature of the barrier between biotas. In addition, distributional data also provide information about regional variation in diversity, which on a global scale can be used as a crude indicator of temperature and thus latitude (see Vine 1973).

All three approaches provide independent sources of geological data that must be accommodated in any palaeogeographical reconstruction. Like all indirect methods of reconstructing the position of continents through time, however, palaeontological data do not provide a single clear and objective solution, but only a set of conditions that have to be met by the palaeogeographical reconstruction that is eventually chosen. That is to say, palaeontological data can only be used to reject certain possibilities, and may be able to accommodate a number of plausible alternatives. In this they are no different from inferences derived from palaeomagnetism, where latitude may be quite tightly constrained but longitude unknown.

Global palaeogeographical reconstructions during the Tertiary and much of the Mesozoic are now well established. This is largely because of the clear and unambiguous evidence of deep-sea magnetic lineaments, which provide a direct record of the opening of oceanic tracts (see Owen 1983). However, such evidence is lacking for earlier periods and palaeogeographies have to be based on more indirect evidence such as palaeobiogeography, facies analysis and polar wandering curves. Direct evidence for plate positions is provided by palaeomagnetic pole positions, but these are as yet too few in number and too scattered in time to provide any detailed picture. Instead they provide only another set of independent constraints on the imagination of those who create plate tectonic reconstructions. The fact that constraints of any kind are relatively few is demonstrated by the plethora of reconstructions that have been suggested for the late Palaeozoic. Palaeogeographical reconstruction for the Palaeozoic of remote regions of the World, such as the Tibetan Plateau, is particularly hazardous. Even in the best current palaeomagnetic reconstruction of this region (Lin & Watts 1988, and this symposium), data from different terranes have been combined into one palaeogeographical map that cover a time span from the Tournasian (Kunlun Terrane) to the end Permian (North

China Block), a period of about 100 Ma. Yet we know that oceans can open and close in this time and continental blocks travel vast distances. Obviously there is a real need for rigorous data from any source that can help to establish relative and absolute positions of plates and set constraints on palaeogeographies in the Palaeozoic. Here I wish to investigate the role palaeontology can play in establishing such constraints.

## 2. The structure of East Asia and late Palaeozoic reconstructions

It is now widely recognized that Asia is formed from a complex of small plates accreted together (see, for example, Zhang et al. 1984; Sengör 1985b; Watson et al. 1987). The boundaries between these plates are defined by linear outcrops of dismembered ophiolite suites, that mark the position of ancient oceanic tracts (figure 1). Several of these cross the Tibetan Plateau (Allègre et al. 1984; Chang Chengfa et al. 1986). Recent work in this area by Chinese, French and British geologists has greatly clarified the evidence concerning when these sutures finally closed (summarized in Yin et al. 1988; Smith & Xu 1988) but the pre-collisional palaeogeography of this region remains poorly understood. In particular, the position and indeed the reality of a large Tethyan ocean in this region postulated in late Palaeozoic Pangean reconstructions is a matter of some contention.

Most workers currently believe that there existed a major ocean separating Eurasia and Gondwana during the late Palaeozoic–early Mesozoic. This is clearly shown, for example, in Smith's (1981) reconstruction of Pangea, which has a large, triangular gap often referred to as 'Tethys' or the 'Tethyan Ocean'.

Early reconstructions such as those of Smith et al. (1973) showed the majority of the Tibetan region and all of China as part of Eurasia, with the Zangbo Suture marking the position of 'Tethys'. As information concerning the structural composition of China and Tibet became better understood, it was realized that the picture was very much more complex. Firstly, it was pointed out that all the known oceanic sediments along the Tethyan belt were Triassic or younger and that there was no field evidence for a Permo-Triassic ocean (see review by Sengör 1984). Whereas some (Hsu & Bernoulli 1978) explained this apparent contradiction away by assuming all late Palaeozoic ocean floor sediments had been subducted, others (Carey 1976; Avias 1978; Crawford 1979, 1982) questioned the very existence of a large Tethyan ocean before the Triassic.

Even among workers who believe that a large late Palaeozoic ocean must have existed there are a bewildering number of alternative hypotheses about precisely where such an ocean was situated (Stocklin 1984). Helmcke (1985), for example, argues that the Zangbo Suture is the only possible site for a major oceanic break and that all regions to the north of this line were already part of Eurasia by the late Palaeozoic. Sengör (1984, 1985a) proposed an alternative interpretation whereby 'Tethys' was inferred to be two oceans separated by a relatively narrow continental strip, which he named the Cimmerian continent. This Cimmerian continent, according to Sengör, formed the northern margin of Gondwanaland during the Upper Palaeozoic but was rifted from Gondwanaland during the Permian and swept across the original Tethys like a 'windshield wiper'. Thus the site of the original late Palaeozoic ocean, which Sengör termed 'Palaeotethys', lies to the north of the Kunlun fold belt system and the North China Block. The younger oceanic tract, 'Neotethys', is represented now by the Zangbo Suture in the Himalayan region.

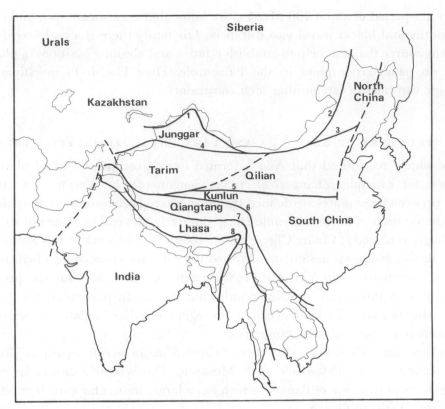

FIGURE 1. Major terranes and sutures discussed in this paper (from Watson *et al.* 1987). Sutures as follows: 1, North Junggar; 2, Tienshan; 3, Hegen; 4, Yinshan; 5, Kunlun–Qinling; 6, Jinsha; 7, Banggong; 8, Zangbo.

Chang Chengfa *et al.* (1986) have suggested that a sequence of small strip-like terranes may have been rifted off Gondwanaland rather than the single Cimmerian continent envisaged by Sengör. Yang & Fan (1983) and Fan (1985), on the basis of palaeontological data, have argued that it is the Banggong Suture, separating the Lhasa and Qiangtang Terranes, that marks the site of 'Palaeotethys'. Others have argued that the Lhasa Block had closer affinities with the Cathaysian region than with Gondwana (Allègre *et al.* 1984; Zhang *et al.* 1985). Wang (1984) believes that the Jinsha Suture should be considered the site of 'Palaeotethys' and Liu (1984 *a,b*) has even suggested that there were four elongate, parallel oceans in the Permian, not one. Watson *et al.* (1987) have argued that the southern margin of Eurasia in the Upper Palaeozoic is marked by the Tienshan–Hegen Suture near the northern margin of China. Smith & Xu (1988) point out that none of the three suture zones that cross the Tibetan Plateau were significant breaks during the late Palaeozoic and propose that global changes in climate explain why some terranes appeared to change their affinities through time. Finally some workers, largely on the basis of palaeomagnetic data, have the various terranes that make up central Asia scattered across the broad oceanic tract of Tethys in the late Palaeozoic (McElhinney *et al.* 1981; Lin *et al.* 1985; Lin & Watts, this symposium).

Nor is the situation much clearer with respect to the North and South China Blocks. Zhang *et al.* (1984) favour a Permian age for the timing of accretion of these two blocks, but Klimetz (1983), Ji & Coney (1985), Sengör (1985 *b*) and Watson *et al.* (1987) all believe the North and South China Blocks were separate until the Triassic, and Mattauer *et al.* (1985) and Laveine *et al.* (1987) argue that these two blocks have been united since the Devonian on structural and palaeontological evidence respectively.

So, although a broad consensus now exists concerning the times of closure for the various sutures (except for the Kunlun–Qinling Suture), there exists a wide spectrum of views on when such sutures opened and how the various continental blocks were arranged before closure. These fall broadly into three camps: those which predict no major Tethyan ocean to be present during the late Palaeozoic (Earth expansionists); those which predict an archipelago of small blocks scattered across a major ocean (mostly palaeomagnetists); and those who would have a large oceanic separation coincident either with one of the Tibetan Suture Zones (see, for example, Helmcke 1985) or to the North of the Tibetan Plateau (Sengör 1984, 1985a).

This paper then sets out to analyse late Palaeozoic faunal distributions rigorously to establish where and when biotic discontinuities are to be found in East Asia. This will then place constraints on possible plate reconstructions. I shall first assemble and present the biogeographical data for the late Palaeozoic, then discuss palaeontological and facies analysis evidence for times of opening and closure of the various sutures, and end with an assessment of how these alternative palaeogeographical reconstructions compare with the palaeontological evidence presented.

## 3. Approach

### (a) Background

The recognition of endemism from fossil distributional data forms the basis for all palaeobiogeographical analyses. Endemism is the restricted occurrence of taxa within geographical regions and can only be defined in relative terms. It is generally assumed that the degree of biotic similarity between areas reflects the prevailing palaeogeographical distribution of barriers to dispersal, but clearly gives no clue as to what constituted these barriers or how distantly separated the biotas were (a narrow land isthmus is just as effective a barrier to dispersal of marine organisms as a wide ocean; see review by Rosen 1988). Various methods exist by which degree of endemism or biotic similarity can be measured (see Rosen & Smith 1988), although almost all workers interested in Asian palaeobiogeography, with the notable exception of Waterhouse & Bonham-Carter (1975), have simply discussed relationships of biotas anecdotally.

Here I have chosen to analyse patterns of endemicity in late Palaeozoic rugose coral faunas. The reasons for this are twofold. Firstly, coral faunas are relatively widespread in the late Palaeozoic of the Tibetan region and have been reported from most tectonic blocks. They have been used previously to designate faunal realms and provinces (Hill 1973; Lin 1984; Luo 1984; Fan 1985), although only in an empirical way, and figure prominently in discussions as to where 'Palaeotethys' was situated. Secondly, in addition to pure distributional data, it is likely that faunal diversity will provide an indication of relative latitude (see Hill 1973), as they appear to have their highest diversity in warm, equatorial latitudes like present-day corals (Rosen 1984).

An initial pilot study of the late Palaeozoic faunas of the Lhasa, Qiangtang and Kunlun Terranes was carried out previously by Smith & Xu (1988) although this was somewhat less rigorous in its approach (ranges of taxa were interpolated in an attempt to overcome sampling problems). Here a wider view is taken of Tibetan coral faunas and faunas from throughout central East Asia are analysed without any interpolation of ranges.

Problems of poor sampling or lack of data for some areas are discussed in Smith & Xu (1988). It is difficult to know whether few corals have been reported from certain areas because of sampling failure or for biogeographical reasons. In some cases, low diversities appear to be genuine, as for example where only a handful of small simple solitary species are known (Siberia, Himalaya). In other cases, where the relatively few records comprise mostly complex solitaries or compound forms known elsewhere from high diversity faunas, one might suspect that poor sampling was the cause. In such cases, these records are best left out of the analysis. Sampling is a particular problem in palaeobiogeographical analysis because it means that, even from relatively well sampled areas, a sizable proportion of the fauna can seem to be endemic (i.e. is unrecorded from elsewhere) simply because insufficiently large samples are available (see Koch 1987). Endemicity at any one site may therefore be more of a reflection of how much collecting has been done rather than a true measure of 'biotic isolation'. For this reason I prefer not to use counts of the number of taxa endemic to one region as a measure of that region's degree of isolation. Rather I have used only methods that calculate similarity on the basis of taxa shared between two or more sample areas.

In palaeogeographical reconstruction we wish to determine how similar the faunas of each region are to one another at specific geological periods. The results are then interpreted as a measure of the degree of biotic contact between regions and palaeogeographical conclusions drawn. This method requires sample areas to be defined *a priori* to any analysis. Clearly, as I am trying to establish data that will test late Palaeozoic plate reconstructions, the boundaries to biotic regions have been chosen to coincide with well-established tectonic lineaments (mostly ophiolitic suture zones). In some cases, where there appears to be a significant faunal change across a single such tectonic terrane (as for example in the Qiangtang Terrane during the early Permian) a finer subdivision of these regions is opted for. The principal tectonic regions adopted here are shown in figure 1 and have been taken from Watson *et al.* (1987).

The analyses presented here are based on generic data because this is the lowest practical level at which to work. As the methods employed rely on the recognition of taxa that are found in more than one sampling locality it is important to carry out the analyses at the right sort of taxonomic level. Few species occur outside their type area and are thus of little use in comparing different biotic regions. Family data provide too low a resolution to identify differences between a number of the small tectonic blocks that are of interest. I have therefore run the analyses at generic level. Where taxonomic monographs predate Hill (1981) generic names have been updated to remove synonymies.

As I feel it is most important to keep biogeographical data independent of any palaeogeography that it sets out to test, results are displayed in the form of dendrograms and venn diagrams, or presented on a present day geographical map. In this way the biogeographical history of the different biotas can be kept separate from any palaeogeographical reconstruction, and used as an independent test.

## (b) *Methods of analysis*

Coral faunas from the Visean, early Lower Permian, late Lower Permian and early Upper Permian were compiled for the areas defined in figure 1. Not all areas had coral faunas reported for these time periods, because of the absence of appropriate rocks or lack of systematic work. However, it was possible to build up faunal lists for many of these areas. From the original faunal lists, presence/absence tables for each time period were constructed that included only those taxa that were recorded from two or more of the sample areas (see figure 4). Total diversity for each area was calculated from the full generic list. Because the taxonomy of complex corals is more reliable (these having more morphological characters on which taxa are established) small, simple solitary genera were generally omitted from the similarity analyses. Generic assignment of small simple solitary corals seemed highly suspect in many cases.

Three forms of numerical analyses of similarity were run with these data. All methods of biogeographical analyses have their advantages and disadvantages (Rosen & Smith 1988) and so it was felt that several different measures of faunal similarity should be employed. Parsimony analysis of endemism, principal coordinates analysis and single linkage cluster analyses were all employed and the results compared for consistency.

## (i) Parsimony analysis of endemism (PAE)

This program calculates a Wagner minimum span tree in a Manhattan metric and then produces a rooted tree in which the distribution of shared endemic taxa are distributed in the most parsimonious hierarchical arrangement. The computer program PAUP (Phylogenetic Analysis Using Parsimony; Swofford 1985) was used (for a discussion of this method see Rosen & Smith 1987). The program's 'acctran' option was used, maximizing the number of reversals (secondary losses from an area, either through sampling failure or extinction) compared to parallelisms (multiple evolution of a taxon in separate areas). The tree was rooted by using the 'Lundberg' option, where total absence of taxa was taken as primitive and the distribution of all shared taxa as derived. This program in effect calculates the closeness of similarity on the basis of shared endemic taxa. The consistency index for each tree gives a measure of how often each variant (taxon) appears in the calculation. A consistency index of 1.0 means that the data is wholly congruent with the result. Less consistent results arise because of apparent loss of taxa (sampling failure or extinction within areas) or poor taxonomy.

In an earlier analysis (Smith & Xu 1988) taxonomic ranges were interpolated between end members along a north–south transect to compensate for the vagaries of sampling. As this presupposes the relative positions of the blocks under scrutiny, no such assumption was made in the analyses presented here and no interpolation of taxa made. (The results of this earlier analysis are fully in keeping with the results presented here suggesting that the assumption of fixed relative position was indeed valid.)

## (ii) Principal Coordinates (PCORDS) Analysis

This program is a form of multivariate analysis. It calculates the percentage similarity, by using Gower's Similarity Index, between the biota of each area and every other area on the basis of the taxa that they share. The results are then converted into a multidimensional array and projections onto the first three axes are plotted. These generally account for the majority of the variation but do not reflect any single variant or suite of variants.

## (iii) Single Linkage Cluster (SLC) Analysis

This calculates the percentage similarity by using presence data only and Gower's Similarity Index. It then calculates a minimum spanning tree from the similarity matrix and produces a single linkage cluster analysis in the form of a dendrogram. Clusters can then be grouped empirically according to levels of similarity if desired.

## 4. RESULTS

### (a) Visean

(i) *Data*

Extensive Visean coral faunas are known from many areas in the Far East associated with shelf carbonate deposits such as those across the Yangtze Craton, the Qilian Mountains and Kazakhstan. Correlation is based largely on coral and brachiopod faunas (Jin & Sun 1981; Yang & Fan 1983; Chen 1984) and is relatively well established where such faunas are diverse. Here I have used the Chinese Upper Lower Carboniferous as an equivalent to the European Visean, although it probably extends into the early Namurian. In regions where fauna is sparse, as in the Lhasa and Himalaya Terranes and Northeast Siberia, precise correlation is not so well established.

Visean coral data were taken from the following sources: Qilian Terrane: She 1983 (Southwest Hoit Taria), Li & Liao 1979 (Qinghai Lake), Academia Sinica *et al.* 1962 (Qilian Shan); Kunlun Terrane: Wang 1983 (south of Golmud); Qiangtang Terrane: Dong & Mu 1984, Fan 1985 (Qamdo District); South Urals: Altmark 1975, Sultanaev *et al.* 1978 (Tataria); Kazakhstan: Keller 1969, Volkova 1941 (Dzhezkazgan); Tien Shan: Shchukina 1975 (northern slopes); NE Siberia: Abramov 1970 (Verkhoyansk); NE China: Guo 1980 (Heilongjiang & Jilin Provinces); SE China: Jiang 1982 (Hunan), Nanjing Institute 1982 (Jiangsu, Zhejiang, Anhui Provinces); Lhasa Terrane: Zhao & Wu 1986 (Xainza District); Himalaya Terrane: Flugel 1966 (W Nepal), Fan 1985 (S Tibet).

The distribution matrix used in the analysis is shown in figure 4, and omits all genera of simple solitary rugose corals and all non-rugose genera.

(ii) *Results*

*Diversity.* When total generic diversity is compared between areas (figure 3) there appears to be a broad central band of high diversity (25–40 recorded genera) running from SE China, through Qilian and Kunlun Terranes to Kazakhstan and the South Ural Mountains. Diversity is less on either side, with both NE China and Qiangtang Terranes having between 10 and 25 genera, whereas in the Lhasa and Himalaya Terranes to the south and in NE Siberia less than five genera are known from each region.

Dividing the faunas into compound forms, solitary forms with complex internal dissepimentation and solitary forms with simple internal partitioning (table 1; figure 2) shows that compound forms form a major component in the high diversity faunas, but are absent (with the exception of *Lithostrotion* in NE Siberia) from the very low diversity faunas. Faunas from both NE China Terrane and Qiangtang Terrane are predominantly composed of compound or complex solitary forms, so that their somewhat lower diversity (compared with SE China and the Kunlun/Qilian regions) could be a sampling artefact. However, it is tempting to interpret the broad symmetry of diversity plots as a latitudinal gradient.

*PAE analysis.* This produced one tree of 58 steps with a consistency index of 0.67, which suggests reasonably internally congruent data (figure 4). The cladogram recognizes two groupings towards the crown, one uniting Kazakhstan and the South Urals, the other uniting SE China, Qilian and Kunlun Terranes with Qilian and Kunlun more closely related than either is to SE China. Below this crown grouping comes a pectinate arrangement of sample areas arranged roughly in order of decreasing diversity.

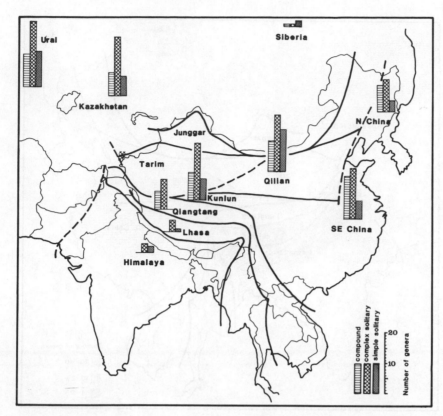

FIGURE 2. Diversity of Visean rugose coral faunas at generic level. Genera have been divided into compound forms, large solitary forms with complex internal dissepimentation and small simple forms without complex dissepimentation.

*PCORDS analysis*. The first two axes of the principal coordinate analysis account for 64 % of the cumulative variation, the first three axes for 76 %. A plot of the first two axes (figure 5) shows a clear differentiation between four clusters: the Qilian–Kunlun–SE China grouping, the Qiangtang–NE China grouping and the Lhasa–Himalaya grouping all differentiated along axis 1, and the Kazakhstan–Ural grouping differentiated from the other three groupings along axis 2. These groupings are largely borne out in other dimensions.

*SLC analysis*. The dendrogram resulting from the single linkage cluster analysis of this data is shown in figure 6. The same four clusters as were identified in the PCORDS analysis are apparent here, namely the Kazakhstan–Ural grouping, SE China–Qilian–Kunlun grouping (Qilian and Kunlun being the closest pair), NE China–Qiangtang grouping, and Lhasa–Himalaya grouping. Unlike the PAE analysis, the Kazakhstan–Ural cluster is separated from the other three groups, which form a second cluster. The NE China–Qiangtang grouping is closer to the SE China–Kunlun–Qilian grouping than to the Lhasa–Himalaya grouping, as was found in the PAE and PCA analyses.

(iii) *Interpretation*

All three analyses give broadly similar patterns, which are best interpreted with reference to the coral diversity gradients highlighted in figures 2 and 3. The most significant feature of these analyses is the marked difference between the coral fauna of Kazakhstan and the Urals on one

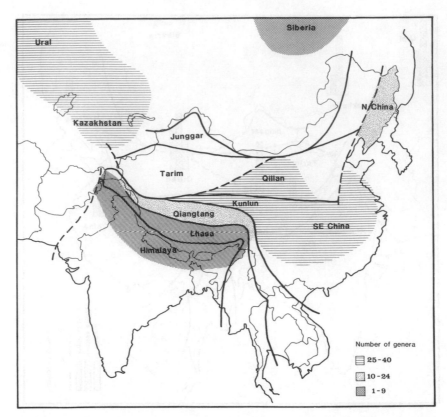

FIGURE 3. Generic diversity of rugose coral faunas in the Visean. A central belt of high diversity faunas runs through the South China Block, the Kunlun–Qilian Terranes and into Kazakhstan and the South Urals interpreted as broadly equatorial.

TABLE 1. DIVERSITY OF VISEAN RUGOSE CORAL GENERA

(Regions as in figure 1. Numbers in parentheses are percentage of total rugose coral fauna.)

|  | 1 | 2 | 3 complex solitary | 4 simple solitary | 5 |
|---|---|---|---|---|---|
|  | total | compound | complex solitary | simple solitary | 2 + 3 |
| Himalaya | 5 | — | 2 (40) | 3 (60) | 2 (40) |
| Lhasa | 5 | — | 4 (80) | 1 (20) | 4 (80) |
| Qiangtang | 16 | 6 (36) | 10 (64) | — | 16 (100) |
| NE China | 24 | 9 (38) | 11 (45) | 4 (17) | 20 (83) |
| SE China | 37 | 14 (37) | 17 (47) | 6 (16) | 31 (84) |
| Qilian | 43 | 10 (22) | 19 (41) | 14 (29) | 29 (63) |
| Kunlun | 35 | 9 (25) | 19 (55) | 7 (20) | 28 (80) |
| Ural | 45 | 11 (24) | 22 (49) | 12 (27) | 33 (73) |
| Kazakhstan | 35 | 8 (23) | 20 (57) | 7 (20) | 28 (80) |
| NE Siberia | 4 | 1 (25) | 1 (25) | 2 (50) | 2 (50) |

hand and the Qilian–Kunlun–SE China region on the other. Both areas have high diversity faunas and were presumably at comparable low latitudes, yet there was clearly some barrier to faunal exchange between these areas in the Lower Carboniferous.

The relative position of the NE China and Qiangtang Terranes is more ambivalent. Both areas have apparent lower diversities than the SE China–Qilian–Kunlun region and in both the PCORDS and SLC analyses are clustered together. In the PAE analysis these regions are close

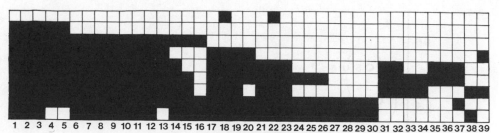

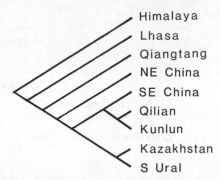

Himalaya
Lhasa
Qiangtang
NE China
SE China
Qilian
Kunlun
Kazakhstan
S Ural

FIGURE 4. Distribution matrix and PAE area cladogram for Visean rugose coral genera, excluding all simple solitary forms (for discussion see text). Genera are as follows: 1, *Lithostrotion*; 2, *Clisiophyllum*; 3, *Dibunophyllum*; 4, *Kueichouphyllum*; 5, *Yuanophyllum*; 6, *Palaeosmilia*; 7, *Diphyphyllum*; 8, *Neoclisiophyllum*; 9, *Arachnolasma*; 10, *Gangamophyllum*; 11, *Lonsdaleia*; 12, *Aulina*; 13, *Heterocaninia*; 14, *Thysanophyllum*; 15, *Koninckophyllum*; 16, *Siphonodendron*; 17, *Corwenia*; 18, *Caninia*; 19, *Orionastraea*; 20, *Siphonophyllia*; 21, *Carcinophyllum*; 22, *Caninophyllum*; 23, *Aulokoninckophyllum*; 24, *Auloclisia*; 25, *Kazakophyllum*; 26, *Cyathoclisia*; 27, *Nemistrium*; 28, *Amygdalophyllum*; 29, *Aulophyllum*; 30, *Arachnolasmella*; 31, *Ekvasophyllum*; 32, '*Lithostrotionella*'; 33, *Cravenia*; 34, *Palastraea*; 35, *Dorlodotia*; 36, *Qinghaiphyllum*; 37, *Carruthersella*; 38, *Stelechophyllum*; 39, *Arachnastraea*.

together but in a pectinate relationship to higher diversity regions, an arrangement characteristic of diversity gradients. On present day geographies the NE China Terrane and the Qiangtang Terrane lie on either side of the belt of high diversity Visean coral faunas (figure 3). Thus this grouping is here interpreted as one uniting slightly higher latitude faunas from which the strictly equatorial elements have been lost, rather than terranes in close proximity. This is precisely what Stehli & Wells (1971) found for Recent coral distributions. They found that coral faunas on either side of a central equatorial belt clustered together. The PCORDS and SLC analyses both demonstrate that the Qiangtang and NE China faunas have stronger similarity with the SE China–Kunlun–Qilian region than with the Ural–Kazakhstan region at this time. Thus the Qiangtang Terrane is inferred to be slightly higher latitude fauna to the south of the SE China–Kunlun–Qilian region, and the NE China Terrane as a slightly higher latitude fauna to the north of the same region. This implies that the Visean palaeoequator traversed the SE China–Kunlun–Qilian region. Also by implication, the Kazakhstan–Ural region must have lain at comparable (broadly tropical) latitudes.

NE Siberia must have been situated at relatively high latitudes during the Visean, outside the tropical belt of broad carbonate shelf deposition and diverse shallow-water coral faunas. The same is true also of the Himalaya and Lhasa Terranes, though a number of complex solitary corals are found in the northern part of the Lhasa Terrane where less extensive limestone sequences are found. The faunas are too impoverished to relate to either the Kazakhstan–Ural region or the Cathaysian region, although both the PCORDS and SLC analyses

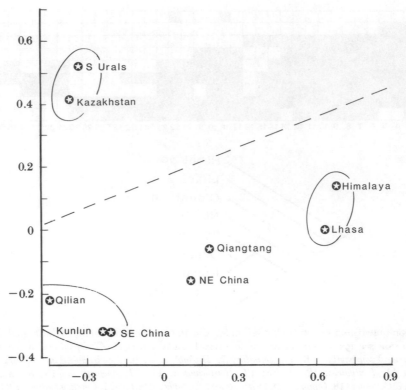

FIGURE 5. First two principal axes of a PCORDS analysis of Visean rugose coral genera,
excluding all simple solitary forms (for discussion see text).

placed the faunas from these areas slightly closer to the Cathaysian fauna than the
Kazakhstan–Ural fauna. Clearly, the Lhasa Terrane lay on the margins of the tropical belt of
high-diversity coral faunas, which today extends to about 35° latitude (Rosen 1984) (although
this is based on scleractinian corals and nothing definitive is known about the latitudinal
controls of Palaeozoic corals).

In summary, Visean coral faunas appear to show a diversity gradient that can be used to
distinguish tropical from temperate regions. There is an unambiguous faunal discontinuity
between the broadly tropical Cathaysian faunas (SE China, Qilian, Kunlun, NE China and
Qiangtang) and the Russian faunas of Kazakhstan and South Urals. There is, however, no
evidence for any faunal discontinuity between the South and North China Blocks at this time.
Himalayan and NE Siberian faunas are interpreted as low-diversity, high-latitude assemblages,
with the Lhasa Terrane occupying a position on the fringes of the tropical belt intermediate
between the Himalaya and Cathaysian regions.

(iv) *Comparison with previous work and other faunal groups*

Hill (1973) provided one of the earliest comprehensive and authoritative accounts of global
coral distribution during the Lower Carboniferous. She divided faunas into three biogeographical realms, North American, Eurasian and Eastern Australian. All of the areas dealt
with here belong to her Eurasian realm, which she further divided into various provinces,
including a Chinese province (Hunan and the Yangse platform, NE China, NW China (Gansu
& Xinqang Province), Tibet, SE Asia, Japan and Western Australia) and a central Asian

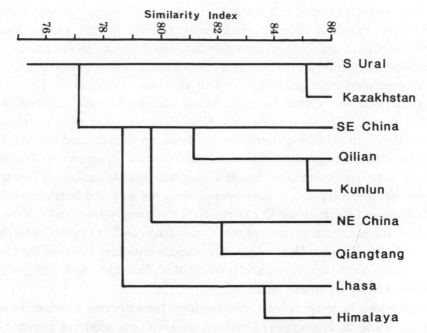

FIGURE 6. Dendrogram from a SLC analysis of Visean rugose genera, excluding all
simple solitary forms (for discussion see text).

province. She noted that in Kazakhstan a significant Chinese element appears towards the end
of the Visean. The analysis presented here supports her separation of Cathaysian and Central
Asian faunas, but she presented no assessment of the relationships of provincial faunas within
her Eurasian Realm. She did, however, discuss the implications of diversity data and equated
high diversity coral faunas with warm, shallow waters.

Later workers have tended to amplify Hill's original observations, though solely through
narrative accounts that lack analytical rigour. Papojan (1977) agreed that Kazakhstan had
closer affinities with the Transcaucasus and European fauna than with China, noting that the
central Asiatic region appeared to have had immigration of coral genera from both east and
west. Similarly Fedorowski (1977) recognized Chinese, Eurasian and NE Siberian realms and
also separated the Chinese coral fauna into a northern and southern province. Minato & Kato
(1977) equated the zone of high-diversity coral faunas running from Turkey through
Kazakhstan, China, Australia and Japan at this time with a 'Tethyan' ocean.

More detailed work on the Lower Carboniferous coral faunas of China by Luo (1984) led
to the recognition of three distinct provinces, although again without using rigorous analytical
techniques. These are the Yangze Province (SE China), The Tienshan–Qilian Province and
the Bayankela–Sanjiang Province (the Qiangtang Terrane of this analysis). Furthermore he
suggested that NE China might be a fourth region. The Tienshan–Qilian Province includes,
in addition to Qilian and Kunlun Terranes, Xinjiang, Gansu and Ningxia Provinces, and the
western part of Nei Mongolia. Fan (1985) separated Visean coral faunas of Tibet into two
groups: a 'Palaeotethyan' province in which relatively high diversity faunas of compound and
complex solitary rugose corals were recorded (Qiangtang Terrane, North of the Banggong
Suture) and a 'Gondwanan'-province with a low diversity fauna containing only small, simple
solitary rugose genera (the Lhasa and Himalaya Terranes).

Smith & Xu (1988) provided a quantitative analysis of Tibetan Visean coral faunas and found a diversity gradient, from high-diversity faunas in the Qilian–Kunlun region, intermediate diversity in the Qiangtang region and low diversity in the Lhasa and especially Himalaya region. They interpreted this as a climatic/latitudinal gradient.

The results presented here generally confirm previous empirical studies of Asiatic coral distribution patterns in the Visean, but make much more explicit the relations between faunal regions. Although my results show Qilian and Kunlun faunas to be more closely related to one another than either is to SE China the differences between 'North' and 'South' China are not marked and the similarities pronounced. More refined work is required to determine whether the differences between North and South China are gradational or abrupt. On current evidence, however, there appears to have been little in the way of a barrier to faunal dispersal between SE China and the Kunlun–Qilian region. Interestingly Luo (1984) found that Visean of Xinjiang and the northern margin of the Tarim Basin to have comparable high-diversity coral faunas to those of Qilian. This suggests that the discontinuity between the Cathaysian and the Middle Asian coral faunas coincides with the Tienshan fold belt, although Luo's observations need to be rigorously analysed.

Other groups whose biogeographical affinities have been discussed include brachiopods and terrestrial flora, Yang & Fan (1983) provided an empirical study of Lower Carboniferous brachiopods from the Tibetan Plateau. They distinguished two faunas; those from the Qamdo region (eastern Qiangtang Terrane) show striking similarity to faunas from western Guizhou (SE China Block), where those from south of the Zangbo Suture show strong affinities with Himalayan faunas of Kashmir. The fauna from northern Xizang (Xainza, Lhasa Terrane) they found to be intermediate, with a mixture of 'Gondwanan' and 'Cathaysian' elements. Quantitative analysis of Visean brachiopod data from the Tibetan Plateau by Smith & Xu (1988) demonstrated a North–South diversity gradient over the region. Highest diversities were found in Qilian–Kunlun–Qiangtang region, intermediate diversities in the Lhasa region and low diversities in the Himalaya region. No endemic taxa at generic level were found between the Himalaya and Lhasa Terranes suggesting that the so-called 'Gondwanan' fauna is simply a depleted equatorial (Cathaysian) fauna.

Lower Carboniferous floras across China have recently been reported by Laveine et al. (1987) who found that the North China and South China Blocks share a common and very characteristic flora during the Visean.

(v) *Conclusions*

Various independent lines of palaeontological evidence lead to the following conclusions.

1. During the Visean there existed a broadly tropical region composed of North and South China Blocks, the Kunlun and Qiangtang Terrane and apparently the Junggar and Tarim Terranes over which faunal and floral exchange was relatively uninhibited.

2. This Cathaysian region was separated by some unspecifiable barrier from Central Asia, which also lay in a broadly tropical latitude.

3. Faunal diversity of both corals and brachiopods decreases southwards towards the Himalayas but no distinct endemic 'Gondwanan' fauna exists. Although the drop-off in diversity is steep, no sharp discontinuity indicating a faunal break, is evident. Thus no significant oceanic barrier can be identified within this region.

4. A similar low-diversity northern fauna is found in NE Siberia.

### (b) *Early Lower Permian (Asselian/Sakmarian)*

(i) *Data*

This is the period of extensive glaciation in the Southern Hemisphere (see Dickens 1977, 1984, 1985) and coral faunas are relatively few and never attain high diversities. Global correlation in the Permian is not good, but here I have correlated Eurasian Asselian/Sakmarian deposits with the Chinese 'Upper Carboniferous' deposits, which have *Pseudoschwagerina* assemblage fusulinids (see Ross 1982*a*; Kanmera *et al.* 1976). In regions where fusulinids are absent correlation has been based on brachiopod assemblages (Waterhouse & Gupta 1979; Yang & Fan 1983; Chen 1984; Jin 1985) and is less certain.

Coral faunas were compiled from the following sources: SE China: Jiang 1982 (Hunan), Nanjing Institute 1982 (Jiangsu, Anhui and Zhejiang Provinces); Qilian Terrane: Li & Liao 1979 (Qilian mountains), Academia Sinica *et al.* 1962 (Qinghai Lake); Nei Mongolia: Bureau of Geology and Northeastern Institute of Geology 1976 (NE corner); Himalaya Terrane: Fan 1985 (S Tibet); Lhasa Terrane: Yang & Fan 1983, Fan 1985; Eastern Qiantang: Dong & Mu 1984, Fan 1985 (Qamdo District); Western Qiangtang: Liang *et al.* 1983 (Ali District); Kazakhstan: Malkovskiy 1975 (NW region); South Urals: Stevens 1975.

(ii) *Results*

*Diversity.* Coral faunas at this time have a low diversity (table 2). Distinguishing between simple solitary rugose corals and compound forms or those large solitaries with complex dissepimentation (figure 7) reveals that faunas with compound forms and complex solitary genera are found in the Urals and Kazakhstan (where corals of the family Durhaminidae occur) and in SE China, Qilian and Eastern Qiangtang (where Waagenophyllidae dominate). Only simple solitary forms are known from the Lhasa and Himalaya Terranes and from Western Qiangtang.

*PAE analysis.* This analysis was based on compound and complex solitary forms only. One tree of branch length 21 and a consistency index of 0.67 was found (figure 8). This distinguished two faunas: a Kazakhstan–Ural grouping and a SE China–Qilian–E Qiangtang grouping. Within the latter, SE China and Qilian share closer similarity than either does to East Qiangtang.

*PCORDS analysis.* This produced a plot in which the first two axes accounted for 82% of the total variation and the first three axes 96.5% of the total variation. A plot of the first two axes (figure 9) shows a clear separation of the Kazakhstan–Ural fauna and the SE China–Qilian–E Qiangtang fauna along the principal axes.

*SLC analysis.* This produced an identical result to the PAE analysis, clearly differentiating between a Kazakhstan–Ural fauna and a SE China–Qilian–E Qiangtang fauna (figure 10). The similarity index suggests that SE China, Qilian and E Qiangtang are separated by only a small difference and that this may not be significant, considering the small sample size on which this analysis is based.

(iii) *Interpretation*

The pattern of coral distribution resembles that seen during the Visean, with a clear demarcation between Central Asian faunas and Cathaysian faunas. Within the Cathaysian region there is a hint that SE China and the Qilian Terrane are slightly more closely related

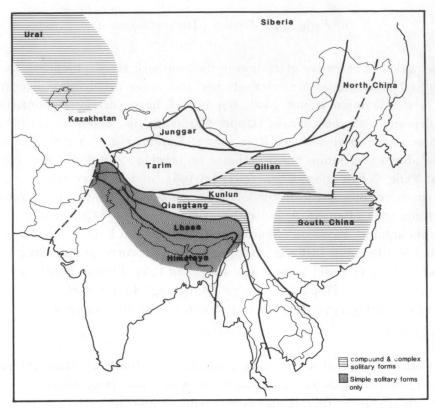

FIGURE 7. Distribution of early Lower Permian (Asselian/Sakmarian) rugose coral genera.

TABLE 2. DIVERSITY OF EARLY LOWER PERMIAN (ASSELIAN/SAKMARIAN)
RUGOSE CORAL GENERA

(Regions as in figure 1. Numbers in parentheses are percentage of total rugose coral fauna.)

|  | 1 total | 2 compound | 3 complex solitary | 4 simple solitary | 5 2+3 |
|---|---|---|---|---|---|
| Qilian | 12 | 3 (25) | 4 (33) | 5 (42) | 7 (58) |
| East Qiangtang | 12 | 7 (58) | 3 (25) | 2 (17) | 10 (93) |
| West Qiangtang | 5 | — | — | 5 (100) | — |
| Lhasa | 1 | — | — | 1 | — |
| SE China | 21 | 9 (43) | 9 (43) | 3 (14) | 18 (86) |
| Kazakhstan | 7 | 6 (86) | 1 (14) | — | 7 (100) |
| Urals | 11 | 7 (64) | 2 (18) | 2 (18) | 9 (82) |
| N Nei Mongolia | 7 | — | — | 7 (100) | — |
| Himalaya | — | — | — | — | — |

than either is to the eastern part of the Qiangtang Terrane, although this may be no more than an artefact of sampling. Both regions are interpreted to be broadly equatorial in position, although unfortunately I have no data on the northern extent of compound or complex solitary corals at this time. The fact that none have been reported from beds of this age in NE China (Guo 1980) may imply that they are absent, but such negative evidence is of questionable validity.

The occurrence of only simple solitary rugose corals in relatively shallow and predominantly

FIGURE 8. Distribution matrix and PAE area cladogram for early Lower Permian rugose coral genera, excluding all simple solitary forms. Genera are as follows: 1, *Tschussovskenia*; 2, *Paralithostrotion*; 3, *Lonsdaleiastraea*; 4, *Bothroclisia–Bothrophyllum*; 5, *Kepingophyllum*; 6, *Lytvophyllum*; 7, *Protowentzelella*; 8, *Stylastraea*; 9, '*Orionastraea*'; 10, *Protolonsdaliastraea*; 11, *Caninia*; 12, *Koninckocarinia*; 13, *Pavastehphyllum*; 14, *Ivanovia*.

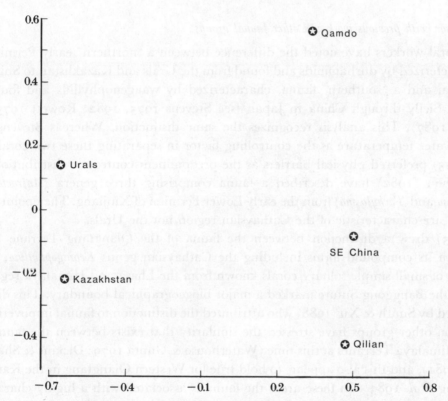

FIGURE 9. First two principal axes of a PCORDS analysis of early Lower Permian (Asselian/Sakmarian) rugose coral genera, excluding all simple solitary forms (for discussion see text).

clastic deposits in the western part of the Qiangtang Terrane, Lhasa Terrane and Himalaya Terrane suggest that these areas lay outside the equatorial zone at this time. In comparison to the Visean distributional pattern, where a few genera of solitary corals with complex dissepimentation extend into the Lhasa Terrane, the early Lower Permian pattern suggests that the equatorial belt was less extensive, as might be expected from the widespread evidence of glaciation throughout Gondwanaland at this time (Dickins 1984).

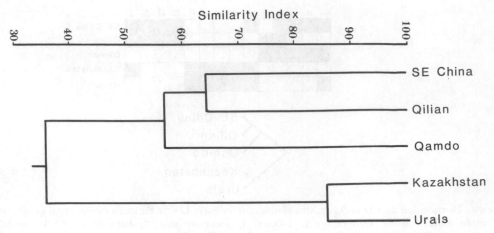

FIGURE 10. Dendrogram from a SLC analysis of early Lower Permian (Asselian/Sakmarian) rugose coral genera, excluding all simple solitary forms (for discussion see text).

(iv) *Comparison with previous work and other faunal groups*

Several coral workers have noted the difference between a 'northern' early Permian coral fauna, characterized by durhaminids and found from the Urals and Kazakhstan to Spitsbergen and America, and a 'southern' fauna, characterized by waagenophyllids and found from Tunisia and Sicily through China to Japan (see Stevens 1975, 1982; Rowett 1975, 1977; Fedorowski 1987). This analysis recognizes the same distinction. Whereas Stevens (1975) argued for water temperature as the controlling factor in separating these two coral faunas, Rowett (1975) preferred physical barriers as the predominent control on distribution.

Wu & Zhou (1982) have described a fauna comprising three genera (*Anfractophyllum*, *Kepingophyllum* and *Tachylasma*) from the early Lower Permian of Xinjiang. They point out that these genera are characteristic of the Cathaysian region not the Urals.

Fan (1985) drew a distinction between the fauna of the Qiangtang Terrane (Qamdo district), with its compound forms including the Cathaysian genus *Kepingophyllum*, and the sparse fauna of small simple solitary corals known from the Lhasa and Himalaya regions. He argued that the Banggong Suture marked a major biogeographical boundary. This difference was also noted by Smith & Xu (1988) who attributed the distinction to faunal impoverishment.

Workers on other groups have stressed the similarity that exists between the fauna of the Lhasa and Himalaya Terranes at this time (Waterhouse & Gupta 1979; Dickins & Shah 1981; Jin & Sun 1981), and this also appears to hold true for Western Qiangtang in the Karakoram region (Liang *et al.* 1983). In these areas the fauna is associated with a highly characteristic mixtite deposit or deposits, interpreted as glacio-marine in origin (see Leeder *et al.* 1988). The characteristic fauna has been termed the 'Stepanoviella Fauna', 'Lytvolasma Fauna', or 'Eurydesma Fauna'. It is dominated by brachiopods, but also includes bivalves, such as *Deltopecten*, fenestellid bryozoans and crinoids. This fauna has been described from the Himalayas, Lhasa Terrane, Western Qiangtang Terrane and from the Tenchong and Baoshan areas of western Yunan and the Menghong Formation of Thailand, Burma and Malaysia (Wang 1983).

Smith & Xu (1988) noted the occurrence of endemic brachiopod elements shared between the Lhasa and Himalaya regions and between Eastern Qiangtang and Qilian regions. They also noted that fusulinids, a group thought to be restricted to warm, shallow waters (see Ross 1982b), occurred only in the Qilian, Kunlun and Eastern Qiangtang Terranes. Gobbett (1973) drew a sharp distinction between the 'Tethyan' fusulinids of the Cathaysian region and the 'Boreal' fusulinids of the Urals.

### (v) Conclusions

The isolation of Cathaysian equatorial faunas from Central Asian equatorial faunas identified in the Visean is still apparent in the early Lower Permian. There is no evidence that SE China, Qilian and Eastern Qiangtang were separated by any significant barrier to dispersal at this time. The Lhasa and Himalaya Terranes and the western part of the Qiangtang Terrane have glacio-marine sediments and a characteristic fauna, lacking complex rugose corals and fusulinids. These regions presumably lay in higher latitudes. There was no significant barrier to dispersal between the Lhasa and Himalaya Terranes. The major faunal break between 'Himalayan' and 'Cathaysian' faunas does not follow tectonic suture lines but crosses the Qiangtang Terrane and thus presumably is the product of a latitudinal/climatic gradient rather than separation of terranes by oceanic distances.

### (c) Late Lower Permian

#### (i) Data

This period covers the Artinskian and Ufimian stages of Russian authors, and the Qixian stage of Chinese authors. I have taken the fusulinid *Parafusulina* Zone (Toriyama 1984), with the *Cancellina* and *Misellina* assemblages as diagnostic. Correlation into areas lacking fusulinid foraminifers is unsatisfactory, but can be done on the basis of brachiopod assemblages (Jin 1985).

Faunal lists were compiled from the following sources: SE China: Jiang 1982 (Hunan), Nanjing Institute 1982 (Jiangsu, Anhui and Zhejiang Provinces); Qilian Terrane: Liu 1984a,b (Qilianshan), Bureau of Geology and Northeastern Institute of Geology 1976 (Inner Mongolia: southern region, Xar Moron River District); Western Qiangtang: Liang et al. 1983 (Ali District); Eastern Qiangtang: Fan 1985, Dong & Mu 1984 (Qamdo District); NE China: Guo 1980 (Heilongjiang and Jilin Provinces); Lhasa Terrane: Wang & Liu 1982, Lin 1984, Zhang et al. 1985 (Xainza District); Himalaya Terrane: Wu et al. 1982, Lin 1984 (southern Tibet); South Urals: Stevens (1975).

Stevens (1975) did not provide an exhausive list of Artinskian genera from the Urals and this area was omitted from the quantitative analyses for this reason.

#### (ii) Results

*Diversity.* At this period the majority of rugose coral genera are simple solitary forms (table 3). High-diversity faunas of compound and complex solitary forms are not recorded. Faunas with four to six genera of compound and complex solitary rugose coral genera are found in SE China, Qiangtang Terrane, NE China and the South Urals (figure 11). NE China has a moderately high diversity of small simple solitary forms and only three compound genera. It may therefore genuinely have had a low diversity of compound and complex solitary forms. The North China Block (Qilian Shan and part of Nei Mongolia) also has only three genera of compound or complex solitary rugose corals recorded, but in the Qilian Terrane this is probably at least partially caused by sampling failure as no simple solitary corals are recorded

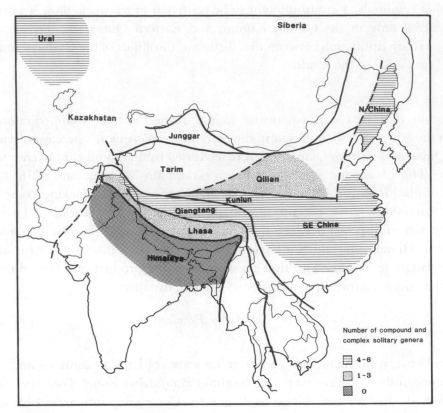

FIGURE 11. Plot of diversity of compound and complex solitary rugose coral genera
in the late Lower Permian (for discussion see text).

TABLE 3. DIVERSITY OF LATE LOWER PERMIAN (QIXIAN) RUGOSE CORAL GENERA

(Regions as in figure 1. Numbers in parentheses are percentage of total rugose coral fauna.)

| | 1 | 2 | 3 | 4 | 5 |
|---|---|---|---|---|---|
| | | | complex | simple | |
| | total | compound | solitary | solitary | 2+3 |
| NE China | 22 | 3 (14) | 2 (9) | 17 (77) | 5 (23) |
| Qilian | 3 | 2 (67) | 1 (33) | | 3 (100) |
| West Qiangtang | 10 | 3 (30) | 1 (10) | 6 (60) | 4 (40) |
| East Qiangtang | 6 | 4 (80) | — | 1 (20) | 4 (80) |
| SE China | 15 | 3 (20) | 2 (13) | 10 (67) | 5 (33) |
| Lhasa | 9 | — | — | 9 (100) | — |
| Himalaya | 9 | — | — | 9 (100) | — |

either. In the Lhasa and Himalaya Terranes moderately diverse faunas of simple solitary corals
are recorded but no compound or complex solitary forms have yet been reported.

*PAE analysis.* Two analyses were run, one with all coral genera, the other with just compound
and complex solitary genera. The analysis of compound and complex solitary generic data
found four equally parsimonious trees each with a branch length of 11 and a consistency index
of 0.64 (figure 12). All identified SE China and West Qiangtang as the most closely related,
with the North China Block (Qilian) as sister group. NE China and Qiangtang were placed
either with the SE China–West Qiangtang cluster or with Qilian. Figure 12 shows one of the
four cladograms. However, the small data set and the multiple solutions suggests that it is

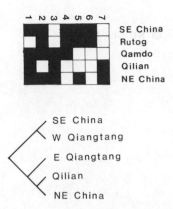

FIGURE 12. Distribution matrix and PAE area cladogram for late Lower Permian rugose coral genera, excluding all simple solitary genera. Genera are as follows: 1, *Yatsengia*; 2, *Polythecalis*; 3, *Wentzellella*; 4, *Wentzellophyllum*; 5, *Paracaninia*; 6, *Thomasiphyllum*; 7, *Metriophyllum*.

impossible to determine the relationships of East Qiangtang and NE China to the other areas without further data.

The analysis run using the entire data set produced even less satisfactory results with nine equally parsimonious trees, each with a consistency index of only 0.5. SE China and West Qiangtang consistently appeared as one group, and Lhasa, Himalaya, NE China and N China Block (Qilian and part of Nei Mongolia) usually appeared as a second group, though not in all trees. The position of East Qiangtang was variable.

*PCORDS analysis*. In the analysis of complex solitary and compound genera alone, the first two axes accounted for 81 % of the total variance and the first three for 92 % of the total variance. The plot on the first two axes (figure 13) shows SE China and West Qiangtang to be separated from other areas along the first axis, and the other areas widely scattered.

An analysis with all coral genera was less satisfactory, with only 56 % of the variation shown on the first two axes and 75 % on the first three axes. West Qiangtang and SE China appear separated from other areas as before. East Qiangtang and Qilian are separated along the first axis from Mongolia, Lhasa, Himalaya and NE China.

*SLC analysis*. As with the other two analyses, West Qiangtang and SE China appear as one cluster separated distinctly from the other areas (figure 14). Qilian and North China (Inner Mongolia) also appear closely similar, but this group, together with NE China and East Qiangtang, form an unresolvable trichotomy.

Analysis of the full data set identified Lhasa and Himalaya Terranes as having very similar fauna, and the East Qiangtang and Qilian Terranes as a second cluster, but left all other relationships unresolved.

### (iii) *Interpretation*

The only consistent feature to appear from these analyses is the similarity between the coral fauna of SE China and Western Qiangtang (Ali District) at this time. The Urals, according to Stevens (1975), have a fauna in which no genera of compound or complex solitary rugose corals are common to those of the Cathaysian region, which suggests that a significant barrier to migration existed between these two areas. The Lhasa and Himalaya Terrane continue to share a similar fauna, basically depleted of taxa. Other regions of China have ambiguous

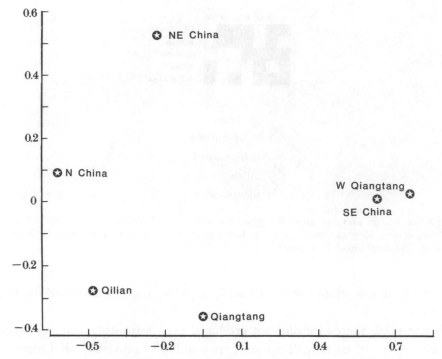

FIGURE 13. First two principal axes of a PCORDS analysis of late Lower Permian rugose coral genera based on compound and complex solitary forms only (for discussion see text).

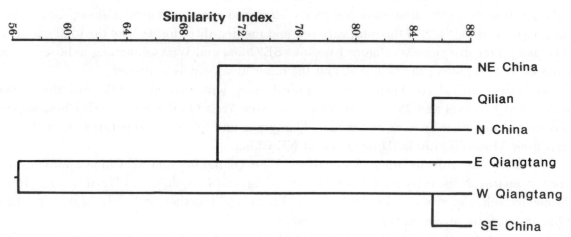

FIGURE 14. Dendrogram from a SLC analysis of late Lower Permian rugose coral genera, excluding all simple solitary forms.

relationships, and it is probable that sampling deficiencies are causing much of the problem. On the available evidence it would be rash to postulate any biogeographical relationships for these regions. The compound genera *Wentzellophyllum* and *Polythecalis* are distributed across Qiangtang and SE China whereas *Yatsengia* is found in both North and South China Blocks, suggesting that these regions maintained at least a reasonable amount of biotic exchange.

(iv) *Comparison with previous work and other groups*

Both Fan (1985) and Lin (1984) have stressed the differences that exist between the coral faunas of the Lhasa and Qiangtang Terranes at this time and have used this as evidence for the Banggong Suture being the site of a major oceanic tract. They have also noted the similarity between the fauna of the Lhasa and Himalaya Terranes. This similarity is borne out by the brachiopod faunas (see Zhang *et al.* 1985; Smith & Xu 1988).

The distribution of fusulinid foraminifers is somewhat different from that of compound rugose corals (Smith & Xu 1988). Fusulinids are most diverse in the Qiangtang Terrane and in SE China (Nanjing Institute 1982), but are well developed in the North China Block (Bureau of Geology and Northeastern Institute of Geology 1976), and extend into NE China where nine genera are reported (Guo 1980). Fusulinids also extend into the Lhasa Terrane at this time, with four genera (Chu 1983). Fusulinid diversity then appears to have been relatively high in the North and South China Blocks, including the Qiangtang Terrane, and to have diminished both northwards in the NE China Terrane and southwards into the Lhasa Terrane. Thus on the basis of fusulinid distributions, the Lhasa Block shares elements with the Cathaysian region, not the Himalayan region. Gobbett (1973) recognized a distinction between the fusulinid fauna of the Urals and that of China and 'Tethys'.

(v) *Conclusions*

Evidence from compound coral and fusulind distribution patterns continues to imply that the Cathaysian region occupied a broadly equatorial latitude, with the highest diversities possibly in the S China Block and Qiangtang Terrane regions. NE China lay on the fringes of this belt, but was still in faunal contact, as it shares a number of coral and fusulinid genera in common. To the south the Lhasa Terrane also lay on the periphery of this high diversity belt, with some genera of fusulinids (but not complex solitary or compound rugose corals) extending from the Cathaysian region. Faunal exchange was extensive between the Lhasa and Himalaya Terranes, which shared a similar brachiopod and simple solitary coral fauna. But the appearance of fusulinids in the Lhasa Terrane may be the first indication of the faunal disjunction between the Lhasa and Himalaya Terranes so prominent in the early Upper Permian.

(d) *Early Upper Permian*

(i) *Data*

Moderately high-diversity coral faunas are known from a number of areas at this time (table 4). Here I have taken early Upper Permian to be represented by the Chinese Maokoan Stage and characterized by the fusulinid foraminiferid *Neoschwagerina–Yabeina* Zones (see Toriyama 1984). This correlates approximately with the American Guadalupian Stage, the Russian Murgabian and Midian Stages and the Kungurian and Kazanian Stages. Correlation into regions from which fusulinids are absent is based on brachiopod assemblages (Jin 1985). In Kashmir this horizon is represented by the lower part of the Zewan Series (Nakazawa *et al.* 1975), whereas in the Salt Range (the western extension of the Lhasa Terrane into northern West Pakistan) it is equivalent to the Middle Productus Limestone or Wargal Limestone (Kummel & Teichert 1970).

Coral faunas were compiled from the following sources: SE China: Jiang 1982 (Hunan), Nanjing Institute 1982 (Jiangsu, Anhui and Zhejiang Provinces); Nei Mongolia: Bureau of

TABLE 4. DIVERSITY OF EARLY UPPER PERMIAN (MAOKOAN) RUGOSE CORAL GENERA

(Regions as in figure 1. Numbers in parentheses are percentage of total rugose coral fauna.)

| | 1 | 2 | 3 | 4 | 5 |
|---|---|---|---|---|---|
| | | | complex | simple | |
| | total | compound | solitary | solitary | 2+3 |
| NE Siberia | 3 | — | — | 3 (100) | — |
| N Nei Mongolia | 11 | 1 (9) | — | 10 (91) | 1 (9) |
| Vladivostok | 17 | 2 (12) | — | 15 (88) | 2 (12) |
| Qilian | 9 | 3 (33) | 2 (22) | 4 (44) | 5 (66) |
| Kunlun | 7 | 2 (29) | 1 (14) | 4 (57) | 3 (43) |
| NE China | 6 | 4 (66) | — | 2 (33) | 4 (66) |
| SE China | 13 | 7 (54) | 1 (8) | 5 (38) | 8 (62) |
| Qiangtang | 8 | 3 (37) | — | 5 (63) | 3 (37) |
| Lhasa | 16 | 10 (62) | 4 (25) | 2 (13) | 14 (87) |
| Himalaya | 18 | — | — | 18 (100) | — |

Geology and Northeastern Institute of Geology 1976; NE Siberia: Sokolov 1960 (Verkhoyansk); Vladivostok: Kotlyar 1978 (South Primorye); NE China: Guo 1980 (Helongjiang & Jiling Provinces); Qilian Terrane: Liu 1984a,b (Tianjun County), Li & Liao 1979 (Qilianshan); Kunlun Terrane: Li & Liao 1979; Eastern Qiantang: Dong & Mu 1984, Fan 1985 (Qamdo District); Western Qiangtang: Liang et al. 1983 (Ali District) Lhasa Terrane: Lin 1984, Zhang et al. 1985 (Xianza District) Heritsch 1937 updated by Minato & Kato 1965 (Salt Range); Himalaya Terrane: Kato 1976 (Kashmir), Flugel 1966 (Nepal), Wang & Liu 1983, Lin 1984 (southern Tibet).

(ii) *Results*

*Diversity.* Highest-diversity faunas of rugose corals are recorded from the Lhasa Terrane (Lin 1984) and SE China (Jiang 1982), where compound and complex solitary forms predominate (table 4). Although some of the apparent diversity in the Lhasa fauna from Xainza probably results from taxonomic oversplitting, when newly proposed genera 'endemic' to that locality are removed, the Lhasa Terrane still retains a coral diversity equivalent to that of SE China. Compound and complex solitary forms are also found in regions to the north, though represented by fewer taxa (figure 15). Five genera are known from beds of this age in the Qilian Terrane, whereas only three are known from NE China Terrane and two from the Vladevostok Terrane. None are reported from NE Siberia, the northern part of Nei Mongolia or the Himalaya Terranes. The Vladivostok, northern Nei Mongolia and Himalaya Terranes all have high diversities of simple solitary rugose corals recorded, suggesting that it is not sampling that is causing the apparent drop in numbers of compound and complex solitary corals in these regions.

Although not included in this analysis, the Trans-Caucasus has a moderate diversity of compound and complex solitary rugose corals composed largely of waagenophyllids (Kotlyar & Stepanov 1984).

*PAE analyses.* Two analyses were run, one with the full coral data, the other with only compound and complex solitary forms (figure 16). For compound and complex solitary forms alone, the PAE analysis found 20 equally parsimonious trees of branch length 14 and with a consistency index of 0.643. The only consistent feature of all 20 trees is the association of SE China with the Lhasa Terrane. A consensus tree of all twenty solutions places the other faunas all together as a basal polychotomy.

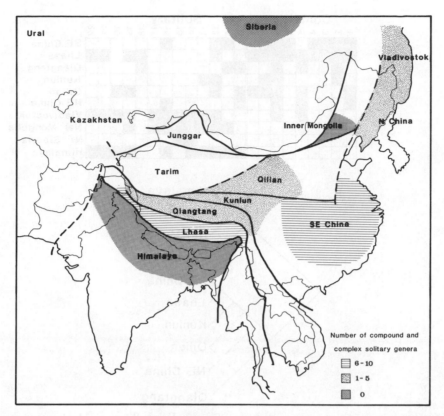

FIGURE 15. Diversity of early Upper Permian (Maokoan) rugose genera, excluding simple solitary forms.

The analysis with the full data set produced only two equally parsimonious solutions of branch length 46. However, the consistency index, at 0.49, was much lower than for the preceding analysis. Both trees recognized two primary clusters, one comprising the Himalaya, Vladivostok, Nei Mongolia and NE Siberia Terranes, the other comprising the SE China, Lhasa, Kunlun, Qilian, NE China and Qiangtang Terranes. Clearly this clustering is caused by the distribution of compound and complex solitary forms. In both solutions SE China and the Lhasa Terrane are grouped together. However, one solution places the Qilian, NE China and Qiangtang Terranes as a discrete cluster, whereas the other places these areas in a pectinate arrangement beneath the Lhasa–SE China cluster. The two alternatives are shown in figure 16.

*PCORDS analyses.* As above, the two analyses were run firstly with only compound and complex solitary forms then afterwards with the total data set. In the first analysis the total variation accounted for by the first two axes is 78.5 % and by the first three axes 90 %. A plot of the first two axes (figure 17) gives a clear distinction between SE China and Lhasa from other terranes along the first axis, whereas a plot of the first and third axes (figure 18) shows distinct clustering of three regions: Lhasa and SE China; Qilian and Kunlun; and NE China, Vladivostok and Qiangtang.

The analysis with the full data produced a less reliable result with the total variation accounted for by the first two axes only 51 % and by the first three axes 66 %. As in the PAE analysis, two groupings can be recognized, one comprising areas dominated by simple solitary genera (NE Siberia, Vladivostok, Nei Mongolia and Himalaya), the other by compound and

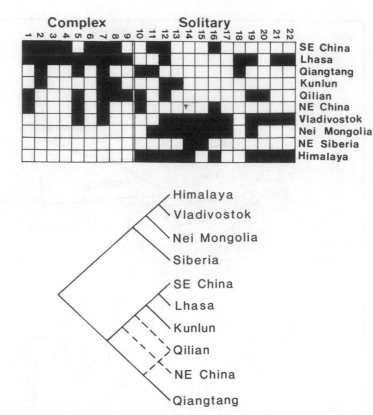

FIGURE 16. Distribution matrix and PAE area cladogram for early Upper Permian (Maokoan) rugose coral genera. Genera are as follows: 1, *Yatsengia*; 2, *Ipciphyllum*; 3, *Pseudohuangia*; 4, *Paraipciphyllum*; 5, *Wentzellella*; 6, *Polythecalis*; 7, *Waagenophyllum*; 8, *Iranophyllum*; 9, *Pavastehphyllum*; 10, *Duplophyllum*; 11, *Allotropiophyllum*; 12, *Amplexocarinia*; 13, *Calophyllum*; 14, *Plerophyllum*; 15, *Soshkineophyllum*; 16, *Lophophyllum*; 17, *Lophocarinophyllum*; 18, *Bradyphyllum*; 19, *Trachylasma*; 20, *Timorphyllum*; 21, *Cyathaxonia*; 22, *Verbeekia*.

complex solitary genera (SE China, Lhasa, Qilian, Kunlun, NE China and Qiangtang). No other clear pattern emerged.

*SLC analyses*. The dendrogram produced from the compound and complex solitary generic data (figure 19) shows two clear clusters, with the Lhasa and SE China Terranes as one and the Qilian, Kunlun, Qiangtang, NE China and Vladivostok Terranes as the other. Within this second cluster NE China, Vladivostok and Qiangtang appear to be closer in similarity to one another. The dendrogram from the entire data set identified few clusters as it had several trichotomies, none of which resembled groupings present in the more restricted data set.

(iii) *Interpretation*

As the taxonomy of small solitary rugose corals is somewhat suspect, more weight is given here to patterns derived from the better established taxa of compound and complex solitary genera. The link between the Lhasa Terrane and SE China appears in almost all analyses, suggesting that the faunas of these two regions were the most similar. The loss of compound and complex solitary forms to the north is more ambiguous because it might reflect sampling. Clearly the fauna of Vladivostok and Nei Mongolia are genuinely impoverished in complex forms since relatively large faunas of simple solitary rugose corals are known from these areas. However, in the Kunlun and Qilian Terranes for example, only small numbers of corals have

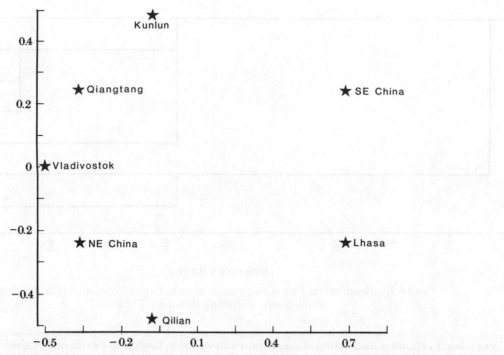

FIGURE 17. Plot of the first two principal axes of a PCORDS analysis of early Upper Permian (Maokoan) rugose coral genera, excluding simple solitary forms.

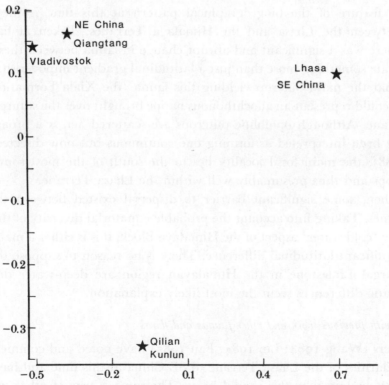

FIGURE 18. Plot of the axes one and three from a PCORDS analysis of early Upper Permian (Maokoan) rugose coral genera, excluding simple solitary forms.

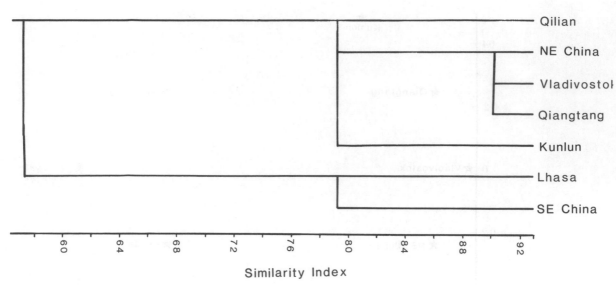

FIGURE 19. Dendrogram form a SLC analysis of early Upper Permian (Maokoan) rugose
coral genera (excluding Himalaya Terrane).

ever been reported and when the composition of these faunas is examined (table 4) compound
and complex solitary corals actually outnumber simple solitary corals. Thus at least part of the
paucity of faunas in these regions is likely to be sampling failure or absence of rocks of the right
facies (both areas have extensive volcanic sequences at this time).

A correlative feature of the biogeographical pattern at this time is the sharp faunal
discontinuity between the Lhasa and the Himalaya Terranes. Taken at face value, this
indicates that there was a significant and abrupt change in fauna between these two regions,
and would indicate something more than just a latitudinal gradient impoverishment. It seems
most unlikely that the main outcrop yielding this fauna (the Xiala Formation of northern
Lhasa Terrane) could represent an allochthonous nappe brought over the suture zone from the
Qiangtang Terrane. Although ophiolitic outcrops are scattered across a broad tract in this
region and have been interpreted as forming one continuous but now disected thrust plane
(Kidd *et al.* 1988), the main fossil locality lies to the south of the most southerly of these
ophiolitic outcrops and thus presumably well within the Lhasa Terrane.

Presumably then, some significant barrier to dispersal existed between the Lhasa and
Himalaya Terranes. Taking into account the probable equatorial diversity of the Lhasa Block
and the distinctly 'cold water' aspect of the Himalaya Block, this is either a marked difference
in depth or a significant latitudinal difference. There is no reason to suppose that formations
such as the Wargal Limestone in the Himalayan region are deep-water deposits and so
latitudinal/climatic differences seem the most likely explanation.

(iv) *Comparison with previous work and other faunas and floras*

Various workers (Wang 1984; Lin 1984; Fan 1985) have noted and commented upon the
close faunal similarities of the Lhasa Terrane to SE China at this time and have constructed
various models to explain why this should be so. The same is true of other groups, namely
brachiopods (Zhan & Wu 1982) and fusulinid foraminifers (Chu 1983), both of which show
strong similarity with SE China faunas during the early Upper Permian. Wu & Liao (1982)

believed that the coral and fusulinid faunas from the Tibetan Plateau north of the Zangbo Suture represented a single 'palaeozoogeographical province'. However, many of the genera of fusulinids and compound corals also occur in the Trans-Caucasus (Minato & Kato 1965; Kotlyar & Stepanov 1984), indicating that this is simply an equatorial fauna. No rigorous analysis of faunal similarity has been carried out between the Cathaysian tropical assemblages and Middle Eastern assemblages to establish whether there was significant exchange taking place longitudinally between these two regions. However, both Ross (1982a) and Gobbett (1973) recognized only two fusulinid 'realms' worldwide in the Upper Permian, an eastern 'Tethyan' one and a western 'Midcontinent–Andean' one, suggesting that there was possibly no major barrier to dispersal at this time within 'Tethys'.

Terrestrial floras of this period have been widely studied (Chaloner & Meyen 1973; Chaloner & Lacey 1973; Plumstead 1973; Zhang & He 1985) and used extensively in the definition of geographical discontinuities (Sengör 1985a,b). Three broad floras have been recognized, named Angaran, Cathaysian and Gondwanan. Chinese authors (Li et al. 1982; Zhang & He 1985) have tended to stress the differences that exist in late Permian floras between North and South China Blocks. However, there exists great similarity between both areas (Laveine et al. 1987) and the significance of the small differences has not yet been established. Because almost all genera are common to the two blocks and the differences are manifest only in the late Permian species, this could be interpreted as the product of an early to mid-Permian vicariance event that separated the two regions after the flora had become established over the whole region.

Two important points emerge from the floral distribution. Firstly all of Cathaysia south of the Tienshan–Hegen Suture and north of the Banggong Suture formed one well defined equatorial region at least until the mid-Permian. Only fragmentary plant material has so far been discovered from the Lhasa Terrane (which has only marine deposits in the Upper Permian) and these are inconclusive as to their biogeographical relationships (Li et al. 1985; Smith & Xu 1988). In the Himalaya region *Glossopteris* flora is found (Singh et al. 1982), whereas to the north of the Tienshan–Hegen Fault Angaran flora is found. Secondly, the barrier between the Cathaysian and Angaran flora is now known to have broken down towards the end of the Permian, with the appearance of mixed affinity floras along the northern part of the North China Block (see Zhang & He 1985). No such mixed Gondwanan/Cathaysian flora is known from the south in this region although it has been reported further west in Saudi Arabia (Hill & El-Khayal 1983).

## (v) Conclusions

The most significant feature of faunal and floral distributions in the early Upper Permian is the marked disjunction between broadly tropical biotas in the Lhasa, Qiangtang, Kunlun, North China and South China Blocks as compared to the higher latitude biotas of the Himalaya Terrane to the south and the Nei Mongolia, Vladivostok, NE Siberia region to the north. Although highest diversities of corals are found in the Lhasa and SE China regions this may reflect their position on the continental shelf, because deposits further north are dominated more by terrigenous and volcanoclastic sequences. The marked difference between the Lhasa and Himalaya faunas at this time argues for a sizeable latitudinal difference between these two terranes. NE China Terrane probably lay on the fringes of this equatorial belt as it has reduced diversities of fusulinid and complex rugose coral genera and increased diversities of small solitary rugose genera. Faunal impoverishment towards the north is more gradual than towards the south and no suture can be identified as marking a sharp drop in diversity. The significance of the minor differences in the flora between North and South China Blocks is not yet clear, but suggests at most that the previously unified Cathaysian flora became differentiated into a north and south flora after the early Permian.

## 5. LATE PALAEOZOIC PALAEOBIOGEOGRAPHY OF ASIA: A SYNTHESIS

By taking a synoptic approach to the evidence presented in the preceding section it is possible to highlight times at which significant changes in faunal distributional patterns occurred. Too often, far-sweeping conclusions about plate tectonic models have been drawn from geographically restricted regions, without consideration of changing global conditions. The end Carboniferous Glaciation must presumably have had a marked effect on patterns of distribution and it is important to try to distinguish between changes in faunal distribution produced by migration of climatic belts and those produced through ocean floor spreading and plate realignment. This cannot be done in any rigorous way as yet, but must rely on narrative interpretation, incorporating data from other sources such as facies analysis.

Figure 20 provides summary Venn diagrams of area relations through the late Palaeozoic based on the analyses presented above. The following points emerge from the distributional patterns of rugose corals and other fauna during the period from the Visean to the end of the Permian.

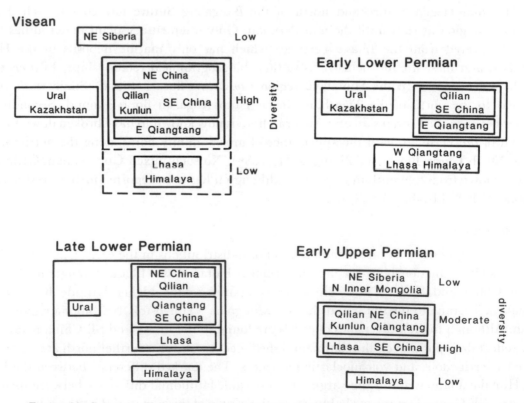

FIGURE 20. Summary Venn diagrams to show the changing pattern of relationship for the biotas of regions within central East Asia, deduced from numerical analyses.

(1) A barrier to marine faunal exchange existed between Central Asia and Cathaysia during the Carboniferous and early Lower Permian. The Tienshan fold belt system marks the approximate boundary between the two faunas throughout this period.

(2) During the late Palaeozoic much of Cathaysia formed a cohesive biotic region open to relatively free faunal exchange. This is true for marine groups such as rugose corals, fusulinids and brachiopods and also for terrestrial flora.

(3) This Cathaysian biotic region lay, broadly speaking, equatorially and extended northwards into NE China and southwards into the Lhasa Terrane. The southern boundary to this region is not sharp but gradual in the Visean. Nor does it coincide with a single suture zone through time. The fact that the Lhasa Terrane can at one time present a 'cold water'-type fauna closely similar to that of the Himalaya Terrane yet shortly after display a fauna with closest similarity to the Qiangtang Terrane and Cathaysia suggests that the primary control on the distribution of the Cathaysian biota is related largely to global climatic conditions. The Lhasa biota shows closest similarity to the Himalaya biota when glaciation is most pronounced in Gondwanaland and Cathaysian (i.e. more tropical) elements appear in this region as global climate warmed. This interpretation is supported by the observation that the Qiangtang Terrane straddled the apparent boundary between 'Cathaysian' and 'Himalayan' biotas during the early Permian. Thus no suture within this region can be taken as the site of a major oceanic barrier in the late Palaeozoic and the whole region represents the northern equatorial extensions of Gondwanaland.

(4) A sharp faunal discontinuity develops between the Lhasa and Himalaya Terranes in the mid-Permian. Before this time the Lhasa Terrane had a biota intermediate between that of the Himalaya and Qiangtang Terranes.

(5) NE China appears to have occupied a position peripheral to the high diversity tropical belt, as its marine fauna shows a pattern of impoverishment that can be correlated with global climatic changes.

(6) There is no palaeontological evidence to suggest that there was a significant difference between the North and South China Blocks until possibly the early Upper Permian, when both coral and floral biotas may start to become differentiated.

(7) SE China maintains throughout this period a high diversity, broadly tropical fauna. However, through the late Palaeozoic it is associated with progressively more southern Terranes to the west. In the Visean and early Lower Permian it has closest ties to the Qilian–Kunlun region. In the late Lower Permian it shows closest similarities to the Qiangtang Terrane, whereas in the early Upper Permian it has closest ties with the Lhasa Terrane. This might be taken to imply a possible northward shift through time of the Kunlun–Qiangtang–Lhasa region vis-à-vis SE China. Conversely it may reflect a progressive terrestrialization to the north caused by uplift in this region.

(8) To the north of the Tienshan–Hegen Suture, marine faunas and floras are distinct from those of the Cathaysian region. However, mixed floras appear along this suture before the end of the Permian, suggesting that the barrier that had existed to biotic exchange was breaking down.

## 6. MAJOR ASIATIC SUTURE ZONES AND THEIR DATING

Palaeontological evidence that is relevant to the dating of suture zones is twofold. Firstly, palaeontological data can establish when a suture zone was open by dating the associated oceanic sediments incorporated into ophiolitic suites along the suture or continental rise sediments and arc-related sediments that are associated. Secondly, palaeontology can provide evidence bracketing the time of closure and plate accretion by dating the latest marine sediments within the suture zone and the molasse that is derived from continental collision. Here I will summarize the relevant evidence for the major suture zones within central East Asia.

### (a) Tienshan–Hegen Suture

Watson et al. (1987) have provided evidence that the western part of this suture was open, but already closing by the early Carboniferous. In support they mention thick arc-related volcanics of Lower Carboniferous age in this region. These authors suggest that closure along this suture was diachronous and occurred progressively from west to east. They deduce a late Carboniferous closure in the Junggar District and a late Permian closure in North East China. The late Permian closure to the east appears to be well substantiated, since it is at this time that the first mixed Angaran–Cathaysian floras are recorded in this region. Angaran floras, however, are found in the Junggar District and are clearly distinguished from the Cathaysian floras of Tarim at least during much of the Permian (Zhang et al. 1985). Collision may therefore have been entirely Permian along this suture zone as suggested by Bally et al. (1980).

### (b) Kunlun–Qinling Suture

The age of closure of this suture is highly contentious. Mattauer et al. (1985), on the basis of extensive field mapping, argued that the Qinling Suture represents a Devonian closure and this was supported by Laveine et al. (1987) on floral grounds. Sengör (1985b), however, has disputed this and favours a Triassic closure, as does Klimetz (1983) on the basis of basin analysis, whereas Zhang et al. (1985) favour a Permian age. Watson et al. (1987) also considered North and South China Blocks to be separated during the Carboniferous and suggested that the Qinling Suture, like the Tienshan–Hegen Suture, represented a diachronous closure; Permian to the east and Triassic to the west.

In all probability both parties are correct. The Qinling fold belt system is an exceedingly complex region and Bally et al. (1980) point out that there are two almost parallel fracture zones superimposed. They suggest that 'Indosinian (Triassic) subduction processes are superimposed on or intersect the preceding Variscan subduction zones described from the same area' (Bally et al. 1980). In the Tibetan Plateau these two zones are apparently more discrete. The Jinsha Suture with its thick Triassic Flysch lies to the south of the Kunlun Shan fold belt. Within the Kunlun Shan, Leeder et al. (1988) have recognized a huge thickness of immature arkosic sandstone of Tournasian age derived from the north. This is best interpreted as molasse arising from the collision of the Kunlun Terrane and North China Block and thus lends support to the Kunlun–Qinling Suture being a Devonian closure. It is probable that the Qinling fold belt also contains a post–early Permian rift suture to account for the floral vicariance event postulated above and the late Permian palaeomagnetic gap of 6° latitude between North and South China Blocks (McElhinney et al. 1981).

### (c) Jinsha Suture

The extensive sequence of contourites and other continental rise sediments that make up the large accretionary prism to the north of the Jinsha Suture (the Bayan Kala Group) appears from fossil evidence to be entirely Triassic in age (Yin *et al.* 1988; Smith & Xu 1988). Andesitic volcanics of the Batang Group, also associated with this suture and interpreted as forming a fore-arc (Pearce & Mei 1988), were active in the early Late Triassic but had ceased by the Norian when a broad carbonate platform covered the area (Leeder *et al.* 1988). Watson *et al.* (1987) derive this sequence of clastics from the Qinling collisional foldbelt, but as argued above this may not be correct and the principal source may have been from the Angara–Cathaysia collisional fold belt. Accretion therefore appears to have taken place during the late Triassic and by the middle Jurassic great thicknesses of molasse derived from the north were being deposited across the Qiangtang Terrane (Leeder *et al.* 1988).

Permian (probably early Permian) rift-type volcanics are known from either side of the suture in both the Qiangtang and Kunlun Terranes (Pearce & Mei 1988), suggesting that this may be when the suture opened.

### (d) Banggong Suture

Flysch sequences associated with this suture appear to be entirely Jurassic in age (Leeder *et al.* 1988; Smith & Xu 1988). However, break up of the broad carbonate platforms that existed in the middle Triassic of the Lhasa Terrane and the development of late Triassic intrabasinal carbonate turbidites (see Leeder *et al.* 1988) may be related to the initiation of this suture. The only direct evidence for the age of the ocean comes from the dating of radiolaria from deep-sea cherts. Li (1986) identified these as Tithonian in age. However, the suture apparently had closed by the end of the Jurassic at the latest (see Girardeau *et al.* 1984).

### (e) Zangbo Suture

This suture incorporates deep-sea cherts that have been dated as Upper Triassic (Wang & Sheng 1982), late Jurassic, early Cretaceous and late Cretaceous as young as Turonian (Wu 1984, 1986; Wu & Li 1982). Deposits with open oceanic planktonic foraminifers are found up to at least the Campanian–Maastrichtian. The latest marine sediments in this zone are Lower Eocene (pre-Lutetian) according to Blondeau *et al.* (1983) or middle Eocene (Lutetian) according to Pan *et al.* (1984).

Sengör (1984) has argued that the Panjal Traps, an extensive sequence of volcanics in Kashmir and adjacent regions, represent rift-related volcanics associated with the opening of this suture. These volcanics are Lower Permian (Artinskian) in age. This would be in close accord with palaeontological data which demonstrates that the biota of the Lhasa and Himalaya Terranes first diverged markedly in the early Upper Permian.

Figure 21 summarizes the dates of closure and the development of molasse for each of these sutures. It appears that of the five suture zones one was closed in the Devonian (Kunlun–Qinling), one was closed or closing in the late Carboniferous or Permian (Tienshan–Hegen), two opened in the late Lower Permian (Jinsha and Zangbo) and one is a wholly Mesozoic back-arc basin (Banggong).

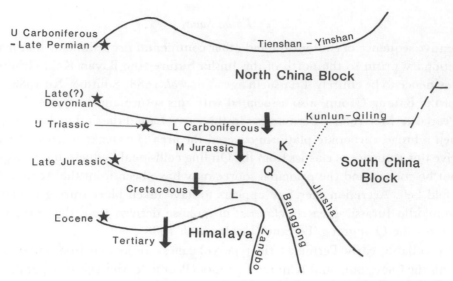

FIGURE 21. Sketch of central East Asia to show the timing of suture closure (asterisked) and the first appearance of molassic sequences (arrows), based on palaeontological evidence. Abbreviations: K, Kunlun Terrane; Q, Qiangtang Terrane; L, Lhasa Terrane.

## 7. COMPARISON WITH ALTERNATIVE LATE PALAEOZOIC PALAEOGEOGRAPHIES

As stated in the Introduction, there are currently three broad alternative models for the palaeogeography of Asia during the late Palaeozoic. These are as follows.

(1) Pangea without a large Tethyan ocean separating Eurasia from Gondwana (Carey 1976; Crawford 1982).

(2) Pangea with a large wedge-shaped Tethyan ocean separating Gondwana from Eurasia but with a scattering of isolated plates, which currently form China and South East Asia, distributed across the ocean (McElhinney et al. 1981; Lin et al. 1985; Lin & Watts, this symposium).

(3) Pangea with a large wedge-shaped Tethyan ocean separating Eurasia and Gondwana. This ocean either coincided with the Zangbo Suture (Helmcke 1985) or with one of the sutures to the north (Wang 1984; Sengör 1984). Thus the 'Cimmerian continent', before break up, was either part of Eurasia, part of Gondwana or divided between the two.

Each model carries a number of implications that can be compared with the available palaeontological evidence and tested for consistency.

In looking for the site of a major late Palaeozoic ocean we would expect to be able to recognize a sharp faunal discontinuity coincident with one of the suture zones. Helmcke (1985) suggested that the early Lower Permian mixtites of the Himalaya and Lhasa Terranes were continental-rise sediments, following Mitchell et al. (1970), and placed 'Palaeotethys' at the Zangbo Suture. This is at variance with palaeontological evidence, which shows that during the early Lower Permian faunas on both sides of this suture were extremely similar (see Jin & Sun 1981; Dickens & Shah 1981). Leeder et al. (1988) also interpret the mixtites as shelf deposits rather than continental-rise deposits on sedimentological and faunal evidence. As discussed above, this suture appears to have been initiated in the early Permian, thus the Zangbo Suture seems improbable as the site of 'Palaeotethys'.

Nor does the Banggong Suture appear likely as the site of a major late Palaeozoic ocean (Allègre et al. 1984; Chang et al. 1986) and no significant faunal discontinuity can be recognized associated with this suture. Indeed, palaeontological evidence suggests that the whole of the Cathaysian region from the Tienshan–Hegen Suture south formed one cohesive biogeographical region during the Carboniferous, lying at low latitudes and connected to the northern part of 'Gondwanaland'. A latitudinal gradient and climatic belt zonation appear to have been the controlling factors influencing the distribution of biotas between Qilian and the Himalayas. The Cathaysian coral fauna was differentiated from comparable diversity coral faunas of Central Asia. Although scattered island terranes need not have posed any barriers to faunal dispersal (as is demonstrated by distributions in today's Indo-Pacific), the absence of any evidence for continental slope or open oceanic sediments of this age associated with any of the proposed island terranes makes this reconstruction less appealing. Clearly negative evidence is unsatisfactory but so is a model that predicts extensive continental break sediments when none are known.

The palaeomagnetic data on which this island-type reconstruction is based indicate a discrepancy (gap) of about 6° latitude between the North and South China Blocks at the end Permian (McElhinney et al. 1981; Zhang & He 1985). But this could easily postdate the early to mid-Permian break up of the region postulated by the other two models. The early Carboniferous palaeoposition of Lin & Watts (1988) for the Kunlun Terrane at ca. 25–25 °S agrees with palaeontological evidence suggesting that the Kunlun–Qilian–SE China region occupied an equatorial position at this time.

Palaeontological evidence then favours an early Permian reconstruction in which the Cimmerian continent formed an integral part of Gondwanaland, as postulated by both Crawford (1974, 1982) and Sengör (1984), and which became rifted along the Zangbo Suture during the early Permian. The distinction between these two models lies in how the northern border of the Cimmerian continent is interpreted.

Rifting along the Jinsha Suture appears to date from the early Permian, to judge from the faunal similarity between the Kunlun and Qiangtang Terranes before that time and the presence of early Permian rift volcanics on either side of the suture. Thus the Jinsha Suture opened at the same time as the Zangbo Suture and both presumably were initiated by the same factors related to the late Palaeozoic break up of the northern platform of Gondwanaland.

So 'Palaeotethys', if it is to be found, must lie to the north of the Tibetan Plateau region. Sengör (1984, 1985b) placed the northern margin of his Cimmerian continent along the northern borders of the Kunlun Terrane and North China Block. On palaeontological grounds, however, the Kunlun, Tarim and Qilian Terranes have very similar biotas in the Carboniferous and early Permian. Furthermore, there appears to be little differences between the biotas of the North and South China Blocks before the Upper Permian. If Mattauer et al. (1985) are correct in interpreting the Kunlun–Qinling Suture as a Devonian closure, as is suggested from faunal similarities across this suture, then the Cimmerian continent must have extended in the early Permian up to the Tienshan–Yinshan Suture. Having the Tienshan–Yinshan Suture as the site of 'Palaeotethys' would then explain why this line defines a biogeographical barrier in the late Palaeozoic for both shallow-water corals and terrestrial plants.

However, there is a difficulty. Watson et al. (1987) attribute a Late Carboniferous closure to the Tienshan Suture. Thus the northern margin of the Cimmerian continent had apparently become accreted to the Central Asian craton before it had rifted from the southern landmass of Gondwanaland. Possibly this accretion dating is wrong, because Permian floras are still quite distinct on either side of the Tienshan Suture and only start to intergradate in the Late

Permian. If we accept an Artinskian age for rifting along the Zangbo Suture and a Dzhulfian age for collision along the Tienshan–Yinshan Suture this implies that any ocean between the Cimmerian and Angaran Blocks was consumed in just 20 Ma, hardly enough time for the size of ocean generally envisaged. The observation that both Cathaysian and Central Asian biotas lay at comparable (broadly tropical) latitudes during the Carboniferous and early Permian would also tend to limit the size of 'Palaeotethys'.

In conclusion, evidence currently points towards the Cimmerian Block (containing most of what is generally referred to as Cathaysia) forming a northward continuation of the Gondwanaland craton into tropical latitudes during the Carboniferous and early Lower Permian. Subsequent rifting led to the fragmentation of this platform during the late Lower Permian along the lines of the Jinsha and Zangbo Sutures, and possibly also the old Devonian Qinling Suture to separate the North and South China Blocks by the late Permian. This Gondwanan shelf extended in the Carboniferous as far north as the Tienshan Suture and occupied a broadly tropical latitude. A 'Palaeotethys' ocean of any magnitude could only have existed to the north of this suture, but currently it is believed that this suture represents a late Carboniferous or Permian closure. There is a clear faunal discontinuity associated with the Tienshan–Yinshan Suture during the Carboniferous and Permian, but both faunas appear to have been situated at relatively low latitudes. These observations could be accommodated in a reconstruction in which a relatively small 'Palaeotethys' formed a broadly equatorial barrier separating northern and southern tropical/subtropical belts.

## References

Abramov, B. S. 1970 *Biostratigraphy of the Carboniferous deposits of Sette-Dabana (S. Verkhoyansk).* (176 pages.) Moscow: Izd-vo Nauka. (In Russian.)

Academia Sinica, Institute of Geology and Palaeontology, Academia Sinica Institute of Geology & Beijing College of Geology 1962 *Geology of Qilian Shan*, vol. 4, *Fossils of Qilian Shan.* Beijing: Science Press.

Allègre *et al.* 1984 *Nature, Lond.* **307**, 17–22.

Altmark, M. S. 1975 *Trudy Inst. Geol. Geofiz. (Novosib.)* **202**, 37–48.

Avias, J. 1978 In *International Symposium on Geodynamics in South-West Pacific, Noumea (New Caledonia)*, pp. 381–386. Paris: Éditions Technip.

Bally, A. W. *et al.* 1980 *US Geol. Surv. Open File Report* no. 80–501, (100 pages.)

Blondeau, A., Bassoullet, J. P., Colchen, M., Han Tonnlin, Marcoux, J. & Mascle, G. 1983 *Terra Cogn.* **3**, 264.

Bureau of Geology, Autonomous Region of Inner Mongolia & Northeastern Institute of Geology 1976 *Palaeontological Atlas of North China (Inner Mongolia Volume)*, **1**, Palaeozoic. (502 pages.) Beijing: Geological Publishing House.

Carey, S. W. 1976 *The expanding Earth.* (488 pages.) Amsterdam: Elsevier.

Chaloner, W. G. & Lacey, W. S. 1973 In *Organisms and continents through time* (ed. N. F. Hughes), *Spec. Papers Palaeont.* **12**, 271–290.

Chaloner, W. G. & Meyen, S. V. 1973 In *Atlas of palaeobiogeography* (ed. A. Hallam), pp. 169–186. Amsterdam: Elsevier.

Chang Chengfa *et al.* 1986 *Nature, Lond.* **323**, 501–507.

Chen Chuzhen 1984 In *Stratigraphy of Xizang (Tibet)*, pp. 311–317. (In Chinese.) Beijing: Scientific Press.

Chu Shuifang 1983 *Contr. Geol. Qinghai-Xizang (Tibet) Plateau* **7**, 136–148. (In Chinese.)

Crawford, A. R. 1974 *Geol. Mag.* **111**, 369–380.

Crawford, A. R. 1979 *J. Petr. Geol.* **2**, 3–9.

Crawford, A. R. 1982 *J. Petr. Geol.* **5**, 149–160.

Dickins, J. M. 1977 *Chayanica Geologica* **3**, 11–21.

Dickins, J. M. 1984 In *Fossils and climate* (ed. P. Brenchley), pp. 317–327. London: J. Wiley & Sons.

Dickins, J. M. 1985 *B.M.R. J. Austral. Geol. Geophys.* **9**, 163–169.

Dickins, J. M. & Shah, C. C. 1981 In *Gondwana five: the fifth international Gondwana symposium, Wellington, New Zealand, 1980* (ed. M. M. Cresswell & P. Vella), pp. 79–83. Rotterdam: A. A. Balkema.

Dong Deyuan & Mu Xinan 1984 In *Stratigraphy of Xizang (Tibet) Plateau*, pp. 237–287. (In Chinese.) Beijing: Scientific Press.

Fan Yingnian 1985 *Contr. Geol. Qinghai-Xizang (Tibet) Plateau* **16**, 87–106. (In Chinese.)

Fedorowski, J. 1977 *Mem. Bur. Rech. geol. min.* **89**, 234–248.

Fedorowski, J. 1986 *Acta palaeont. pol.* **31**, 253–275.

Flugel, H. 1966 *Jb. geol. Bundesanst., Wien* **12**, 101–120.

Girardeau, J., Marcoux, J. Allègre, C. J., Bassoullet, J. P., Tang Youking, Xiao Xuchang, Zao Yougong & Wang Xibin 1984 *Nature, Lond.* **307**, 27–31.

Gobbett, D. J. 1973 In *Atlas of palaeobiogeography* (ed. A. Hallam), pp. 151–158. Amsterdam: Elsevier.

Guo Shengzhe 1980 In *Palaeontological atlas of North East China, Palaeozoic volume* (ed. Shenyang Institute of Geology and Mineral Resources), pp. 106–152. Beijing: Geological Publishing House.

Helmcke, D. 1985 *Geol. Rdsch.* **74**, 215–228.

Heritsch, F. 1937 *Akad. Wiss. Wien Math-naturwiss.* **146**, 1–16.

Hill, C. A. & El-Khayal, A. A. 1983 *Bull. Br. Mus. nat. Hist. A* **37**, 105–112.

Hill, D. 1973 In *Atlas of palaeobiogeography* (ed. A. Hallam), pp. 133–142. Amsterdam: Elsevier.

Hill, D. 1981 In *Treatise on Invertebrate Paleontology, part F: Coelenterata. Supplement 1, Rugosa and Tabulata* (2 vols) (ed. C. Teichert). (762 pages.) Lawrence, Kansas: Geological Society of America and University of Kansas Press.

Hsu, K. J. & Bernoulli, D. 1978 In *Initial reports of the Deep Sea Drilling project* no. 42/1, pp. 943–949.

Ji, X. & Coney, P. J. 1985 In *Tectonostratigraphic terranes of the circum-Pacific region* (ed. D. G. Howell), pp. 349–362. Houston, Texas: Circum-Pacific Council for Energy and Mineral Resources.

Jiang Shuigen 1982 In *Palaeontological atlas of Hunan Province* (ed. Hunan Provincial Geological Bureau), pp. 81–162. Beijing: Scientific Press. (In Chinese.)

Jin Yugan 1985 *Palaeont. Cathayana* **2**, 19–72.

Jin Yugan & Sun Dongli 1981 In *Palaeontology of Xizang* vol. 3, pp. 127–168. Beijing: Science Press. (In Chinese.)

Kanmera, K., Isu, K. & Toriyama, R. 1976 *Geol. Palaeont. S.E. Asia* **17**, 129–154.

Kato, M. 1976 *J. Fac. Sci. Hokkaido Univ., Ser. 4*, **17**, 357–364.

Keller, N. B. 1969 *Palaont. Zh.* **4**, 90–99.

Kidd, W. S. F, Pan Yusheng, Chang Chengfa, M. P. Coward, J. F. Dewey, A. Gansser, P. Molnar, R. M. Shackleton & Sun Yiyin 1988 *Phil. Trans. R. Soc. Lond. A.* (In the press.)

Klimetz, M. P. 1983 *Tectonics* **2**, 129–166.

Koch, C. 1987 *Paleobiology* **13**, 100–107.

Kotlyar, G. V. 1978 In *The Upper Palaeozoic of northeastern Asia* (ed. L. I. Popeko), pp. 5–23. Publ. Izd. Akad. Nauk. Dal'nevost. nauchn tsentr. (In Russian.)

Kotlyar, G. V. & Stepanov, D. L. (eds) 1984 *Trans A. P. Karpinsky all Union Order of Lenin Geol. Res. Inst.* no. 286. (280 pages.) (In Russian.)

Kummel, B. & Teichert, C. (eds) 1970 *Stratigraphic boundary problems: Permian and Triassic of West Pakistan.* Lawrence, Kansas: University of Kansas Press. (474 pages.)

Laveine, J. P., Lemoigne, Y., Li Xingxue, Wu Xiuyuan, Zhang Shanzhen, Zhao Xiuhu, Zhu Weiqing & Zhu Jianan 1987 *C.r. Acad. Sci., Paris* (II) **304**, 391–394.

Leeder, M., Smith, A. B. & Yin Jixiang 1988 *Phil. Trans. R. Soc. Lond. A.* (In the press.)

Li Hongsheng 1986 *Acta micropal. sin.* **3**, 297–315.

Li Xingxue, Wu Yiming & Fu Zaibin 1985 *Acta palaeont. sin.* **24**, 140–170.

Li Xingxue, Yao Zhaoqi & Deng Longhuo 1982 In *Palaeontology of Xizang* vol. **5**, pp. 17–44. Beijing: Scientific Press. (In Chinese.)

Li Zhangcha & Liao Weihua 1979 In *Palaeontological atlas of Northwest China (Qinghai Volume)* (ed. Nanjing Institute of Palaeontology, Academia Sinica and Qinghai Institute of Geosciences). Beijing: Geological Publishing House.

Liang Dingyi, Nei Zetong, Guo Tieying, Zhang Yizhi, Xu Baiwen & Wang Weipin 1983 *Earth Sci. J. Wuhan Coll. Geol.* **19**, 9–28. (In Chinese.)

Lin Baoyu 1984 In *Mission Franco-Chinoise au Tibet* (ed. J. L. Mercier & Li Guangcen), pp. 77–107. Paris: Éditions du Centre National de la Recherche Scientifique.

Liu Baotan 1984a *Contr. Geol. Qinghai-Xizang (Tibet) Plateau* **14**, 125–136.

Liu Baotan 1984b *Contr. Geol. Qinghai-Xizang (Tibet) Plateau* **15**, 101–112. (In Chinese.)

Luo Jinding 1984 *Palaeontogr. am.* **54**, 427–432.

Lin Jinlu, Fuller, M. & Zhang Wenyou 1985 *Nature, Lond.* **313**, 444–449.

Lin Jinlu & Watts, D. 1988 *Phil Trans. R. Soc. Lond. A.* (In the press.)

Mattauer, M., Mathe, P., Malavieille, J., Tapponier, P., Maluski, H., Xu Zhinqin, Lu Yilun & Tang Yaoqin 1985 *Nature, Lond.* **317**, 496–500.

McElhinney, M. W., Embleton, J. J., Ma, Xh. & Zhang Zk. 1981 *Nature, Lond.* **293**, 212–216.

Malkovskiy, F. S. 1975 *Trudy Inst. Geol. Geofiz. (Novosib.)* **202**, 195–197.

Minato, M. & Kato, M. 1965 *J. Fac. Sci. Hokkaido Univ. Ser. 4*, **12**, 1–241.

Minato, M. & Kato, M. 1977 *Mem. Bur. Rech. geol. min.* **89**, 228–233.

Mitchell, A. H. G., Yong, B. & Jantaranipa, W. 1970 *Geol. Mag.* **107**, 411–428.

Nakazawa, K., Kapoor, H. M., Ishii, K. I., Bando, Y., Okimura, Y. & Tokuoka, T. 1975 *Mem. Fac. Sci. Kyoto Univ., Series of Geol. Min.* **42**, 1–106.

Nanjing Institute of Geology and Mineral Resources (ed.) 1982 *Palaeontological Atlas of East China: Late Palaeozoic.* Beijing: Geological Publishing House. (In Chinese.)

Owen, H. G. 1983 *Atlas of continental displacement: 200 million years to the present.* (159 pages.) Cambridge University Press.

Pan Guitang, Jiao Shupei, Wang Peisheng, Xu Yaorong & Xiang Tianxiu 1984 *Contr. Geol. Qinghai-Xizang (Tibet) Plateau* **17**, 183–196. (In Chinese.)

Papojan, A. S. 1977 *Mem. Bur. Rech. geol. min.* **89**, 197–202.

Pearce, J. A. & Mei Houjun 1988 *Phil. Trans. R. Soc. Lond.* A. (In the press.)

Plumstead, E. P. 1973 In *Atlas of Palaeobiogeography* (ed. A. Hallam), pp. 187–205. Amsterdam: Elsevier.

Rosen, B. R. 1984 In *Fossils and climate* (ed. P. J. Brenchley) *Geol. J. Spec. Issue* no. **11**, pp. 201–262. Chichester: Wiley.

Rosen, B. R. 1988 In *Biogeographical analysis* (ed. A. Myers & P. Giller), ch. 14. London: Chapman & Hall.

Rosen, B. R. & Smith, A. B. 1987 In *Gondwana and Tethys* (ed. M. G. Audley-Charles & A. Hallam), pp. 275–306. Oxford University Press.

Ross C. A. 1982*a* In *Foraminifera – Notes for a short course* (ed. T. W. Broadhead). Univ. Tennessee Dept. Geol. Sci. Studies in Geology no. 6, pp. 163–176.

Ross, C. A. 1982*b* *Proc. Third N. American Paleont. Convention* (ed. B. Mamet & M. Copeland), pp. 441–445. Montreal: Université de Montreal and Geological Survey of Canada.

Rowett, C. L. 1975 *Trudy Inst. Geol. Geofiz. (Novosibirsk)* **202**, 205–211.

Rowett, C. L. 1977 *Mem. Bur. Rech. geol. min.* **89**, 190–196.

Sengör, A. M. C. 1984 *Spec. Pap. geol. Soc. Am.* **195**, 1–82.

Sengör, A. M. C. 1985*a* *Geol. Rdsch.* **72**, 181–213.

Sengör, A. M. C. 1985*b* *Nature, Lond.* **318**, 16–17.

Shchukina, V. Y. 1975 *Trudy Inst. Geol. Geofiz. (Novosib.)* **202**, 180–185.

She Xide 1983 *Contr. Geol. Qinghai-Xizang (Tibet) Plateau* **2**, 187–206. (In Chinese.)

Singh, G., Maithy, K. & Bose, M. N. 1982 *Palaeobotanist* **30**, 185–232.

Smith, A. B. & Xu Juntao 1988 *Phil. Trans. R. Soc. Lond.* A. (In the press.)

Smith, A. G. 1981 *Geol. Rdsch.* **70**, 91–127.

Smith, A. G., Briden, J. C. & Dewey, G. E. 1973 In *Organisms and continents through time* (ed. N. F. Hughes), Spec. Pap. Palaeont. no. 12, pp. 1–42.

Sokolov, B. S. 1960 *neft. nauchno-issled. geol.-razv. Inst. Trudy vses.* **2**, 38–77.

Stehli, F. G. & Wells, J. W. 1971 *Syst. Zool.* **20**, 115–126.

Stevens, C. H. 1975 *Trudy Inst. Geol. Geofiz. (Novosib.)* **202**, 197–204.

Stevens, C. H. 1982 *Bull. geol. Soc. Am.* **93**, 798–803.

Stocklin, J. 1984 *Reports 27th Intern. Geol. Congr.* vol. 5, pp. 65–84.

Sultanaev, A. A. *et al.* 1978 *Key sections and faunas of the Visean and Namurian stages of the Central and Southern Urals,* 131 pp. Leningrad: Nedra Leningradskoe Otdelerie.

Swofford, D. L. 1985 *PAUP: Phylogenetic analysis using parsimony.* Computor program and instruction manual distributed by the author, Illinois Natural History Survey, Champaign, Illinois.

Toriyama, R. 1984 *Geol. Palaeont. S.E. Asia* **25**, 137–147.

Vine, F. J. 1973 In *Organisms and continents through time* (ed. N. F. Hughes), Spec. Pap. Palaeont. no. 12, pp. 61–77.

Volkova, M. S. 1941 *Mater. Geol. polezn. Iskop. Kazakhstan* **11**, 1–120. (In Russian.)

Wang Naiwen 1984 In *Mission Franco-Chinoise au Tibet, 1980* (ed. J. L. Mercier & Li Guangcen), pp. 33–54. Paris: Éditions du Centre national de la Recherche Scientifique.

Wang Yizhao 1983 *Contr. Geol. Qinghai-Xizang (Tibet) Plateau* **11**, 71–77. (In Chinese.)

Wang Yujiang & Sheng Jingzhang 1982 In *Palaeontology of Xizang* no. 4, pp. 81–96. Beijing: Scientific Press. (In Chinese.)

Wang Zengji 1983 *Contr. Geol. Qinghai-Xizang (Tibet) Plateau* **2**, 207–225. (In Chinese.)

Wang Zhengji & Liu Shikun 1982 *Contr. Geol. Qinghai-Xizang (Tibet) Plateau* **7**, 59–85. (In Chinese.)

Waterhouse, J. B. & Bonham-Carter, G. F. 1975 *Can. J. Earth Sci.* **12**, 1085–1146.

Waterhouse, J. B. & Gupta, V. J. 1979 *J. geol. Soc. India* **20**, 461–464.

Watson, M. P., Hayward, A. B., Parkinson, D. N. & Zhang Zm 1987 *Mar. Petr. Geol.* **4**, 205–225.

Wu Haoru 1984 *Scientia Geol. Sinica* **1**, 26–33. (In Chinese.)

Wu Haoru 1986 *Acta micropal. sin.* **3**, 347–360.

Wu Haoru & Li Hongsheng 1982 *Acta palaeont. sin.* **21**, 64–71. (In Chinese.)

Wu Wangshi, Liao Weihua & Zhao Jiaming 1982 In *Palaeontology of Xizang* no. 4, pp. 107–145. Beijing: Science Press. (In Chinese.)

Wu Wangshi & Zhou Kangjie 1982 *Bull. Nanjing Inst. Geol. Palaeont. Acad. Sinica* **4**, 213–239. (In Chinese.)

Yang Shiping & Fan Yingnian 1983 *Contr. Geol. Qinghai-Xizang (Tibet) Plateau* **11**, 265–289. (In Chinese.)

Yin Jixiang, Xu Juntao, Liu Chengjie & Li Huan 1988 *Phil. Trans. R. Soc. Lond.* A. (In the press.)
Zhan Lipie & Wu Rangyong 1982 *Contr. Geol. Qinghai-Xizang (Tibet) Plateau* **7**, 103–121.
Zhang Zhengui, Chen Jirong & Yu Hongjin 1985 *Contr. Geol. Qinghai-Xizang (Tibet) Plateau* **16**, 117–137.
Zhang Shanzhen & He Yuanliang 1985 *Palaeont. Cathayana* **2**, 77–86.
Zhang Zm, Liou Jg & Coleman, R. G. 1984 *Bull. geol. Soc. Am.* **95**, 295–312.
Zhao Jiaming & Wu Wangshi 1986 *Bull. Nanjing Inst. Geol. Palaeont. Acad. Sinica* **10**, 169–194.

## Discussion

M. Colchen (*Lab. Géologie stratigraphique et structurale, UFR Sciences,* 86022 *Poitiers Cédex, France*). What is the palaeogeographic distribution of the Upper Permian fusilinid *Colaniella parva* biozone, faunal associations known only in the exotic blocks of the ophiolite mélange of Ladakh and Oman and not in the sediments of the North Indian platform. In my opinion, it should be located near to the northern margin of the Indian continent.

A. B. Smith. My analyses extend only up to the early Upper Permian and do not include the latest Djulfian *C. parva* biozone that Dr Colchen refers to. Until a rigorous, quantitative analysis is carried out comparing the similarity of the Ladakh exotics fauna with that from other regions, any comments would be no more than speculation. Absence of a fauna is a dangerous criterion to use in biogeography because it can be the result of sampling failure as well as biogeographical control.

*Phil. Trans. R. Soc. Lond.* A **326**, 229–255 (1988)   [ 229 ]

*Printed in Great Britain*

# The geochemical and tectonic evolution of the central Karakoram, North Pakistan

By A. J. Rex,[1] M. P. Searle,[1] (the late) R. Tirrul,[2] M. B. Crawford,[1] D. J. Prior,[3] D. C. Rex,[3] and A. Barnicoat[4]

[1] *Department of Geology, University of Leicester, Leicester LE1 7RH, U.K.*
[2] *Geological Survey of Canada, 588 Booth Street, Ottawa, Ontario, Canada K1A 0E4*
[3] *Department of Earth Sciences, University of Leeds, Leeds LS2 9JT, U.K.*
[4] *Department of Geology, University College of Wales, Aberystwyth, Dyfed SY23 3DB U.K.*

The Main Karakoram Thrust (MKT) separates the Karakoram Plate from the accreted Kohistan–Ladakh Terranes and Indian Plate to the south. Within the central Karakoram three geologically distinct zones are recognized: from south to north (i) the Karakoram metamorphic complex, (ii) the Karakoram batholith and (iii) the northern Karakoram sedimentary terrane.

Magmatic episodes of Jurassic and mid-Upper Cretaceous age are recognized before India–Asia collision at *ca.* 50–45 Ma. Both reveal subduction-related petrographic and geochemical signatures typical of Andean-type settings. Associated with the Jurassic event was a low-pressure metamorphism (M1). Synchronous with the mid-Upper Cretaceous episode was the passive accretion of the Kohistan–Ladakh terrane to the Karakoram and closure of the Shyok Suture Zone (SSZ). The main collision between the Indian and Asian Plates resulted in crustal thickening beneath the Karakoram and development of Barrovian metamorphism (M2). Early post-collisional plutons dated at 36–34 Ma cross-cut regional syn-metamorphic foliations and constrain a maximum age on peak M2 conditions. Uplift of the Karakoram metamorphic complex in response to continued crustal thickening was synchronous with culmination collapse along the inferred Karakoram Batholith Lineament (KBL). A combination of thermal re-equilibration of thickened continental crust and the proposed addition of an enriched mantle component promoted dehydration, partial melting and generation of the Baltoro Plutonic Unit (BPU). It was subsequently emplaced as a hot, dry magma into an extensional mid-crustal environment. A contact aureole (M3) was imposed on the low-grade sediments along the northern margin, whereas isograds in uplifted metamorphic rocks to the south were thermally domed with *in situ* migmatization.

## Introduction

The inter-relations of crustal melting and metamorphic processes within continent–continent collision zones have been widely addressed (England & Thompson 1984, 1986). Thickening of continental crust comprises two end-member geometries: underthrusting of one crustal sheet beneath another (single thrust model) and homogeneous thickening of the entire lithosphere. Thermal relaxation and regional metamorphism then occur synchronous to thinning of the crust by erosion. More rapid exhumation may also occur through differential uplift of fault-bounded blocks. The pressure–temperature–time ($P$–$T$–$t$) paths followed by rocks in this environment (England & Thompson 1984) suggests that considerable melting of the crust is inevitable, the amount dependent upon water availability. Where there is a delay of some

20–30 Ma between crustal thickening and the onset of extension, large volumes of melt are expected (England & Thompson 1986). The High Himalayan granites, products of thermal re-equilibration after India–Asia collision, were water-saturated melts that migrated short distances through the crust. However, crustal melting under water-saturated conditions is generally considered to be the exception, the rule being that crustal melting occurs under fluid-absent conditions (see Clemens & Vielzeuf 1987).

In this contribution we document the metamorphic and magmatic history of the Karakoram Range, north Pakistan. The well-constrained tectonic framework allows both pre- and post-collisional processes to be studied. The magmatic history is emphasized, especially the tectonic implications that arise through interpretation of geochemical and isotopic data (Pearce et al. 1984; Harris et al. 1986). It is shown that geochemical constraints, when integrated with other evidence, provide important insights into the development of the Karakoram.

## TECTONIC FRAMEWORK

The mountain ranges of the Karakoram extend from the Afghanistan–Pakistan border eastwards, along the northern frontiers of Pakistan and India with the Chinese province of Xinjiang, to western Tibet. Two regions of the central Karakoram have been studied in detail and represent the basis to the tectonic interpretations presented here. The first region is the Hunza Valley and satellite valleys and the second is the Baltoro–Shigar–Hushe area (figure 1).

Early reconnaissance studies of the Hunza region were made by Desio & Martina (1972) with more recent studies of specific geological zones including Le Fort et al. (1983), Prior (1987) and Debon et al. (1987). Early reconnaissance work in the Baltoro glacier region includes Lydekker (1883), Auden (1935, 1938) and Schneider (1957). Geological maps were published on the basis of the Italian expeditions to the Karakoram and Hindu Kush in 1953 and 1954 (Zanettin 1964; Desio & Zanettin 1970). The Hushe valley was described by Desio & Mancini (1974) and the Biafo–Hispar area by Desio et al. (1985). Recent studies in the Baltoro and Hushe regions include Bertrand & Debon (1986), Debon et al. (1986a), Brookfield (1980, 1981), Brookfield & Reynolds (1981), Reynolds et al. (1983) and Searle et al. (1988). A more regional synthesis on the evolution of the Karakoram was presented by Desio (1979). A synthesis of the present-day crustal structure of the Karakoram inferred and interpreted from geophysical studies is presented by Molnar (this symposium).

### Karakoram Plate and its boundaries

The Karakoram Plate extends from Afghanistan, through northern Pakistan and northern Ladakh to western Tibet (Desio 1964, 1980). Its boundaries are defined as the Rushan–Pshart Suture Zone within the southern Pamirs (Shvolman 1978) to the north, and the Shyok Suture Zone (SSZ), also referred to as the Northern Suture (Pudsey et al. 1985; Pudsey 1986; Coward et al. 1982, 1986, 1987), to the south. The western limit of the Karakoram Plate is not precisely known but it extends into the Hindu Kush Range of eastern Afghanistan (Tapponnier et al. 1981). Along the southern boundary of the central Karakoram, the SSZ structures have been reactivated by a major late Tertiary breakback thrust, the Main Karakoram Thrust (Searle et al. 1988). The MKT is largely responsible for the uplift of the mid- and lower-crustal rocks of the southern Karakoram. Within the Karakoram Plate, a major dextral strike-slip fault, the Karakoram Fault, shows approximately 150 km of right-lateral offset (Molnar &

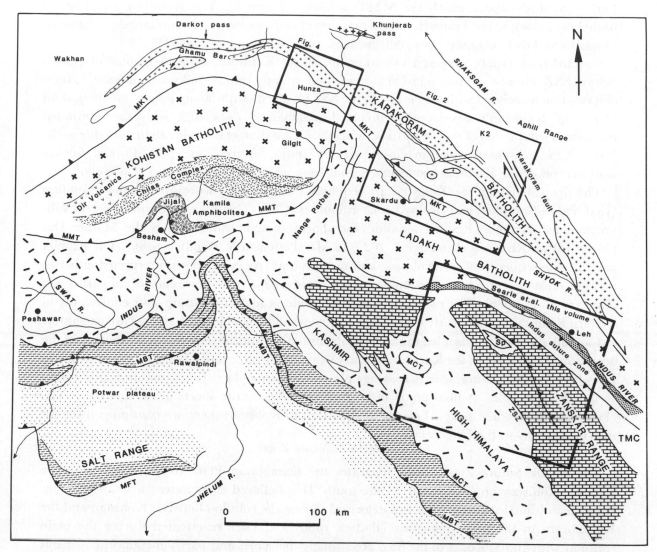

FIGURE 1. Simplified geological map of the western Himalaya–Karakoram showing the central Karakoram region discussed in this paper and the Ladakh–Zanskar Himalaya described by Searle *et al.* (this symposium). Abbreviations: MKT, Main Karakoram Thrust; MMT, Main Mantle Thrust; MCT, Main Central Thrust; MBT, Main Boundary Thrust; MFT, Main Frontal Thrust; Sp, Spontang ophiolite.

Tapponnier 1975). The stratigraphy and structure of the area remains poorly known and this paper is concerned mainly with the south-central part of the Karakoram Plate from the Hunza Valley in the west to the Hushe Valley in the east (figure 1).

## Kohistan and Ladakh Terrane

The Kohistan–Ladakh Terrane, bounded to the north by the SSZ and to the south by the Main Mantle Thrust (MMT), is widely interpreted as a late Jurassic–Cretaceous island-arc. The Kohistan Terrane was first interpreted thus by Tahirkheli *et al.* (1979) and Bard *et al.* (1980). A synthesis of the geological history of Kohistan and its relations to regional collision tectonics is presented by Coward *et al.* (1986, 1987). A review of the Ladakh Terrane is presented by Searle *et al.* (this symposium). The two terranes are separated by the Nanga

Parbat Syntaxis about which the MMT is folded (figure 1). The Kohistan and Ladakh batholiths belong to the Trans-Himalaya belt that extends eastwards to the Gangdese batholith in southern Tibet (Gansser 1964; Allègre *et al.* 1984; Debon *et al.* 1986*b*).

Coward *et al.* (1986) suggest a two-stage history for Kohistan; an earlier phase of growth before SSZ closure in the mid-Cretaceous and a later phase following closure. Major deformation associated with the closure of the SSZ involved south-facing isoclinal folding. Both phases of Kohistan's development involved magmatism associated with the continued northward subduction of the Tethyan oceanic plate. Cessation of oceanic subduction due to the India–Asia collision resulted in closure of the Indus Suture Zone (MMT) and further deformation of the Kohistan sequence (crustal 'pop-up' structure; Coward *et al.* 1986).

The deepest part of the Kohistan arc sequence consists of high-pressure garnet granulites (Jijal–Patan Complex) and two-pyroxene granulites (Chilas Complex), interpreted as the sub-arc magma chamber. Related tonalitic and dioritic magmatism includes the Matum Das pluton with a Rb–Sr isochron age of $102 \pm 12$ Ma (Petterson & Windley 1985). The Kamila amphibolite sequence of metavolcanic and metaplutonic rocks has been interpreted either as oceanic crust precursory to the Kohistan arc (Bard 1983) or as the relics of an earlier arc (Coward *et al.* 1986). The overlying Chalt Volcanics and Yasin Group and equivalents in western Kohistan include early Cretaceous sediments and volcanics with typical arc chemistry.

Calc-alkaline magmatism led to the development of an Andean-type batholith following the accretion of Kohistan to the Karakoram and closure of the SSZ. Early basic dykes immediately post-date the SSZ closure, with main-stage plutonism taking place from uppermost Cretaceous to late Eocene times. A final phase of layered aplite-pegmatite sheets are Oligocene in age (Petterson & Windley 1985). Episodic magmatism in Kohistan therefore continued from the mid-Cretaceous to the Oligocene.

### The Shyok Suture Zone

The Shyok Suture Zone (SSZ) separates the Karakoram Plate to the north from the Ladakh–Kohistan arc/microplate to the south. It is believed to represent a small back-arc basin depositional sequence that was deformed during the collision between Kohistan and the Karakoram in the mid-Cretaceous (Pudsey 1986) and later reactivated during the main collision with India. Rocks of the SSZ are strongly deformed and partly metamorphosed but consist of a mélange 150 m–4 km wide, containing blocks of volcanics, sediments and serpentinite in a slate matrix. Most of the blocks are derived from the Kohistan island-arc sequence to the south (Pudsey *et al.* 1985; Pudsey 1986).

Evidence for mid-Cretaceous closure of the SSZ includes: (*a*) Cretaceous volcanic rocks and Aptian–Albian limestones (Yasin Formation) are the youngest marine rocks in the SSZ (Pudsey 1986); (*b*) post-tectonic intrusions yield radiometric ages of 111–62 Ma, and a diorite dyke that cuts deformed volcanics within the SSZ has an $^{39}$Ar/$^{40}$Ar age of 75 Ma (Petterson & Windley 1985); (*c*) the earlier deformation phase in Kohistan is mid-Cretaceous (Coward *et al.* 1986); and (*d*) along the SSZ in northern Ladakh, Tertiary continental molasse similar to the Indus molasse of the Indus Suture Zone unconformably overlie the Mesozoic oceanic sediments (Thakur & Misra 1984).

## The central Karakoram

Within the central Karakoram three geologically distinct zones are recognized.

### (i) The Karakoram metamorphic complex

The metamorphosed sediments and igneous rocks north of the MKT and SSZ, and south of the Karakoram batholith (the Karakoram metamorphic complex) have been divided into a number of lithological units (Desio 1964, 1979; Bertrand & Debon 1986; Searle et al. 1988). These units include dominantly metasedimentary sequences (Dumordo, Ganschen and Hunza schist units), felsic gneisses (Dassu gneiss) and migmatites and a prominent mélange that includes ultramafic rock (Panmah ultramafic unit).

### (ii) The Karakoram batholith

The Karakoram batholith is now recognized as a composite magmatic belt following the reconnaissance work of Desio et al. (1964), Desio & Zanettin (1970) and Desio & Martina (1972). Recent geochemical, isotopic and geochronological determinations in the central Karakoram have resulted from studies along the Hunza Valley (LeFort et al. 1983; Debon et al. 1987; this paper) and in the Baltoro–Hushe region (Bertrand & Debon 1986; Searle et al. 1987, 1988). Magmatism evolved from pre-collisional, subduction-related to post-collisional types (Debon et al. 1987; Searle et al. 1988), with the prevailing tectonic environment strongly influencing petrogenetic processes and magma characteristics.

Episodic magmatism over a period of at least 80 Ma has been confined to the Karakoram batholith, which extends for over 600 km but reaches maximum widths of only 30 km. Such localization of magmatism implies that this narrow belt constitutes a major crustal lineament, which we refer to here as the Karakoram Batholith Lineament (KBL).

### (iii) The northern Karakoram Terrane

Carboniferous to Jurassic (or possibly Lower Cretaceous) sediments constituting the Broad Peak and Gasherbrum Ranges north of the BPU (figure 2) are described by Desio (1964, 1979), Desio & Zanettin (1970) and Searle et al. (1988). The Gasherbrum sedimentary succession includes Carboniferous black shales that are overlain by a thick Permian and Mesozoic carbonate sequence with interbedded conglomerate, shale and tuffaceous horizons. Quartz diorites of the Broad Peak Porphyry unit intrude the sediments in the Broad Peak and Gasherbrum IV region (Desio & Zanettin 1970; Searle et al. 1988). The K2 area comprises interbanded ortho- and para-gneisses interpreted as mid-crustal rocks uplifted along the hangingwall of a major structural culmination (Searle et al. 1988). A sample of K2 Gneiss yielded a U–Pb date of $115 \pm 3$ Ma (R. Parrish, personal communication) and two other samples have K–Ar (Hbl.) ages of 111 and $94 \pm 3$ Ma (figure 2). Cross-cutting leucogranitic pegmatites have K–Ar ages ranging between 70 to 58 Ma (Searle et al. 1988).

## THE PRE-COLLISIONAL EVOLUTION OF THE KARAKORAM

Before India–Asia collision, Carboniferous to Lower Cretaceous sedimentary and volcanic sequences were intruded by Jurassic and mid-Upper Cretaceous batholiths. Pre-collision magmatism is characterized by calc-alkaline to mildly subalkaline intrusives of dioritic to

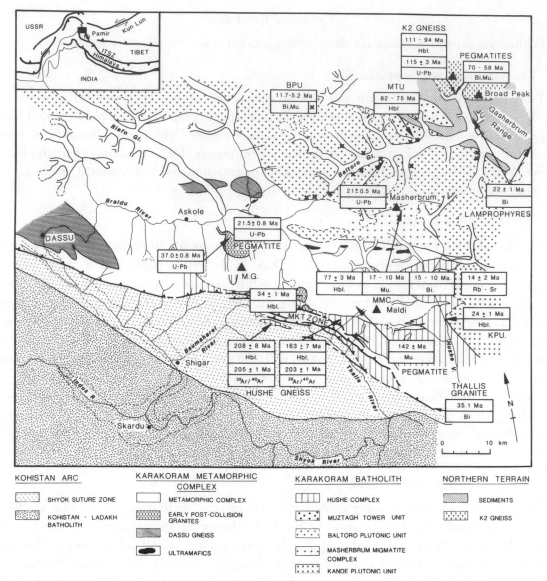

FIGURE 2. Simplified geology of the Baltoro–Hushe–Shigar region of the central Karakoram with radiometric
age determinations for magmatic units. Hornblende (Hbl.), biotite (Bi.) and muscovite (Mu.) dates are all
determined by K–Ar; data are listed fully in Searle *et al.* (1988). The black box in the inset map delineates the
area within a regional context.

granodioritic composition. Mafic inclusions are typically observed and the granitoids are
deformed and often recrystallized. Observations are consistent with reconstructions involving
northward subduction of oceanic plate beneath the southern margin of the Karakoram Plate
and voluminous subduction-related magmatism of Andean type. Magmas were preferentially
emplaced into upper crustal levels along the Karakoram Batholith Lineament.

## The Hushe complex

The oldest component of the Karakoram batholith, the Hushe complex, crops out south of
the BPU along the Hushe Valley (figure 2). It is dominated by foliated orthogneisses

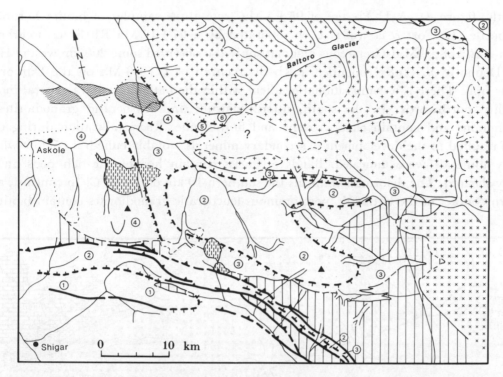

FIGURE 3. Map of the Baltoro–Hushe–Shigar region showing the distribution of metamorphic grades within the Karakoram metamorphic complex and sedimentary rocks of the northern Karakoram Terrane north of the Karakoram batholith. 1, very low grade, chlorite + white mica; 2, low grade, musc + Chl ± Chlt ± Bi ± Gnt; 3, staurolite grade; 4, kyanite grade; 5, sillimanite grade; 6, sillimanite grade with migmatites. Legend as for figure 2.

interlayered with minor amphibolites and felsic gneisses of granodioritic to monzogranitic composition. Jurassic ages of $208 \pm 8$ Ma and $163 \pm 7$ Ma (K–Ar; Hbl) pre-date cross-cutting monzogranitic pegmatite dykes with ages of $142 \pm 6$ Ma (K–Ar; Mu) and are supported by $^{39}Ar/^{40}Ar$ step heating analyses on two amphiboles showing minimum ages of 165 Ma and integrated ages of $205 \pm 1.4$ Ma and $203 \pm 0.6$ Ma. Although typically foliated and lacking any primary mineralogy and textures, the igneous character of the Hushe complex compels us to include it as an early component of the Karakoram batholith. The outcrop distribution of the Hushe complex suggests that it has been tectonically fragmented along south-directed thrusts within the MKT zone. The Hushe complex is spatially and temporally associated with a widespread low-pressure metamorphism (M1) in the area NE of the Thalle Valley (figure 3). Andalusite, staurolite and garnet-bearing pelitic assemblages (figure 6) are thought to represent the oldest metamorphic rocks in the area, associated with the Jurassic magmatic event (Searle *et al.* 1988).

### The Hunza Plutonic Complex

The Hunza Valley segment of the Karakoram batholith is termed the Hunza Plutonic Complex (HPC). It encompasses at least three magmatic episodes and includes intricate, often diffusive, intrusions as well as well-defined leucogranitic bodies. Within the approximately 15 km wide HPC, semi-discrete zones of distinctive magmatism and broad north–south compositional gradients can be recognized.

The oldest component of the HPC, into which all younger magmas intrude, is a calc-alkaline batholith (figure 4). A U–Pb age of $95 \pm 5$ Ma from three samples of biotite–hornblende granodiorite (Le Fort *et al.* 1983) is corroborated by a $97 \pm 17$ Ma Rb–Sr errorchron age (Debon *et al.* 1987). Mid-Cretaceous magmatism is also observed some 200 km west of Hunza where Debon *et al.* (1987) obtained a Rb–Sr isochron of $111 \pm 6$ Ma on the Darkot Pass plutonic unit. There is a compositional gradient within the batholith with a general increase in acidity northwards. In the south, quartz diorites and mesocratic granodiorites are metaluminous and contain clinopyroxene and hornblende. Recrystallization during deformation resulted in the development of secondary minerals, notably euhedral epidote whereas primary minerals were annealed. Bands of gneiss contain hornblende-rich pods and are strongly sheared. The deformation affects the southern 10 km of the HPC. In contrast, at the northern contact of the batholith, peraluminous leucocratic granodiorites contain biotite. In

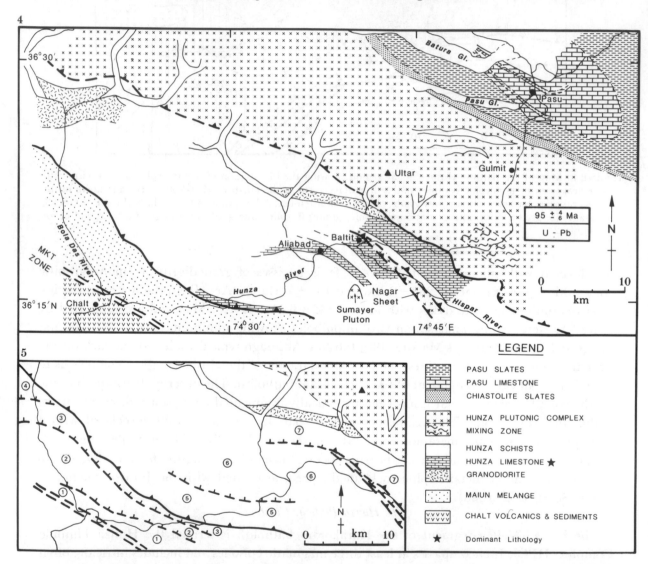

FIGURE 4. Geological map of the Hunza Valley region of the central Karakoram, with reliable radiometric data (from Le Fort *et al.* 1983).

FIGURE 5. Distribution of metamorphic grades in the Hunza Valley region. 1, Chtd + Bi + Chl; 2, Chtd + Gnt + Chl; 3, Gnt + Chl + Bi; 4, Gnt + Chl + Bi; 5, staurolite grade; 6, sillimanite grade; 7, migmatites.

the contact zone between the meta- and peraluminous granitoids they intercalate on scales between centimetres and sheets of 100 m or more. We therefore favour the intrusion of two separate plutons rather than the proposed magmatic differentiation model of Debon *et al.* (1987).

Within the deformed part of the HPC, approximately 1 km north of the southern contact, is a conspicuous mixing zone 2 km in width. Quartz diorites and granodiorites are associated with acidic and basic gneisses, plutons of acidic granodiorites and boudined leucogranitic sweats of pegmatite. Complex magma mixing processes and incipient migmatization were intimately involved in the development of the mixing zone. Aplitic sheets trending N–S emanate from the mixing zone and decrease in abundance away from each side of it. They are composed of biotite monzogranite with small inclusions of granodiorite.

Two generations of leucogranitic dykes can be recognized. The earlier ones are parallel to the foliation in the HPC, striking approximately east–west, and occur throughout the batholith. These are cut, folded and displaced by all other structures. The later ones are larger, more continuous sheets that transect all other structures including the southern contact of the HPC. They only occur in the deformed part of the batholith. Both igneous and metasedimentary inclusions are hosted by the early granitoids of the HPC and are confined to the marginal zones of the batholith.

The deformed southern and central part of the HPC has a distinct metamorphic foliation orientated about 135°/56° NE. It becomes indistinct northwards before changing to a southerly dip of about 100°/35° SSW. The two foliations can be interpreted as two discrete generations. The south-dipping structures to the north are associated with an intrusive contact parallel to the regional foliation of the country rocks (Pasu Slates). This contact is vertical or dips steeply south but not NE (Debon *et al.* 1987). The northward-dipping structures to the south, however, are parallel to the southern contact of the HPC, which shows clear evidence for post-emplacement thrusting. Therefore we suggest that the deformation and recrystallization, including the northward-dipping foliation, resulted from southward thrusting along the southern contact of the HPC, this foliation thus overprinting the primary intrusive foliation.

A second pre-collision magmatism in the Baltoro region is represented by the late Cretaceous Muztagh Tower Unit (MTU) (figure 2). Sheared biotite-bearing and hornblende-bearing tonalitic to granodioritic gneisses have K–Ar (Hbl.) ages of 82–75 ± 3 Ma (Searle *et al.* 1988). On the basis of compositional and age similarities, it is suggested that the MTU metagranitoids are the along-strike equivalents of the mid-Cretaceous intrusives of the HPC. Similarly, the K2 gneiss (figure 2) has a U–Pb zircon age of 115 ± 3 Ma and is interpreted as belonging to the Karakoram batholith (Searle *et al.* 1988; R. Parrish, personal communication).

### The geochemistry of the mid-Cretaceous magmatism

Both the Muztagh Tower Unit (MTU) and the Cretaceous component of the Hunza Plutonic Complex (HPC) display distinct calc-alkaline, subduction-related signatures typical of Andean-type tectonic settings. Representative analyses for these units are presented in table 1, although compositional ranges are very broad emphasizing the complex magmatic assemblages in the HPC. Granodioritic compositions are dominant within both of these units, typically with 60–70 % (by mass) $SiO_2$. On the normalized trace-element variation diagrams (spidergrams; figure 8), the relative enrichments of Rb, Ba, Th, U and K (the large-ion lithophile elements, LILES) and the negative Nb anomaly are characteristic of subduction-related magmas (Brown *et al.* 1984; Thompson *et al.* 1984). Absolute trace-element abundances are also compatible with typical continental subduction-related granitoids. The rare-earth element (REE) patterns (figure 9) indicate typical plagioclase + hornblende-dominated fractionation within the HPC to produce evolved patterns enriched in all trivalent REEs relative to the less evolved patterns, but with the progressive development of the characteristic negative europium (Eu) anomaly.

The pre-collision magmatic component of the Karakoram batholith is isotopically heterogeneous. Recent data indicate initial $^{87}Sr/^{86}Sr$ ratios of 0.705 to 0.713 on the HPC taking the 95 Ma zircon age as a model age. The least radiogenic end member is consistent with an initial $^{87}Sr/^{86}Sr$ ratio of 0.7044 for the Darkot Pass plutonic unit ($111 \pm 6$ Ma; Debon *et al.* 1987). These values are typical of subduction-related magmas in continental environments. The range of initial $^{87}Sr/^{86}Sr$ ratios is not consistent with derivation from a single, homogeneous magma, even assuming that the model age of 95 Ma is applicable to all samples. The variation may result from differences in source composition or contamination. U–Pb ages on the Hushe and K2 gneiss suggest that both are deformed constituents of the Karakoram batholith that have inheritance and incorporation of older crustal material (R. Parrish, personal communication).

*Evolution*

The magmatic evolution of the Karakoram before India–Eurasia collision is partially paralleled in the Kohistan–Ladakh Terranes (figure 10). Mid-Cretaceous subduction-associated magmatism is observed both in the Karakoram batholith and the Kohistan–Ladakh batholiths. The contemporaneous activity is possibly related to a common northward-directed subduction zone. An alternative model (Debon *et al.* 1987) relates synchronous mid-Cretaceous magmatism to subduction of a 'southern Neotethys' south of Kohistan–Ladakh and parallel subduction of a 'northern Neotethys' south of the Karakoram. However, if the Kohistan–Ladakh Terranes have been correctly identified as island-arcs, with suture zones such as the SSZ representing ephemeral back-arc basins, such a model is considered untenable.

Accretion of the Kohistan–Ladakh Terranes with the Karakoram Plate at *ca.* 100–90 Ma (figure 10) is associated with observed folding and cleavage development in Kohistan (Coward *et al.* 1986). The Albian–Aptian Yasin Group limestones along the northern margin of Kohistan represent the latest marine sedimentation before accretion. The passive docking of the back-arc basin resulted in upright folding but no large-scale overthrusting. The SSZ has, however, been reactivated by breakback thrusting and strike-slip faulting after main ISZ closure (Coward *et al.* 1986).

## POST-COLLISIONAL EVOLUTION OF THE KARAKORAM

Timing of main collision between the Indian and the Ladakh–Kohistan–Karakoram Blocks along the ISZ is well constrained by palaeomagnetic (Klootwijk 1979; Patriat & Achache 1984), structural and sedimentological (Searle *et al.* 1987) data at *ca.* 50 Ma. East of the Nanga Parbat Syntaxis in Ladakh, the ISZ contains oceanic sediments and volcanics of Mesozoic–Lower Eocene age after which continental molasse deposition dominated. West of Nanga Parbat the ISZ (Main Mantle Thrust) places rocks of the Kohistan island-arc

TABLE 1. CHEMICAL ANALYSES OF REPRESENTATIVE ROCK TYPES FROM THE MAJOR PLUTONIC COMPLEXES AND UNITS OF THE CENTRAL KARAKORAM BATHOLITH

| | 1 B16 | 2 B9 | 3 B15 | 4 R42 | 5 R33 | 6 R35 | 7 R68 | 8 R11 | 9 SG2 | 10 SLV20 | 11 |
|---|---|---|---|---|---|---|---|---|---|---|---|
| $SiO_2$ | 57.8 | 63.3 | 68.8 | 64.7 | 69.5 | 74.2 | 70.1 | 52.9 | 74.7 | 74.9 | 73.65 |
| $TiO_2$ | 0.7 | 0.8 | 0.7 | 0.7 | 0.4 | 0.1 | BD | 1.6 | 0.1 | BD | 0.1 |
| $Al_2O_3$ | 14.8 | 16.5 | 15.4 | 15.0 | 14.7 | 14.3 | 13.9 | 10.3 | 14.8 | 14.9 | 14.9 |
| $Fe_2O_3$ | 7.2 | 5.4 | 4.4 | 5.4 | 2.1 | 1.0 | 1.1 | 6.9 | 0.9 | 0.5 | 0.8 |
| $MnO$ | 0.1 | 0.1 | 0.1 | 0.1 | BD | BD | 0.4 | 0.1 | BD | 0.1 | BD |
| $MgO$ | 5.1 | 2.7 | 1.3 | 2.4 | 0.7 | 0.1 | BD | 11.1 | BD | BD | 0.1 |
| $CaO$ | 7.4 | 5.5 | 3.7 | 5.0 | 1.7 | 0.8 | BD | 6.0 | 0.8 | 0.9 | 0.5 |
| $Na_2O$ | 3.0 | 3.0 | 2.8 | 2.9 | 4.0 | 3.9 | 4.3 | 1.1 | 4.2 | 4.7 | 4.05 |
| $K_2O$ | 3.0 | 2.0 | 2.7 | 3.2 | 5.0 | 4.6 | 4.1 | 7.9 | 4.8 | 4.6 | 4.6 |
| $P_2O_5$ | 0.2 | 0.2 | 0.2 | 0.1 | 0.2 | 0.1 | 0.1 | 1.7 | 0.1 | BD | 0.1 |
| Ni | 25 | 14 | 5 | 9 | 10 | 3 | 1 | 403 | 1 | BD | < 10 |
| Cr | 179 | 51 | 19 | 60 | 23 | 16 | 12 | 580 | BD | 9 | < 10 |
| V | 157 | 118 | 53 | 107 | 32 | 5 | 1.5 | 135 | BD | BD | 27 |
| Zn | 68 | 44 | 66 | 65 | 54 | 33 | 16 | 153 | 38 | 5 | ND |
| Pb | 24 | 19 | 21 | 26 | 92 | 46 | 22 | 106 | 60 | 82 | ND |
| Cu | 33 | 4 | 19 | ND | 4 | 5 | 4 | ND | 2 | 2 | ND |
| Ga | 18 | 15 | 29 | 21 | 20 | 22 | 27 | 19 | 26 | 16 | ND |
| Rb | 113 | 78 | 101 | 126 | 234 | 272 | 273 | 350 | 383 | 168 | 287 |
| Sr | 509 | 387 | 267 | 347 | 796 | 156 | 9 | 1221 | 41 | 86 | 76 |
| Y | 22 | 19 | 20 | 24 | 16 | 14 | 13 | 34 | 24 | 17 | ND |
| Zr | 97 | 148 | 206 | 182 | 244 | 71 | 30 | 692 | 40 | 17 | 43 |
| Nb | 12 | 10 | 14 | 14 | 22 | 21 | 21 | 46 | 16 | 6 | ND |
| Ba | 537 | 696 | 653 | 658 | 2317 | 442 | 10 | 3679 | 95 | 178 | 213 |
| La | 17 | 47 | 27 | 24 | 66* | 20* | 6* | 114 | 10 | 2 | 6 |
| Ce | 34 | 84 | 56 | 46 | 141* | 45* | 20* | 266 | 18 | 4 | 11 |
| Nd | 18 | 25.5 | 24 | 23 | 57* | 18* | 13.5* | 136 | 9 | 2 | 5 |
| Th | 18 | 34 | 14 | 14 | 62 | 13 | 8 | 8 | 13 | 10 | 6 |
| U | 3 | 5 | 2.5 | 3 | 7 | 9 | 9 | 7 | 17 | 9 | 9 |

Locations: 1, Hunza Plutonic Complex; south contact; 2, Hunza Plutonic Complex; granodiorite, N. contact; 3, Hunza Plutonic Complex; granodiorite; 4, Muztagh Tower Unit; Hbl–Bi gneiss; 5, Baltoro Plutonic Unit; monzogranite; 6, Baltoro Plutonic Unit; leucogranite; 7, Baltoro Plutonic Unit; leucodyke; 8, Lamprophyre; Baltoro glacier; 9, Sumayar Pluton, Hunza; leucogranite; 10, Hunza Plutonic Complex; leucogranite dyke; 11, Average Manaslu granite, Nepal (Le Fort et al. 1987).

ND, not determined; BD, below detection limit; *, INAA data.
All internal data analysed by XRF techniques.

southwards on to the Indian Plate gneisses. After collision at 50 Ma, thrusting and crustal thickening propagated both south of the ISZ in the Zanskar and High Himalayan Ranges (Searle 1986; Searle *et al.*, this symposium) and north of the ISZ in the Karakoram. The regional metamorphism (M2) in the central Karakoram is related to post-collisional crustal thickening, largely by thrust stacking and/or homogeneous thickening (figure 11). An early post-collisional magmatism of Eocene–Oligocene age post-dates peak M2 metamorphism. Following a second magmatic gap late post-collisional magmatism occurred in response to major crustal thickening beneath the Karakoram.

### Regional metamorphism, M2

The Karakoram metamorphic complex in the Hunza region is represented by two main units, the Hunza schists and the Maiun mélange, which are bounded to the south by the MKT and along the north by the Karakoram batholith (figure 4). In general terms, metamorphic grade in pelites and marbles increases northwards, and a pervasive foliation dips consistently NNE between 40 and 90°. The foliation is sub-parallel to bedding and shear sense indicators show thrusting towards the SSW. The two-dimensional isograd pattern shown in figure 5 is poorly constrained in three dimensions; it is suspected, but not proven that the metamorphism along the Hunza Valley is inverted (Coward *et al.* 1986). Mineralogical changes, usually parallel to foliation and bedding correlate with changes in bulk rock composition.

The phyllite matrix of the mélange includes chlorite–biotite–chloritoid and garnet-bearing assemblages. The highest grade assemblages in the mélange are andalusite-bearing and occur adjacent to the thrust contact with the Hunza schists. In pelitic lithologies within the Hunza schists, staurolite + garnet assemblages pass northwards into sillimanite + garnet assemblages and then into migmatites. In the more Ca-rich pelites and marbles, hornblende-bearing assemblages pass northwards into diopside-bearing marbles and close to the Karakoram batholith, plagioclase is replaced by scapolite. Maximum metamorphic conditions given by pelitic rocks close to the Karakoram batholith are 670 °C and 5 kbar† (Broughton *et al.* 1985). Snowball garnet and staurolite porphyroblast textures indicate that metamorphism was syn-tectonic (Powell & Vernon 1979; Prior 1987).

K–Ar biotite ages (D. C. Rex, unpublished data; Coward *et al.* 1986) young towards the region of highest metamorphic grade (figure 7). The andalusite hornfels of the Maiun mélange results from the tectonic uplift and juxtaposition of hotter garnet–staurolite schists along a major shear zone. The andalusite-forming reaction (garnet + chlorite = andalusite + biotite) occurs at about 400 °C, indicating that thrust uplift occurred close to the biotite K–Ar closure temperature. The biotite K–Ar ages can, in this case, be explained by syn- and post-metamorphic uplift on SSW-directed thrusts and shear zones that brought deeper and hotter rocks up in the north. These took longer to cool through the K–Ar closure temperature and give younger ages. Fission track ages do not vary across major tectonic contacts in the Hunza Valley (figure 7), suggesting that differential uplift occurring at moderately high temperatures (300–400 °C) had ceased by the time the area had cooled to 100–200 °C and uplift may have been occurring on one fault only. We correlate the main metamorphism in the Hunza Valley with M2 in the Baltoro region at *ca.* 45–36 Ma.

† 1 kbar = $10^8$ Pa.

The northward-dipping foliations within the southern part of the HPC are parallel to the southern contact of the HPC, which shows clear evidence for post-metamorphic south-verging thrusting. We suggest that the deformation and recrystallization within the HPC, including the northward-dipping foliation, resulted from southward thrusting along the southern contact of the HPC, this foliation thus overprinting the primary intrusive foliation. The mixing zone developed at this time and localized melting resulted in the injection of aplitic sheet intrusions emanating from the mixing zone probably during the late Eocene–early Oligocene. Several north-dipping shear zones within the Hunza schists are probably related to this phase of crustal shortening. Syn-metamorphic deformation in the staurolite–garnet zones south of Aliabad (figure 5) therefore must be earlier than the post-metamorphic thrusting along the southern margin of the batholith to the north. This northward propagation of thrusting supports the contention that major crustal thickening processes in the Himalaya–Karakoram orogenic belt occurred by breakback thrusting propagating in the hangingwall of earlier thrusts (Searle *et al.*, this symposium; Hodges *et al.*, this symposium). Post-collisional magmatism in Hunza resulted in multi-stage leucogranite dyke injection and the emplacement of leucogranitic sheets and plutons at Sumayer and Nagar (figure 4).

In the Baltoro region, the main regional metamorphism is a high pressure-temperature Barrovian sequence, characterized by the assemblage kyanite–staurolite–biotite–garnet–muscovite–plagioclase–quartz (figures 3 and 6) with sillimanite–muscovite and sillimanite–K-feldspar assemblages locally developed (Searle *et al.* 1988). Within the Karakoram metamorphic complex there are isolated satellite plutons of early post-collision age (figure 2). The Mango Gusar two-mica granite has a U–Pb zircon date of $37.0\pm0.8$ Ma (R. Parrish, personal communication) and the Chingkang-la pyroxene–hornblende–biotite granite has a K–Ar (Hbl.) age of $34\pm1$ Ma (Searle *et al.* 1988).

### The Baltoro Plutonic Unit

The Baltoro Plutonic Unit (BPU) is a batholith about 20 km wide and 80 km long (figure 2). It is restricted in composition to granitic and leucogranitic lithologies that are considered to be consanguinous. A well-defined U–Pb date of $21\pm0.5$ Ma on zircons from monzogranite samples reflects the crystallization age of the granite whereas monazite ages of 19–17 Ma reflect cooling through U–Pb closure temperatures of 650–700 °C (Parrish & Tirrul 1988). Eleven K–Ar (Bi.) dates reported by Searle *et al.* (1988) range between 12 and $5\pm0.5$ Ma on BPU granites and concur with an average age of $8.8\pm0.3$ Ma from three whole-rock–plagioclase–biotite internal isochrons (Debon *et al.* 1986). We interpret these dates as recording uplift and cooling events. Furthermore, the younging of K–Ar dates towards the margins of the BPU suggests fluid interaction effects between the granite and wallrock. The crystallization age of 21 Ma clearly refers the BPU to a late post-collisional intracontinental magmatism.

Although the BPU is compositionally restricted, three lithological groups are apparent. The monzogranites are the least evolved and contain plagioclase + K-feldspar + biotite + quartz ± muscovite, the latter being secondary; amphibole is very rare. Minor phases include apatite, sphene, zircon, allanite, monazite, thorite, tourmaline and opaques. The second group includes the two-mica ± garnet leucogranites and leucogranitic aplite and pegmatite dykes constitute the third group.

The southern margin of the BPU, south of the Baltoro glacier, is a migmatitic complex up

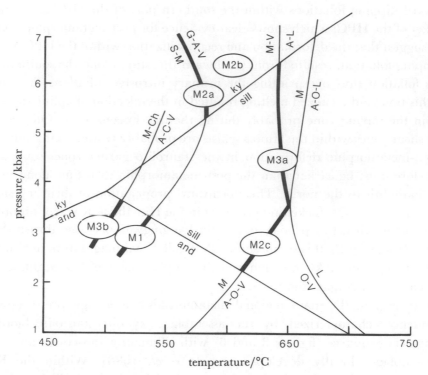

FIGURE 6. Pressure–temperature grid showing key assemblages observed in the Baltoro–Hushe Karakoram (from Searle *et al.* 1988). All assemblages contain biotite, hornblende and quartz. A, andalusite, kyanite or sillimanite; C, cordierite; Ch. chlorite; G, garnet; L, granitic liquid; M, muscovite; O, orthoclase; S, staurolite; V, vapour. (1 kbar = 10⁸ Pa.)

to 10 km wide: the Masherbrum Migmatite Complex (MMC). Paragneisses, including calc-silicate marbles, orthogneiss, strongly deformed migmatite and penetrating granitic sheets are cross-cut by complex leucogranitic dyke swarms. A Rb–Sr whole-rock isochron on these yields an age of 14.1 ± 2 Ma with an $^{87}Sr/^{86}Sr$ initial ratio of 0.7084 (Searle *et al.* 1988). K–Ar (Bi., Mu.) determinations on various pegmatites, leucogranites and gneisses yield ages between 17 and 10 Ma. One K–Ar (Hbl.) age of 77 Ma on a schist is consistent with the interpretation that older components of the Karakoram batholith are represented as protolithic material in the MMC.

Spatially and temporally associated with the BPU are biotite–minette (lamprophyre) dykes. Three K–Ar (Bi.) dates (figure 2) of 24–22 ± 1 Ma suggest their emplacement immediately before the BPU at 21 ± 1 Ma. The dykes form a 'shadow-zone' about the BPU.

The intrusion of the Baltoro plutonic unit at 21 Ma superimposed a contact metamorphism (M3) on the low-grade sediments along the northern margin of the BPU (figure 3). Andalusite in the aureole at Mitre Peak constrains the pressure to less than 3.75 kbar (figure 6). The M3 imprint along the southern margin of the BPU is more controversial. Searle *et al.* (1988) noted an increase in temperature towards the granite contact within the M2 sequence with the successive disappearance of staurolite, the appearance of sillimanite and the development of granitic melt pods within migmatitic terrain. This increase in temperature could be attributed to the thermal effect of the granite superimposed on regional M2 metamorphism (see

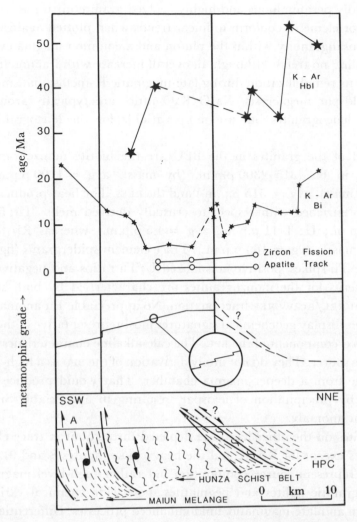

FIGURE 7. Schematic cross section through the Hunza Valley showing major tectonic boundaries, extent and general attitude of foliation and shear sense interpreted from macrostructural and some microstructural evidence. Metamorphic grade increases northwards and is shown schematically. Close-up of profile is to show that changes in metamorphic grade relate primarily to minor thrusts rather than a palaeo-geotherm. The upper part of the diagram shows K–Ar data of D. C. Rex (unpublished, and in Coward *et al.* 1986). Fission track data from Zeitler (1985). HBL., hornblende; BI., biotite; ZIR., zircon; AP., apatite.

Discussion). A leucogranite dyke cross-cutting this sequence approximately 10 km SW of the BPU has a U–Pb zircon age of $21.5 \pm 0.8$ Ma (figure 2; R. Parrish, personal communication). Dyke emplacement within the migmatite terrane was therefore synchronous with emplacement of the BPU.

### Geochemistry of the post-collisional Baltoro Plutonic Unit

The Baltoro Plutonic Unit is the largest and best documented post-collisional granite within the Karakoram batholith. Field observations and comprehensive geochemical data on over 50 representative samples allow lithological classification and subsequent modelling of geochemical evolution. Table 1 lists representative analyses for the different lithologies constituting the BPU.

The BPU is a mildly peraluminous and highly evolved granite intrusion with 68–75% (by mass) $SiO_2$. All major elements conform to linear trends when plotted against a fractionation index, indicating consanguinuity within the pluton and a common magma evolution. $K_2O$ is exceptional in revealing no trend, although an overall increase with fractionation is observed. This is attributed to its remobilization during late-magmatic K-metasomatism associated with the growth of K-feldspar megacrysts. $Na_2O/K_2O$ ratios are typically around one for the monzogranites and leucogranites but are more variable for the leucogranitic aplites and pegmatites.

The least evolved of the granites in the BPU are the biotite monzogranites. These are unusually enriched in Ba (315–2300 p.p.m. (by mass), avg. = 1100 p.p.m., $n = 28$), Sr (100–800 p.p.m (by mass), avg. = 515 p.p.m.) and the REEs. The heat-producing elements, K, Th and U, are also enriched with respect to crustally derived melts (Th: 5–85 p.p.m. (by mass), avg. = 23 p.p.m.; U: 1–11 p.p.m., avg. = 4.5 p.p.m.) whereas Rb tends to be low (90–260 p.p.m. (by mass); avg. = 160 p.p.m.). Trace element spidergrams (figure 8) illustrate pronounced positive Th spikes, positive normalized U/Th ratios and negative spikes for Nb, P and Ti. REE patterns for the monzogranites are characterised by both strong light REE enrichment and light REE/heavy REE fractionation. No appreciable Eu anomaly is developed.

The monzogranites display geochemical signatures most representative of the magma source being the least evolved component of the BPU. The calc-alkaline characteristics of the BPU must be inherited from the source. They do not imply derivation of the BPU as a high-level, physically separated fractionate from a deeper magma chamber. That would produce a fractionation pattern dominated by precipitation of feldspar, resulting in appreciable Sr depletion and development of a Eu anomaly.

All the leucogranites on the other hand are depleted throughout in trace elements relative to the monzogranites except for Rb, K and the heavy REEs (figures 8 and 9). This depletion does not appear to represent source composition but rather high-level magmatic processes. Likewise, the leucogranitic aplites and pegmatites show wide chemical variation reflecting extreme fractionation and late-magmatic, fluid-enhanced processes. Differentiation of the melt through some fractionation process can account for the observed elemental depletions during magma evolution. In particular, Ba and Sr are removed in feldspar and biotite and P, (Zr) and Ti by apatite, (zircon) and sphene respectively. The spectacular depletion in the light REES from monzogranites to evolved differentiates, whereas HREE abundances remain fairly stable, is attributed to monazite (& allanite?) control, although feldspar will also be effective. Monazite (and thorite?) fractionation is also responsible for changing the normalized U/Th ratio from less than 1 in the monzogranites to more than 1 in the leucogranites. Uranium thus behaves more incompatibly than Th throughout magma evolution (partition coefficient data for quantitative geochemical modelling from Arth 1976, Hanson 1978 and Henderson (ed.) 1984).

Initial $^{87}Sr/^{86}Sr$ ratios for the BPU vary between 0.7072 and 0.7128. Preliminary data on whole rocks indicate that oxygen isotope compositions are uniformly heavy, in the range 9.6–11.2 ‰. The isotopic characteristics of the BPU are considered to reflect the isotopic composition of the source region except where exchange has taken place close to the granite contact. Strontium and oxygen isotopic compositions are appreciably lower than those of the High Himalayan post-collisional leucogranites that exceed 0.730 (at 20 Ma) and 11.0, respectively (Deniel et al. 1987; Le Fort et al. 1987).

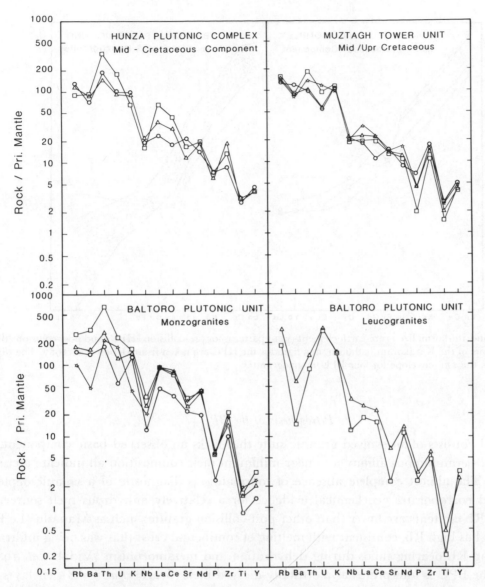

FIGURE 8. Primordial-mantle normalized trace element variation diagrams (spidergrams) for pre-collision (HPC and MTU) and post-collision (BPU) components of the Karakoram batholith.

The lamprophyres that are spatially and temporally associated with the BPU are extraordinarily enriched in most trace elements (table 1) with overall geochemical signatures displaying some similarities to the BPU. Compared to other late tectonic to post-tectonic calc-alkaline lamprophyres, however (Rock 1984; P. Henney, personal communication), they show similar elemental abundances. Of particular interest are the high concentrations of $K_2O$, Sr, Ba, Th, U and the light REEs, just the elements that are anomalously enriched in the BPU granites. A primary continental mantle lithosphere source is envisaged for the lamprophyres; concentrations of refractory elements such as V, Ni, Cr preclude a crustal derivation (Rock 1984; Foley et al. 1987). However, a component derived from subducted continental crust is possible (see Discussion).

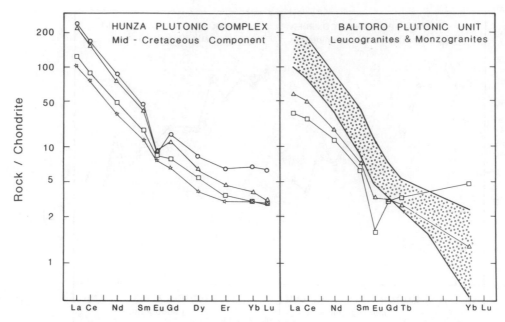

FIGURE 9. Chondrite-normalized rare-earth element (REE) patterns for pre-collision (HPC) and post-collision (BPU) components of the Karakoram batholith. REE data for the HPC are taken from Debon *et al.* 1987. The stipple represents the REE envelope for four BPU monzogranites.

## Petrogenesis of the BPU

The BPU consists of an evolved granitic suite that lacks an observed basic component. Its mineralogy, peraluminous affinity and near-minimum melt composition all indicate a crustal derivation. The almost complete absence of tourmaline is diagnostic of a volatile-depleted magma and corroborates geochemical evidence for a relatively anhydrous melt source. In particular, Rb contents are lower than other post-collision granites such as Manaslu (Le Fort 1981). This has high Rb, consistent with melting of continental crust that was being infiltrated by migrating Rb-bearing fluids during dehydration and metamorphism (Vidal *et al.* 1982). Although the BPU is interpreted as a crustally-derived melt (Searle *et al.* 1988), enclaves are rare and confined to the pluton margins (xenoliths). However, rare xenocrystic sillimanite within monzogranite, breaking down to produce K-feldspar and muscovite under K-rich fluid fluxing, implies a metasedimentary component to the source. In addition, a large inherited component to the zircons suggests a protolith heterogeneous in primary age and hence of a hybrid source. These source ages range from 600–500 Ma to older than 1700 Ma (Parrish & Tirrul 1988).

Further evidence suggests that a major source component for the BPU was dominantly quartzo-feldspathic. Likely bulk compositions include immature quartzo-feldspathic meta-pelites, volcaniclastics and orthogneisses. These are generally relatively depleted in Rb, thus yielding a melt with low Rb and unradiogenic initial $^{87}Sr/^{86}Sr$ ratios. They would also be expected to provide moderate $\delta^{18}O$ values. At high (granulite facies) $P–T$ conditions, dehydration melting (mainly biotite) in the source would produce a granitic melt, deficient in included restite (Clemens & Wall 1981; Clemens & Vielzeuf 1987). Retention of garnet in

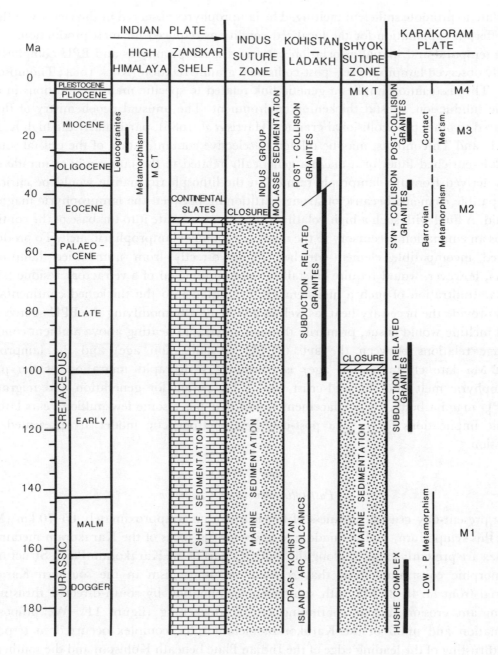

FIGURE 10. Simplified space-time chart based on radiometric ages for Kohistan–Ladakh and the Karakoram presented in Searle *et al.* (1988).

the residue may help to explain the heavy REE depletion in BPU monzogranites. The amount of melt required before a magma forms and begins to migrate (the critical melt fraction: CMF) is dependent mainly upon the temperature and water content of the melt and the restite content of the magma. Although the CMF is in much debate at deep crustal levels (see Wickham 1987; Clemens & Vielzeuf 1987; Rutter & Wyllie 1988), it is considered that, in fluid-absent conditions, the introduction of hot material in or adjacent to the potential melting site is a likely candidate to promote sufficient melting. The lamprophyres observed in the vicinity of the BPU may offer an explanation for the localized conditions required for melt production.

The temporal and spatial association between the lamprophyres and BPU complies with a globally observed lamprophyre – post-collisional granite relation (Rock 1984; Thompson *et al.* 1984). This association implies a genetic link related to specific mantle conditions in a post-oceanic subduction, crustal thickening environment. The unusual geochemistry of the BPU compared to other post-collisional granites (Harris *et al.* 1986), particularly the high K, Ba, Sr, LREE, U and Th contents, may be related to selective contamination of the crustal source by a volatile-enriched fluid, or a magma genetically related to the lamprophyre mantle source. Fluids derived from the lamprophyre melt in the lithospheric mantle would be enriched in incompatible elements because of strong partitioning between the lamprophyric magma and the fluid. A fluid with such a high volatile content may migrate into the base of the continental crust as an emanation precursory to the main extraction of lamprophyric melt. To produce the required incompatible element-enriched liquid directly from unmodified lamprophyric magma, however, would require crystallization and removal of a refractory residue from the magma. Infiltration of such a high-temperature liquid into the thickened continental crust would provide the necessary heat, as well as geochemically modifying the BPU source region. Rapid melting would ensue, primarily through very rapid heating above ambient conditions. The age relations between the BPU (21 Ma crystallization age) and the lamprophyres (24–22 Ma late crystallization age) is consistent with rapid migration of the primary lamprophyric melt into the mid-crust, and a lag period for generation and migration of the BPU magma body. Its emplacement would take place some few million years later. The tectonic implications for such a post-collisional petrogenetic model are presented in the Discussion.

## DISCUSSION

### Post-collisional tectonic evolution

The present-day crustal thickness of the Karakoram is approximately 65–70 km (Molnar 1984; this symposium) and yet middle to lower crustal rocks of the Karakoram metamorphic complex are presently exposed along the southern part of the Karakoram Plate. Structural and metamorphic constraints show that regional metamorphism in the southern Karakoram occurred from *ca.* 45–36 Ma with subsequent exhumation by south-directed thrusting and concomitant erosion during continuing crustal thickening (figure 11). We propose that deformation and uplift of the Karakoram metamorphic complex occurred in response to underthrusting of the leading edge of the Indian Plate beneath Kohistan and the southern part of the Karakoram (figure 11). Coward & Butler (1985) and Coward *et al.* (1986, 1987) record 470 km of shortening from the MMT to the Pakistan foreland in Indian Plate rocks, and argue for large-scale underthrusting of Indian lower crust northwards as far as the Pamirs. From shortening estimates of the Zanskar, Ladakh and Kashmir Himalaya, Searle *et al.* (this symposium) argue for underthrusting of Indian lower crust northwards as far as the Karakoram batholith but not beneath the Pamirs and Tibet.

The timing of metamorphism and deformation is constrained by undeformed granite intrusions that cross-cut the syn-metamorphic foliation, and have ages of 36–34 Ma (Searle *et al.* 1988). Along the Hunza Valley, syn-metamorphic south-directed thrusting, which may

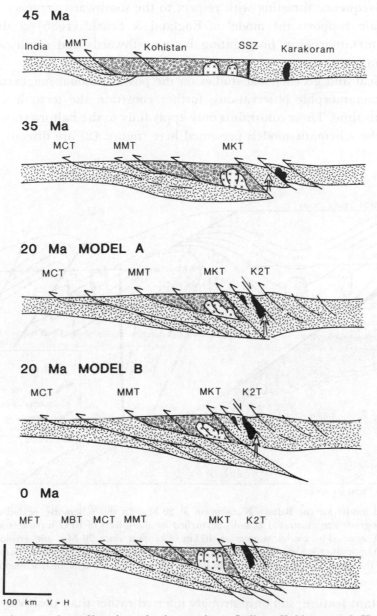

FIGURE 11. Schematic crustal profiles through the northern Indian, Kohistan and Karakoram Plates to illustrate the post-collisional evolution of the Karakoram. At 20 Ma two models are presented, model A being simple homogeneous thickening and model B crust–mantle imbrication. Abbreviations: K2T, K2 Thrust; MKT, Main Karakoram Thrust; MMT, Main Mantle Thrust; MCT, Main Central Thrust; MBT, Main Boundary Thrust; MFT, Main Frontal Thrust.

have been responsible for inverting the metamorphic isograds within the Hunza schists, propagated northwards to post-metamorphic thrusting along the southern margin of the Karakoram batholith. M2 metamorphism in the Karakoram is broadly synchronous with High Himalayan metamorphism to the south (Oligocene-Miocene; see Searle *et al.*, this symposium). We can therefore demonstrate that crustal thickening and regional meta-morphism were occurring both in the Himalaya and the Karakoram following closure of the

ISZ. The out-of-sequence thrusting with respect to the southward vergence observed in the Hunza Karakoram supports the model of England & Searle (1986) of thrusting, crustal thickening and metamorphism propagating both northwards and southwards of the ISZ following collision.

Geochronological and geochemical studies on the post-collisional magmatism in the early Miocene, with metamorphic observations, further constrain the tectonic evolution of the Karakoram at this time. These constraints only apply fully to the Baltoro transect through the Karakoram, so the schematic models presented here (figure 12) are directly relevant to this area.

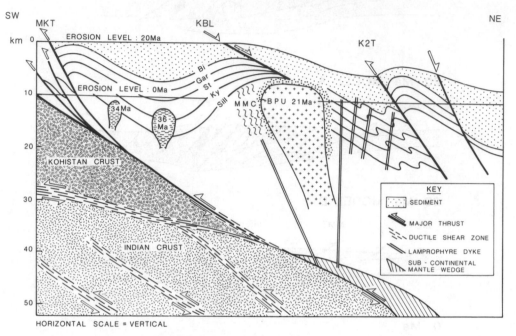

FIGURE 12. Thermal model for the Baltoro Karakoram at 20 Ma. In this schematic, scaled crustal section, M2 metamorphic isograds are illustrated as both perturbed by the effects of BPU intrusion at 21 Ma and offset along the MKT zone. The model assumes *ca.* 10 km of erosion since 20 Ma, and erosion levels are shown schematically. Major ductile shear zones bound Kohistan, Indian and Karakoram crustal blocks and basal detachments of the latter two are considered to flatten with depth. See text for discussion.

A most important feature, still only strongly inferred rather than positively observed, is the Karakoram Batholith Lineament. As stated earlier, this has been the site of magmatic activity at least since the mid-Cretaceous (or even the Jurassic if the Hushe complex is a tectonically fragmented component) until the mid-Miocene. The KBL was reactivated during and for some time before the mid-Miocene as a normal fault, downthrowing north. We suspect that culmination-collapse was initiated in response to crustal thickening beneath the Karakoram and resultant uplift of the Karakoram metamorphic complex. The mid-Miocene intrusion of the BPU into mid-crustal levels, estimated from the M3 aureole assemblages at around 12–15 km (Searle *et al.* 1988), is compatible with an extensional régime in the mid-upper crust along the lineament. Magma was tapped along deep crustal structures, probably ductile shear zones, and directed into the mid-crust along the footwall of the KBL.

Extension along the KBL before and during emplacement of the BPU is required to juxtapose medium- to high-grade metamorphic terrain south of the Karakoram batholith against low-grade sediments to the north, both superimposed by M3. This is seen both in the Hunza section and in the Baltoro. Considerable uplift of the southern crustal block is therefore required; it is proposed that movement occurred along the MKT to the south and the KBL to the north. If steady-state uplift took place following 10 Ma of crustal thickening after main collision at 50–45 Ma, an estimated uplift rate of 1 mm $a^{-1}$ is required before intrusion of the BPU. The uplift of the M2 sequence then complies well with the early post-collisional granites, intruded at 36–34 Ma, cross-cutting the regional M2 fabric (figure 12). Barrovian metamorphism would therefore post-date crustal thickening after collision and pre-date the lowest Oligocene granites (therefore, *ca.* 40–36 Ma).

### Post-collisional thermal evolution

With post-M2 displacement along the KBL before intrusion of the BPU, low-temperature sedimentary rocks to the north were juxtaposed against hotter medium- to high-grade metamorphic rocks to the south. The tectonic juxtaposing of rocks of different metamorphic grade but otherwise identical characteristics will result in the downwarping of isotherm surfaces on the upthrow side and vice versa, with lateral heat transfer due to thermal disequilibration (Harte & Dempster 1987). The lower thermal conductivity of sedimentary rocks over crystalline rocks, gneisses and granites would promote a conductivity contrast, envisaged by Pinet & Jaupart (1987) to be intimately associated with the petrogenesis of the High Himalayan granites and their localization at corresponding structural positions. Sedimentary rocks north of the KBL would thus have acted as a barrier to lateral heat transfer (Jaupart & Provost 1985).

It is therefore significant that the BPU should be emplaced along the footwall of the KBL where a horizontal temperature maximum is localized, as this would be a further inhibitor to magma cooling. The BPU could thus migrate higher in the crust because of the low water content of the melt as a consequence of biotite-dominated dehydration partial melting, to the KBL acting as a magma conduit, to an extensional régime in the mid-upper crust and to the localized temperature anomaly imposed by the thermal contrast effect.

We suggest that the combination of emplacement of the BPU and the thermal effect across the KBL was responsible for the extensive M3 thermal metamorphism in the Karakoram metamorphic complex. Assemblages including garnet + biotite + muscovite + quartz + plagioclase + sillimanite + melt are developed within migmatitic terrane (Searle *et al.* 1988) so migmatization is considered an *in situ* development, with the associated leucosome melts genetically unrelated to the BPU melt.

The prevailing crustal framework (figure 11) has important implications for the source region of the BPU and the lamprophyres. Essentially, a simple homogeneous thickening model (figure 11; 20 Ma model A) for the deep structure of the Karakoram would necessitate vertical migration of melt through an estimated 50–60 km of continental crust. However, the crust–mantle imbrication model (figure 11; 20 Ma model B), whereby a mantle wedge separated Karakoram crust from Indian crust, would permit melting near to the base of the Karakoram crust to produce the BPU magma, the mantle wedge constituting a likely source for the lamprophyres. The relatively young, hot continental lithospheric mantle may have been enriched through devolatilization and dehydration in the subducting Indian crust. The resultant melts would only have to migrate a maximum of 30 km through the Karakoram crust.

### The Karakoram: an analogue to the High Himalaya?

Finally, we note a parallel in several temporal and structural processes between the Karakoram and the High Himalaya. The Barrovian metamorphism observed in both regions, within structurally bound blocks, is probably coeval and related to crustal thickening propagating both northwards and southwards from the ISZ. Uplift and exposure of the high-grade metamorphic terrane occurred along the MKT north of the ISZ and along the Main Central Thrust (MCT) to the south; both structures acting as south-directed thrusts. In response to crustal thickening and uplift of crustal blocks, dorsal culmination collapse occurred at high levels along the inferred KBL in the Karakoram (Searle *et al.* 1988) and along the Zanskar Shear Zone north of the High Himalaya (Burg *et al.* 1984*a*, *b*; Burchfield & Royden 1985; Searle 1986; Hodges *et al.*, this symposium). Juxtaposition of high-grade metamorphic terrane against Tethyan sediments in Zanskar is estimated to have involved a vertical displacement of *ca.* 19 km (Herren 1987).

A significant difference between the Karakoram and the High Himalaya is the composition and source of the leucogranites and their genesis. The anatectic, often 'minimum-melt', leucogranites of the High Himalaya were generated more or less *in situ*, and a number of petrogenetic models have been proposed (see Hodges *et al.*, this symposium). The Miocene post-collisional granites in the Karakoram, however, were generated in a deeper, more anhydrous region of the crust. The melt zone corresponds more or less to our estimate of the northernmost limit of underplating of the Indian crust. Our petrogenetic interpretation of the BPU is primarily based on petrographic, geochemical and isotopic evidence. Tectonic, thermal and structural considerations have also been integrated to produce an overall schematic model (figure 12). In particular, the lower initial $^{87}Sr/^{86}Sr$ ratios and $\delta^{18}O$ values for the BPU relative to leucogranites of the High Himalaya suggest a source fundamentally different from the mainly metasedimentary compositions that comprise the Karakoram metamorphic complex and the Central Crystalline Complex. Further, the absence of volatiles together with evidence for dehydration melting is consistent with a deep crustal source where contamination by a mantle-derived component is a realistic proposition.

A. J. R. and M. P. S. acknowledge NERC grant GR3/4242 to B. F. Windley and also a Royal Society grant (1985) and additional funding from the Mount Everest Foundation. D. C. R. acknowledges NERC grant GR/5932; M. B. C., grant GT4/85/GS/47; and D. J. P., NERC grant GT/83/GS/117. We thank Brian Windley, Paul Hoffman, Marc St Onge, Patrick Le Fort and Paul Henney for discussions, Randy Parrish for permission to quote unpublished U–Pb dates and Andy Saunders for critically reviewing earlier versions of the manuscript. Our thanks to Sue Button for producing an excellent set of slides for the symposium and diagrams in this paper.

### REFERENCES

Allègre, C. J. *et al.* 1984 *Nature, Lond.* **307**, 17–22.
Arth, J. G. 1976 *J. Res. U.S. geol. Surv.* **4**, 41–48.
Auden, J. B. 1935 *Rec. geol. Surv. India* **69**, 123–167.
Auden, J. B. 1938 In *The Shaksgam Expedition, 1938* (ed. E. Shipton) Geographical Journal vol. 91, pp. 335–336.
Bard, J. P. 1983 *Earth planet. Sci. Lett.* **65**, 133–144.

Bard, J. P., Maluski, H., Matte, P. & Proust, F. 1980 *Geol. Bull. Peshawar Univ.* (*Spec. Issue*) **13**, 87–94.

Bertrand, J. M. & Debon, F. 1986 *C. r. Acad. Sci., Paris* **303**, 1611–1614.

Brookfield, M. E. 1980 *AEEi Acad. Naz. Lincei Memorie* **59**, 248–253.

Brookfield, M. E. 1981 In *Metamorphic tectonites of the Himalaya* (ed. P. S. Saklani), pp. 1–14. New Delhi: Today and Tomorrow Publishers.

Brookfield, M. E. & Reynolds, P. H. 1981 *Earth planet. Sci. Lett.* **55**, 157–162.

Broughton, R. D. Windley, B. F. & Jan, M. Q. 1985 *Geol. Bull. Peshawar Univ.* **18**, 119–136.

Brown, G. C., Thorpe, R. S. & Webb, P. C. 1984 *J. geol. Soc. Lond.* **141**, 411–426.

Burchfield, C. & Royden, L. 1985 *Geology* **13**, 679–682.

Burg, J.-P., Guiraud, M., Chen, G. M. & Li, G. C. 1984*a* *Earth planet. Sci. Lett.* **69**, 391–400.

Burg, J.-P., Brunel, M., Gapais, D., Chen, G. M. & Liu, G. H. 1984*b* *J. struct. Geol.* **6**, 535–542.

Clemens, J. D. & Vielzeuf, D. 1987 *Earth planet. Sci. Lett.* **86**, 287–306.

Clemens, J. D. & Wall, V. J. 1981 *Can. Mineralogist* **19**, 111–131.

Coward, M. P. & Butler, R. H. W. 1985 *Geology* **13**, 417–420.

Coward, M. P., Jan, M. Q., Tarney, J., Thirlwall, M. & Windley, B. F. 1982 *J. geol. Soc. Lond.* **139**, 299–308.

Coward, M. P., Windley, B. F., Broughton, R., Luff, I. W., Petterson, M. G., Pudsey, C., Rex, D. & Khan, M. A. 1986 In *Collision tectonics* (eds. M. P. Coward & A. Ries), Special Publication of the Geological Society of London no. 19, pp. 203–219.

Coward, M. P., Butler, R. H. W., Asif Khan, M. & Knipe, R. J. 1987 *J. geol. Soc. Lond.* **144**, 377–391.

Debon, F., Zimmermann, J. L. & Bertrand, J. M. 1986*a* *C. r. Acad. Sci., Paris* **303**, 463–468.

Debon, F., Le Fort, P., Sheppard, S. M. F. & Sonet, J. 1986*b* *J. Petr.* **27**, 219–250.

Debon, F. Le Fort, P., Dautel, D., Sonet, J. & Zimmerman, J. L. 1987 *Lithos* **20**, 19–40.

Deniel, C., Vidal, P., Fernandez, A., Le Fort, P. & Peucat, J. 1987 *Contr. Miner. Petr.* **96**, 78–92.

Desio, A. 1964 Geological tentative map of the western Karakoram, scale 1:500,000. Institute of Geology, Milan University.

Desio, A. 1979 In *Geodynamics of Pakistan* (ed. A. Farah & K. A. De Jong), pp. 111–124. Quetta: Geological Survey of Pakistan.

Desio, A., 1980 *Geology of the Upper Shaksgam Valley, Northeast Karakoram, Xinjiang.* (196 pages.) Holland: Brill-Leiden.

Desio, A. & Mancini, E. G. 1974 *Atti Accad. naz. Lincei Memorie* **12**, 79–100.

Desio, A. & Martina, E. 1972 *Boll. Soc. geol. Ital.* **91**, 283–314.

Desio, A. & Zanettin, B. 1970 *Geology of the Baltoro Basin.* (308 pages.) Leiden: Brill.

Desio, A., Tangiorgi, E. & Ferrara, G. 1964 *Report of XXII Sess. India, Intern. Geol. Congr. Vol.* 11, pp. 479–496.

Desio, A., Martina, E., Spadea, P. & Notarpietro, A. 1985 *Atti Accad. naz. Lincei Memorie* **18**, 3–53.

England, P. C. & Searle, M. P. 1986 *Tectonics* **5**, 1–14.

England, P. C. & Thompson, A. B. 1984 *J. Petr.* **25**, 894–928.

England, P. C. & Thompson, A. B. 1986 In *Collision tectonics* (ed. M. P. Coward & A. C. Ries), special publication of the Geological Society of London no. 19, pp. 67 82.

Foley, S. F., Venturelli, G., Green, D. H. & Toscani, L. 1987 *Earth Sci. Rev.* **24**, 81–134.

Gansser, A. 1964 Geology of the Himalayas. (298 pages.) London: J. Wiley.

Hanson, G. N. 1978 *Earth planet. Sci. Lett.* **38**, 26–43.

Harris, N. B. W., Pearce, J. A. & Tindle, A. G. 1986 In *Collision tectonics* (ed. M. P. Coward & A. Ries), Special Publication of the Geological Society of London no. 19, pp. 67–82.

Harte, B. & Dempster, T. J. 1987 *Phil. Trans. R. Soc. Lond.* A **321**, 105–127.

Henderson, P. (ed.) 1984 *Rare Earth Element Geochemistry. Developments in Geochemistry*, 2. (510 pages.) Amsterdam: Elsevier.

Herren, E. 1987 *Geology* **15**, 409–413.

Jaupart, C. & Provost, A. 1985 *Earth planet. Sci. Lett.* **73**, 385–397.

Klootwijk, C. J. 1979 In *Structural geology of the Himalaya* (ed. P. S. Saklani), pp. 307–360. New Delhi: Today and Tomorrow Publishers.

Le Fort, P. 1981 *J. geophys. Res.* **86**, 10545–10568.

Le Fort, P., Michard, A., Sonet, J. & Zimmerman, J. L. 1983 In *Granites of the Himalayas, Karakoram and Hindu Kush* (ed. F. A. Shams), pp. 377–387. Institute of Geology, Punjab University, Lahore.

Le Fort, P., Cuney, M., Deniel, C., France-Lanord, C., Sheppard, S. M. F., Upreti, B. N. & Vidal, P. 1987 *Tectonophysics* **134**, 39–57.

Lydekker, R. 1883 *Mem. geol. Surv. India* **22**, 108–122.

Molnar, P. 1984 *A. Rev. Earth planet. Sci.* **12**, 489–518.

Molnar, P. & Tapponnier, P. 1975 *Science, Wash.* **189**, 419–426.

Parrish, R. & Tirrul, R. 1988 *Geology*. (Submitted.)

Patriat, P. & Achache, J. 1984 *Nature, Lond.* **311**, 615–621.

Pearce, J. A., Harris, N. B. W. & Tindle, A. G. 1984 *J. Petr.* **25**, 956–983.

Petterson, M. G. & Windley, B. F. 1985 *Earth planet. Sci. Lett.* **74**, 45–57.

Pinet, C. & Jaupart, C. 1987 *Earth planet. Sci. Lett.* **84**, 87–99.

Powell, C. McA. & Vernon, R. H. 1979 *Tectonophysics*, **54**, 25–43.
Prior, D. J. 1987 *J. metamorph. Geol.* **5**, 27–39.
Pudsey, C. J. 1986 *Geol. Mag.* **123**, 405–423.
Pudsey, C. J., Coward, M. P., Luff, I. W., Shackleton, R. M., Windley, B. F. & Jan, M. Q. 1985 *Trans. R. Soc. Edinb.* **76**, 463–479.
Reynolds, R. H., Brookfield, M. E. & McNutt, R. H. 1983 *Geol. Rdsch.* **72**, 981–1004.
Rock, N. M. S. 1984 *Trans. R. Soc. Edinb.* **74**, 193–227.
Rutter, M. J. & Wyllie, P. J. 1988 *Nature, Lond.* **331**, 159, 160.
Schneider, H. J. 1957 *Geol. Rdsch.* **46**, 426–476.
Searle, M. P. 1986 *J. struct. Geol.* **8**, 923–936.
Searle, M. P., Windley, B. F., Coward, M. P., Cooper, D. J. W., Rex, A. J., Rex, D., Tingdong, L., Xuchang, X., Jan, M. Q., Thakur, V. C. & Kumar, S. 1987 *Bull. geol. Soc. Am.* **98**, 678–701.
Searle, M. P., Rex, A. J., Tirrul, R., Rex, D. C. & Barnicoat, A. 1988 *Bull. geol. Soc. Am.* (In the press.)
Shvolman, V. A. 1978 *Himalayan Geol.* **8**, 369–378.
Tahirkheli, R. A. K., Mattauer, M., Proust, F. & Tapponier, P. 1979 In *Geodynamics of Pakistan* (ed. A. Farah & K. A. De Jong,) pp. 125–130. Quetta: Geol. Surv. Pakistan.
Tapponnier, P., Mattauer, M., Proust, F. & Carsaigneau, C. 1981 *Earth planet. Sci. Lett.* **52**, 355–371.
Thakur, V. C. & Misra, D. K. 1984 *Tectonophysics* **101**, 207–220.
Thompson, R. N., Morrison, M. A., Hendry, G. L. & Parry, S. J. 1984 *Phil. Trans. R. Soc. Lond.* A **310**, 549–590.
Vidal, P., Cocherie, A. & Le Fort, P. 1982 *Geochim. cosmochim. Acta* **64**, 2274–2292.
Wickham, S. M. 1987 *J. geol. Soc. Lond.* **144**, 281–297.
Zanettin, B. 1964 *Geology and petrology of the Haramosh-Mango Gusar area.* Leiden: Brill.
Zeitler, P. K. 1985 *Tectonics* **4**, 127–151.

*Note added in proof* (3 *March* 1988). Here we discuss the BPU in relation to petrogenetic models for other post-collisional granites, especially those in the Himalayan belt. The model of Le Fort (1981) for the generation and emplacement of the Manaslu granite is now widely applied to other post-collisional granites of the High Himalaya. It differs from that proposed here for the BPU, consistent with the outstanding differences in chemical, mineralogical and isotopic characteristics. Therefore it is important to emphasize that although the High Himalayan petrogenetic model is convincing, it should not be accepted as a general model for all melts derived by partial melting of crustal sources (see Clemens & Vielzeuf 1987).

## Discussion

J.-M. BERTRAND (*Centre de Recherches Pétrographiques et Géochimiques, France*). The tectonic evolution of the Karakoram presented by the Leicester group fits well with our observations on the lower Braldu Valley and along two N–S sections (Hoh Lunghma and Panmah Valleys). However, I would like to comment briefly on two points of disagreement.

(1) The discrimination between several lithological groups or formations and especially the reference to the Ganschen and Dumordo Formations first described by Desio (1964) is very doubtful as the same rock associations may be found in most of them; for example the mafic and ultramafic units exist both in the Ganschen Formation and Dumordo Formation. If several distinctive formations have to be distinguished among the Karakoram gneisses, they have to be assessed upon very detailed sections.

(2) From small-scale and large-scale structures and from the recorded metamorphic evolution, we tend to divide the M2 metamorphic event into two distinct events, our D1 and D2 (Bertrand & Debon 1986); M1 has not been recognized on the area studied. D2 corresponds to superimposed low-pressure mineral assemblages and to large-scale recumbent folding. In contrast to the clear separation between D1 and D2, outlined by emplacement of

post-D1 pre-D2 leucogranitic bodies, a continuous evolution between D2 and a dome-forming D3 event is assumed, the latter being followed shortly by the emplacement of the Baltoro granite. In our opinion, the D2–D3 low-pressure metamorphism, which is not restricted to the margins of the Baltoro granite, cannot be interpreted as contact metamorphism but is, more likely, a dome-related thermal anomaly, the last consequence of which being probably the emplacement of the Baltoro granite. Such an interpretation is supported by the complete similarity in tectonic and metamorphic evolution between the Dassu dome and the high-grade gneisses and migmatites occurring just south of the contact of the Baltoro granite.

*Reference*

Bertrand, J. M. & Debon, F.  1986  *C. r. Acad. Sci., Paris* (II) **17**, 1611–1614.

A. J. Rex and M. P. Searle. We agree with Dr Bertrand that there is insufficient detail to distinguish the Ganschen and Dumordo Formations in any stratigraphic sense, hence their omission from our maps and their loose definition as units. The mafic–ultramafic blocks are exotic to both units, with which they are in tectonic contact.

The M1 metamorphic event is only regionally present around the Hushe and Thalle Valleys, east of the area studied by Bertrand & Debon (1986). We have documented an increase in temperature of *ca.* 75 K from Bardumal to Paiju approaching the contact of the BPU along the Baltoro glacier (Searle *et al.* 1988), which we have modelled here as the thermal upwarping of pre-36 Ma M2 isograds around the 21 Ma M3 aureole isotherms of the BPU. This temperature increase is restricted to the margin of the BPU and although it may be consequential and hence part of regional M2, we suggest that it is a thermal overprint after initial uplift of the M2 isograds. The 'dome-shaped thermal anomalies' around the Dassu and Panmah domes are, in our opinion, formed by post-metamorphic folding of isograds due to thrust culmination.

*Phil. Trans. R. Soc. Lond.* A **326**, 257–280 (1988)     [ 257 ]

*Printed in Great Britain*

# Metamorphic constraints on the thermal evolution of the central Himalayan Orogen

By K. V. Hodges, Mary S. Hubbard and D. S. Silverberg

*Department of Earth, Atmospheric, and Planetary Sciences, Massachusetts Institute of Technology,
Cambridge, Massachusetts 02139, U.S.A.*

Recent studies that integrate conventional thermobarometry of pelitic mineral assemblages with thermodynamic modeling of garnet zoning reveal complex Tertiary $P-T$ paths for the Greater Himalayan metamorphic sequence in the central Himalaya. Viewed in light of our current understanding of the structural evolution of the Himalaya, these data provide insights into the relations between tectonic and thermal processes during orogenesis. In this paper, we present an interpretive model for tectonothermal evolution of the Greater Himalaya in the central part of the range. This model involves: (1) middle Eocene–early Oligocene burial to depths of more than 30 km during the early stages of collision between India and Asia; (2) early–late Oligocene uplift and cooling; (3) late Oligocene heating and renewed burial synchronous with the early stages of anatectic melting and leucogranite plutonism; (4) latest Oligocene–middle Miocene rapid uplift and continued leucogranite production associated with ramping on the structurally lower Main Central Thrust and tectonic denudation on structurally higher low-angle detachment systems; and (5) middle Miocene–Recent rapid cooling during the final stages of uplift to the surface.

## Introduction

Although the current thermal structure of the lithosphere can be inferred from geophysical measurements, exhumed igneous and metamorphic rocks provide our only direct evidence of the thermal evolution of the deeper levels of ancient orogenic belts. Numerical experiments indicate that tectonic processes control the thermal evolution of mountain belts, and thus the pressure–temperature $(P-T)$ paths pursued by metamorphic rocks within orogens (Oxburgh & Turcotte 1974; England & Richardson 1977; England & Thompson 1984). Clearly, if we can reconstruct the $P-T$ history of a metamorphic terrane, then we should be able to use this information to assess the relative importance of different heat-transfer mechanisms associated with tectonic activity.

In theory, the Himalayan Orogen provides one of the world's great laboratories for the study of thermal processes because (1) the metamorphic core of the belt is immense (more than 100 000 km² of exposure); (2) it contains abundant assemblages appropriate for $P-T$ studies; and (3) extreme relief and the moderate northward dip of metamorphic units yield exposures of crustal sections that commonly exceed 10 km in thickness. Although there have been many studies of the distribution of metamorphic assemblages in the Himalaya and some attempts to estimate the peak $P-T$ conditions of metamorphism by comparing observed assemblages with experimentally constrained petrogenetic grids (see Windley (1983), Le Fort (1986) and Pêcher & Le Fort (1986) for reviews), there have been very few quantitative studies of the $P-T$ evolution of Himalayan metamorphic terranes. In 1985, we began studying the thermal history of three transects through the metamorphic core of the central Himalaya: the Dudh Kosi–Hongu–Hinku section of eastern Nepal, the Burhi Gandaki–Darondi section of central Nepal, and the Alaknanda–Dhauli section of north–central India. Our results, combined with those of Brunel & Kienast (1986) in eastern Nepal and Le Fort *et al.* (1987) in central Nepal, document a complex thermal history spanning much of Tertiary time, and they indicate a close relation between tectonic processes and thermal evolution. In this paper, we review the available data from the central Himalaya and discuss their implications for the relations between metamorphism and tectonics in the central Himalayan Orogen.

## TECTONIC SETTING OF THE CENTRAL HIMALAYA

The product of the Eocene collision between India and Eurasia and subsequent intraplate deformation, the Himalaya can be divided into six tectonic zones running parallel to the length of the orogen. From north to south, these are: (1) the Transhimalayan Zone; (2) the Indus–Tsangpo Suture Zone; (3) the Tibetan Sedimentary Zone; (4) the Greater Himalayan Metamorphic Sequence; (5) the Lesser Himalayan Nappe Sequence; and (6) the Subhimalayan Zone (figure 1). The Transhimalayan Zone is dominated by calc-alkaline batholiths, which range in age from roughly 110 to 40 Ma (Brookfield & Reynolds 1981; Honegger *et al.* 1982; Maluski *et al.* 1982; Schärer *et al.* 1986), and which are thought to have been produced during northward subduction of Tethys beneath the southern margin of Eurasia before India–Eurasia collision. The Indus–Tsangpo Suture Zone consists of Mesozoic

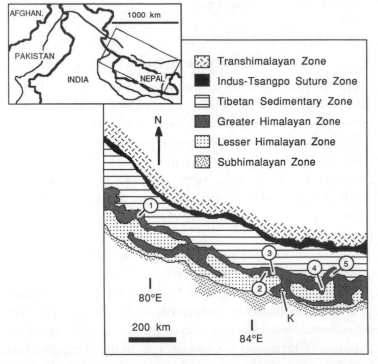

FIGURE 1. Generalized tectonic map of the central Himalaya after Le Fort (1975) and Pêcher & Le Fort (1986). Box in inset map shows the approximate boundaries of the central Himalaya. Circled numbers refer to *P–T* studies reviewed in the text: (1) eastern Garhwal (Hodges & Silverberg 1988); (2) west–central Nepal (Le Fort *et al.* 1986); (3) east–central Nepal (Hodges *et al.* 1988); (4) Everest region (Hubbard 1988); and (5) Makalu region (Brunel & Kienast 1986). K indicates location of Kathmandu.

ophiolites, arc volcanic rocks, and flysch marking the initial zone of collision (Gansser 1964; Bally *et al.* 1980). Various lines of geological and geophysical evidence, most of them indirect, imply that this collision began between 40 and 50 Ma (see Molnar 1984, for a review). South of the suture, the Tibetan Sedimentary Zone includes the miogeoclinal succession developed along the passive northern margin of India from Cambrian–Eocene(?) time (Gansser 1964; Le Fort 1975). The zone is structurally complex, exhibiting S-vergent recumbent folds and thrusts, N-vergent 'back-folds' and 'back-thrusts', and extensional structures with a variety of orientations, all of which developed during several late Cretaceous–Holocene(?) events (Le Fort 1975; Searle 1983; Burg & Chen 1984). For many years, unmetamorphosed rocks of the Tibetan sedimentary sequence were thought to rest unconformably on the metamorphosed 'Precambrian basement' of the Greater Himalayan Metamorphic Sequence (Gansser 1964). Recent mapping of the base of the Tibetan sedimentary sequence near the eastern Nepal–Tibet and Bhutan–Tibet borders (Burg *et al.* 1984; Burchfiel *et al.* 1986; Burchfiel, K. V. Hodges & L. H. Royden, unpublished data), as well as in Ladakh, India (Herren 1987), demonstrates that the contact in these areas is a N-dipping, low-angle normal fault zone of probable Miocene age. Similar observations in central Nepal (Caby *et al.* 1983) and Garhwal, India (Valdiya 1986), indicate that a major structural discontinuity may characterize the Tibetan Zone–Greater Himalayan Zone contact in many segments of the orogen.

Crystalline rocks of the Greater Himalayan Zone occur both in a continuous belt, which roughly coincides with the physiographic Greater Himalaya, and in klippen and half-klippen extending into the physiographic Lesser Himalaya. The zone is characterized by amphibolite facies pelitic to psammitic schists and gneisses, calc-silicate marbles, quartzites, amphibolites, and coarse orthogneisses (Le Fort 1975). Detailed structural studies of the Greater Himalayan Zone (Pêcher 1978; Brunel 1983) indicate polyphase deformational histories, but few workers have demonstrated the existence of major structural discontinuities within the sequence. In many areas, multiple generations of two-mica granitic dikes and sills invade the Greater Himalayan metamorphic rocks, and a series of Upper Oligocene–Miocene leucogranite plutons (e.g. Makalu, Manaslu, Bhagirathi-Badrinath and Nanga Parbat) lie near the top of the zone (Le Fort 1975, 1981). The Greater and Lesser Himalayan Zones are separated by the Main Central Thrust (MCT), a structurally complex zone up to 10 km thick (Pêcher 1978; Brunel 1986). The Lesser Himalayan nappe sequence consists primarily of low-grade metasedimentary rocks with metavolcanic intervals, but the structurally highest portions of some sections contain amphibolite facies assemblages (Le Fort 1975; Valdiya 1980). Unlike the Greater Himalaya, some of the least-metamorphosed strata in the Lesser Himalaya have yielded fossils ranging in age from Upper Precambrian to Lower Eocene (see Stöcklin 1980, for a succinct review). Detailed studies of the Lesser Himalaya in Garhwal (Valdiya 1980, 1981) and in western Nepal (Frank & Fuchs 1970) have demonstrated that the zone is structurally complex, but poor exposure generally limits the quality of mapping in the physiographic Lesser Himalaya and, thus, our understanding of structures. The base of the Lesser Himalayan Zone is defined as the Main Boundary Thrust (MBT) zone, a series of N-dipping faults with a cumulative displacement of demonstrably tens (Heim & Gansser 1939; Stöcklin 1980) and probably hundreds of kilometres (Powell & Conaghan 1973; Molnar 1984). The Miocene–Pleistocene Siwalik molasse (Gansser 1964; Johnson *et al.* 1979), which constitutes the bulk of the Subhimalayan Zone, forms the footwall of the MBT.

Despite the abundance of mesoscopic and macroscopic structures within the six tectonic zones of the Himalaya, most Himalayan geologists believe that much of the convergence within the range was accommodated along the Indus–Tsangpo Suture, the Main Central Thrust, and the Main Boundary Thrust. Various lines of structural, stratigraphic, and geochronologic evidence imply that the suture behaved as an intercontinental subduction zone by Eocene time, accommodating the initial collision of India and Eurasia. With continued convergence, the MCT and MBT formed as intracontinental subduction zones, permitting large-scale imbrication of the downgoing Indian Plate. Although the MBT cuts Plio-Pleistocene molasse units and must be very young, the age of the MCT is less certain. In some areas (e.g. Garhwal) (Valdiya 1980), the age of fossil assemblages exposed beneath klippen of MCT zone rocks in the physiographic Lesser Himalaya require the fault zone to be post-early Eocene. The most often cited argument for the age of the MCT depends on the assumption of a genetic relationship between the thrust and the Upper Oligocene–Miocene leucogranites of the Greater Himalaya (Le Fort 1975). If the south-vergent MCT has a late Oligocene–Miocene age, it could be part of a kinematically complicated group of structures that include north-directed back-thrusts and low-angle normal faults in the southern Tibet (Burg & Chen 1984). Burchfiel & Royden (1985) have suggested that the extensional structures in this group accommodated gravitational collapse of the Miocene topographic front as it was being built by movement on the MCT. By analogy with the western Alps (Milnes 1978), it is tempting to ascribe the present steep dip of the Indus–Tsangpo Suture, as well as the back-thrusts and spectacular back-folds south of the suture, to a ramp in the MCT where it projects beneath southern Tibet. Lyon-Caen & Molnar (1983) attributed the broadly antiformal nature of the Lesser Himalaya, and the presence of erosional remnants of the Greater Himalayan metamorphic sequence within the physiographic Lesser Himalaya, to a similar ramp structure in the MBT.

## INVERTED METAMORPHISM IN THE CENTRAL HIMALAYA

Most models of the thermal history of the Himalaya have focused on one of the most distinctive and controversial characteristics of the metamorphic core of the orogen: 'inverted metamorphism'. Throughout the central and eastern portions of the mountain belt, the basal Greater Himalayan Zone contains mineral assemblages characteristic of intermediate $P-T$ metamorphism (kyanite + staurolite grade). The grade of metamorphism increases structurally upward to upper amphibolite facies (sillimanite ± cordierite) at the structural level occupied by the Upper Oligocene–Miocene leucogranite plutons (Gansser 1964; Thakur 1976). This apparently inverted metamorphic gradient is mimicked in some sections through the Lesser Himalaya; in central Nepal, the metamorphic grade ranges systematically from chlorite roughly 10 km below the principal structural discontinuity in the MCT zone to kyanite + staurolite at the thrust (Pêcher 1978). Pêcher (1978) and Caby et al. (1983) concluded that there is no apparent metamorphic discontinuity associated with the MCT in several central Nepal sections. This view has been disputed by Stöcklin (1980) who found a distinct metamorphic discontinuity across the Mahabarat Thrust, which he considered to be part of the MCT beneath an erosional outliner of the Greater Himalayan sequence (the Kathmandu nappe). In eastern Nepal (Brunel & Kienast 1986) and in Garhwal (Heim & Gansser 1939), the MCT has been mapped as a major metamorphic break, separating kyanite grade rocks in the hanging wall from chlorite grade rocks in the footwall.

Many mechanisms have been invoked to explain inverted metamorphism in the Greater and

Lesser Himalaya (Le Fort 1975, 1981; Thakur, 1980). These include: (1) thermal perturbations and local inversions of the geothermal gradient associated with the intrusion of the Greater Himalayan leucogranites; (2) recumbent folding and/or thrust imbrication of pre-existing 'normal' metamorphic sequences; (3) shear heating along the MCT; and (4) conductive heating of the Lesser Himalaya and concomitant cooling of the basal Greater Himalaya as a result of 'hotter-over-colder' thrusting. In recent years the last of these possibilities has become widely accepted. Such a model (figure 2) is especially appealing because it explains the observed isograd distribution and also provides a mechanism for producing the Greater Himalayan leucogranites: shear heating along the MCT and/or the release of Lesser Himalayan volatiles through prograde metamorphism might trigger anatectic melting near the base of the Greater Himalayan Zone (Le Fort 1975, 1981).

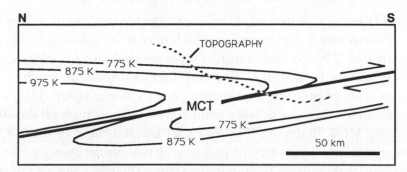

FIGURE 2. The Le Fort (1975) model for the development of inverted isograds in the Greater Himalaya.

If the Le Fort (1975, 1981) model is correct, then numerical models of the thermal effects of thrust faulting and subsequent uplift (Oxburgh & Turcotte 1974; England & Richardson 1977) suggest that metamorphic temperatures near the base of the Greater Himalayan sequence initially should have decreased during heating of the Lesser Himalayan footwall and subsequently increased as the rock column attempted to attain thermal equilibrium. If an influx of fluids from the devolatilizing footwall triggered the development of syntectonic metamorphic assemblages near the base of the hanging wall, and if an appropriate combination of rapid uplift and low radioactive heat production in the footwall prevented substantial re-equilibration of these assemblages during transport to the surface, then thermobarometric studies of the Greater Himalaya could yield direct evidence of the magnitude and length-scale of the temperature inversion caused by thrusting.

## AVAILABLE CONSTRAINTS IN THE $P-T$ EVOLUTION OF THE CENTRAL HIMALAYA

Within the central Himalaya, there have been five quantitative metamorphic studies that provide general insights into the thermal evolution of the Greater Himalaya and specifically yield some of the information necessary to evaluate the model of Le Fort (1975, 1981). From northwest to southeast (figure 1), the sampling areas for these studies were: (1) eastern Garhwal, India (Hodges & Silverberg 1988); (2) west–central Nepal (Le Fort et al. 1986); (3) east–central Nepal (Hodges et al. 1988; Hodges et al., in preparation); (4) the Everest region, eastern Nepal (Hubbard 1988); and (5) the Makalu region, eastern Nepal (Brunel & Kienast 1986). A detailed discussion of the petrologic techniques used in these studies is well beyond the scope of this paper. Suffice it to say that three well-calibrated pelitic thermobarometers provided the bulk of the $P-T$ data: the garnet–biotite geothermometer (Ferry & Spear 1978), the garnet–plagioclase–aluminum silicate–quartz geobarometer (Newton & Haselton 1981), and the garnet–muscovite–biotite–plagioclase geobarometer (Ghent & Stout 1983; Hodges & Crowley 1985). Those readers interested in the accuracy and precision limits of these thermobarometers should refer to Hodges & Crowley (1985) and Hodges & McKenna (1987). In addition, the studies by Hodges & Silverberg (1988), Hodges *et al.* (in preparation) and Hubbard (1988) include attempts to reconstruct the $P-T$ paths followed by individual samples through a combination of Gibbs's method modelling of garnet zoning (Spear & Selverstone 1983) and garnet inclusion thermobarometry (St-Onge 1987).

### Eastern Garhwal, India (Hodges & Silverberg 1988)

In the Garhwal Himalaya, northwest of the Nanda Devi massif (figure 1), the Main Central Thrust Zone is a complicated duplex system, which may be as much at 10 km thick in some sections (Valdiya 1980). The footwall consists of unmetamorphosed to weakly metamorphosed sedimentary rocks of late Precambrian–Eocene(?) age, whereas the hanging wall includes amphibolite facies rocks of probable Precambrian age (Valdiya 1980). The Alaknanda and Dhauli Valleys of eastern Garhwal provide natural cross sections through the 10–12 km thick hanging wall of the MCT (figure 3). The lower 3.1 km and the upper 6.9–8.9 km of these sections are composed predominantly of pelitic and psammitic gneisses; the intermediate 3.8–5.8 km of section is dominated by impure quartzites. Petrographic examination revealed that the two different pelitic–psammitic packages corresponded to two distinctive textural suites. Suite I includes samples collected within 3.1 km structurally above the MCT. These are

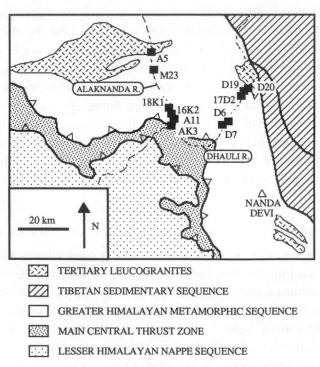

FIGURE 3. Generalized tectonic map of the Alaknanda–Dhauli area, Garhwal, showing sample localities. From Hodges & Silverberg (1988).

characterized by the assemblage: quartz ± muscovite + biotite + plagioclase + garnet ± kyanite. Garnets in Suite I samples are subhedral to anhedral, and were clearly prekinematic with respect to a prominent shear foliation defined by muscovite + biotite ± secondary chlorite. Because Hodges & Silverberg (1988) interpreted this foliation to have been related to movement on the MCT, they believed that these garnets grew before development of the thrust. Suite II samples, collected further than 6.0 km structurally above the thrust zone, contain quartz + biotite + muscovite + plagioclase + garnet ± microcline ± sillimanite. Garnets in these samples are subhedral and synkinematic with respect to the dominant muscovite + biotite + fibrolitic sillimanite schistosity in the rocks. Hodges & Silverberg (1988) inferred that this amphibolite facies schistosity developed significantly before the greenschist facies shear foliation observed in the structurally lower Suite I samples. Despite the clear textural distinctions between the Suite I and Suite II samples, existing geologic maps of the area (Valdiya 1979) indicate no major structural discontinuities between the upper and lower pelitic to psammitic sequences.

Conventional rim thermobarometry, garnet inclusion thermobarometry, and Gibbs's method modelling of garnet zoning yield $P-T$ paths for the Suite I samples (AK3, A11, D7, D6, 16K2 and 18K1), which indicate nearly 15 km of uplift subsequent to tectonic burial at depths of at least 36 km. The Suite II samples give very different $P-T$ paths that indicate an 80–90 K temperature increase and 5–7 km of burial. The apparent differences in Suite I and Suite II thermal histories, despite the lack of post-metamorphic structural discontinuity between the two samples suites, suggest that the different assemblages grew during two distinct metamorphic events: an early high $P$–high $T$ event (M1), and a subsequent moderate $P$–high $T$ event (M2). Preliminary Ar–Ar data for hornblende from the basal part of the sequence (P. Zeitler 1987, unpublished data) suggest a pre-late Eocene age for the first metamorphic event. Hodges & Silverberg (1988) believed that this event was associated with the early stages of India–Asia collision. M2 effects are strongest in the upper portions of the Greater Himalayan metamorphic sequence, near large leucogranite plutons (figure 3) and within a zone of intense migmatization. These relations suggest that the second metamorphic event was clearly related to leucogranite magmatism. In Garhwal, available geochronologic data (Seitz et al. 1976; Stern et al. 1988) do not closely constrain the age of these granites, but we infer a late Oligocene(?)–Miocene age by analogy with other Greater Himalayan leucogranites that have been dated more precisely (Schärer 1984; Deniel 1985; Schärer et al. 1986).

## West–central Nepal (Le Fort et al. 1986)

Le Fort et al. (1986) presented conventional rim thermobarometric data for samples collected in the Kali Gandaki drainage south of Annapurna (figure 1). The samples represent the lower 3 km of the Greater Himalayan sequence and include one sample collected 200 m below the MCT zone (figure 5). In general, apparent temperatures increase downward between 3 and 1 km above the thrust; one sample indicates a moderate decrease in temperature just above the MCT (figure 6a). Le Fort et al. (1986) interpreted the low garnet–biotite temperature for the sample collected just above the thrust as indicative of conductive cooling of the hanging wall during thrust emplacement (figure 2). The thermobarometric pressure gradient is roughly twice that of a normal lithostat (figure 6b), indicating substantial re-equilibration of the samples during uplift.

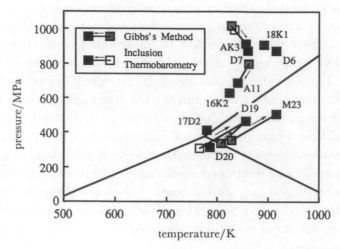

FIGURE 4. $P-T$ paths for the samples shown in figure 3. Arrows indicate core to rim $P-T$ trajectories calculated by using Gibbs's method (core $P-T$ indicated by shaded boxes, rim $P-T$ indicated by solid boxes) and inclusion thermobarometry (core $P-T$ indicated by open boxes, rim $P-T$ indicated by solid boxes). From Hodges & Silverberg (1988).

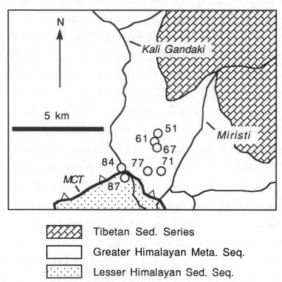

FIGURE 5. Simplified tectonic map of the Kali Gandaki–Miristi area, west–central Nepal, showing sample localities. After Le Fort *et al.* (1986).

### East-central Nepal (*Hodges et al.* 1988)

Hodges *et al.* (1988) studied a suite of pelitic samples from the Darondi and Burhi Gandaki drainages southeast of Manaslu (figure 1). These samples represent a 12 km structural cross section of the Greater Himalayan sequence (figure 7). Caby *et al.* (1983) and Pêcher & Le Fort (1986) documented textural evidence for two prograde metamorphic events in the area: an early, high $P$–high $T$ phase, and a later, intermediate $P$–high $T$ phase. Thermobarometric data indicate that the second event was sufficiently widespread and intense that no record of the high $P$–high $T$ event was recorded by mineral chemistry. There is excellent correspondence between the thermobarometric pressure gradient and a nominal lithostatic gradient (*ca.* 27 MPa km$^{-1}$; figure 8$a$). This result is not predicted by thermal models of simple burial

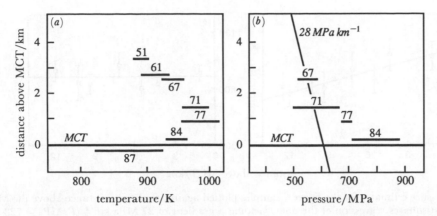

FIGURE 6. (a) Calculated temperatures for figure 5 samples plotted against structural distance from the MCT. (b) Calculated pressures against structural distance from the MCT. Reference lithostatic gradient (28 MPa km⁻¹) shown for reference. After Le Fort *et al.* (1986).

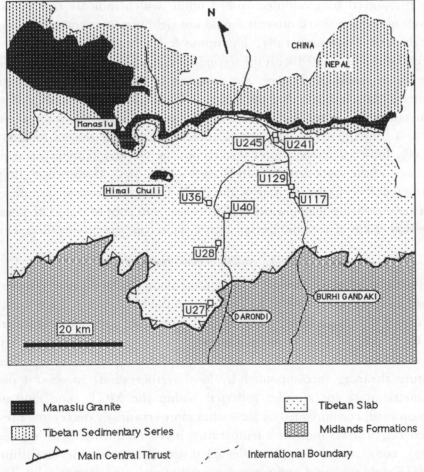

FIGURE 7. Simplified tectonic map of the Burhi Gandaki–Darondi region, east–central Nepal, showing sample localities. After Hodges *et al.* (1988).

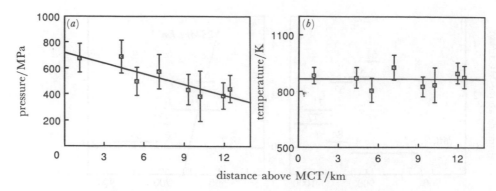

FIGURE 8. (a) Pressure estimates for the figure 7 samples plotted against structural distance above the MCT. Line indicates least-squares regression of the data, yielding a gradient of 27 MPa km$^{-1}$ ($P$/MPa = 723–27$z$/km). (b) Temperature estimates for the Figure 7 samples plotted against structural distance. Line shows mean $T = 870$ K. From Hodges *et al.* (1988).

metamorphism followed by erosion-controlled uplift, which indicate that rocks at different structural levels in compressional orogens should not yield pressures corresponding to a normal lithostat (England & Thompson 1984; Thompson & England 1984).Thus, the data suggest that the thermal pulse associated with the second metamorphic phase was short lived and was followed by rapid cooling. Despite the reasonable lithostatic pressure gradient, the thermobarometric data indicate that the entire 12 km section was roughly isothermal during the second metamorphic event (figure 8b). This surprising result was interpreted by Hodges *et al.* (1988) as a consequence of widespread anatexis during the second metamorphic event, which effectively buffered temperatures throughout the Greater Himalayan sequence.

As an extension of the work described in Hodges *et al.* (1988), Hodges *et al.* (in preparation) have reconstructed $P$–$T$ paths for four of the Burhi Gandaki–Darondi samples. All of these samples indicate increasing temperature and pressure from core to rim. The pressure increase implies more than 4 km of tectonic burial during the second metamorphic event.

*Everest region, eastern Nepal (Hubbard* 1988)

Hubbard (1988) studied metamorphic conditions along the Dudh Kosi, Hinku, and Hongu drainages south of Mt Everest (figure 1). In this area the MCT is a 3–5 km thick zone, which separates kyanite ± sillimanite grade hanging wall rocks from garnet or biotite grade footwall rocks (figure 9). Within the zone, strain was markedly inhomogeneous. Although narrow, high-tempreature mylonite zones are distributed throughout the sequence, some evidence exists for low-tempreature shearing (accompanied by local retrogression) in several discrete zones. Thermobarometric data for samples collected within the MCT zone indicate increasing temperatures upward, accompanied by somewhat more erratically decreasing pressures (figure 10a, b). In general, the data indicate a tempreature inversion during thrusting, consistent with Le Fort's (1975, 1981) model. Samples collected at various structural levels within the hanging wall of the MCT yield pressures and temperatures that vary unsystematically. This behaviour was interpreted as a consequence of either: (1) patchy, late-stage heating in the gneiss sequence associated with the injection of leucogranites; (2) post-metamorphic faulting, for which there is some field evidence: or (3) some combination of these. $P$–$T$ paths calculated for MCT zone samples by using Gibbs's method modelling and inclusion thermobarometry vary from roughly

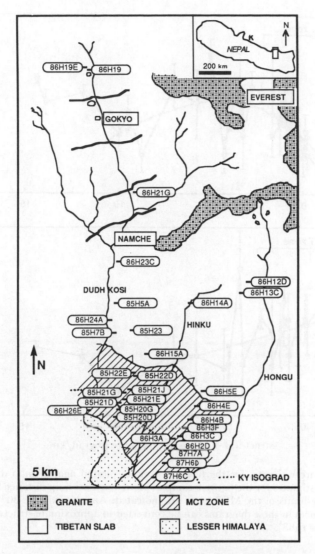

FIGURE 9. Simplified tectonic map of the Everest region, eastern Nepal, showing
sample localities. From Hubbard (1988).

isobaric cooling to roughly isothermal decompression, indicating a complicated thermal history
for rock units within the zone before final mineral rim equilibration.

*Makalu region, eastern Nepal (Brunel & Kienast 1986)*

Brunel & Kienast (1986) analysed a collection of pelitic samples from the Greater
Himalayan gneisses of eastern Nepal, near Makalu (figure 1). Although most of the samples
came from the main outcrop belt of the gneisses in the upper Barun Valley, some were collected
from erosional outliers that are part of the Kathmandu Klippe. Together the samples represent
the lowermost 3 km of the Greater Himalayan sequence. Two phases of prograde
metamorphism were identified petrographically: an early phase, during which kyanite ±
staurolite assemblages were produced in the lower part of the section; and a later phase,
during which sillimanite ± cordierite assemblages crystallized in the upper part of the section.
Garnet–biotite and garnet–plagioclase–aluminum silicate–quartz thermobarometry were used

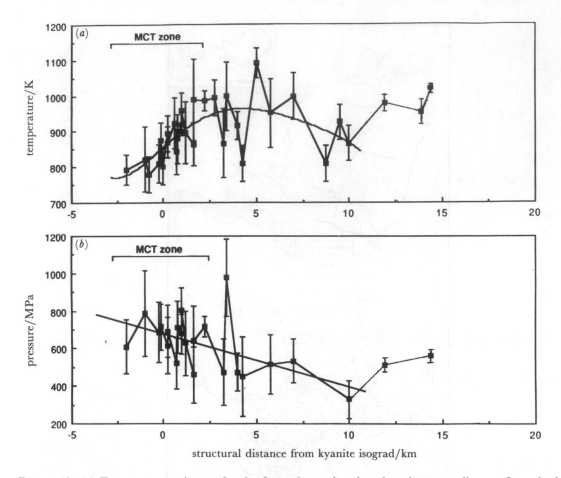

FIGURE 10. (*a*) Temperature estimates for the figure 9 samples plotted against map distance from the kyanite
isograd. (*b*) Pressure estimates for the figure 9 samples plotted against map distance from the kyanite isograd.
In both diagrams, the limits of the MCT zone are indicated. An approximate 30 °N dip of the MCT zone
means that map distances in these diagrams can be converted to approximate structural distances by dividing
by 2. From Hubbard (1988).

to estimate conditions of 825–925 K and 600–900 MPa (equivalent to depths of 22–33 km) for
the first event near the base of the slab, and 785–990 K and 350–500 MPa (13–19 km) for the
second event near the Makalu leucogranite pluton (figure 11). There is no systematic relation
between structural level and calculated *P–T* conditions for either event. Based on metamorphic
textures and the thermobarometric data, Brunel & Kienast (1986) inferred *P–T* paths that
involved: (1) high-pressure metamorphism of the entire Greater Himalayan sequence; (2)
decompression of the basal part of the sequence between the two metamorphic events; and (3)
selective heating of the upper part of the sequence, associated with intrusion of the Makalu
granite, during decompression.

### CONTROLLING FACTORS IN THE THERMAL EVOLUTION OF THE HIMALAYA

Although two of the studies outlined above (Hubbard 1988; Le Fort *et al.* 1986) generally
support the thermal model of Le Fort (1975, 1981), it seems clear from the data that the
Greater Himalayan sequence in Garhwal and Nepal experienced at least two prograde

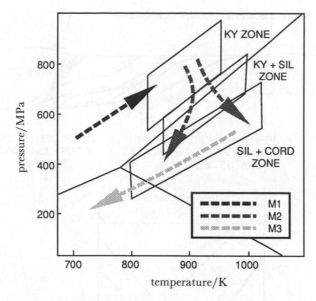

FIGURE 11. *P–T* trajectories inferred by Brunel & Kienast (1988) from *P–T* data for the Makalu area, eastern Nepal. Rhombic fields indicate range of calculated *P* and *T* for kyanite, kyanite + sillimanite, and sillimanite + cordierite zone samples. After Brunel & Kienast (1986).

metamorphic events in Tertiary time. This observation implies that the Tertiary thermal structure of the Himalayan metamorphic core was influenced by a variety of tectonic processes in addition to movement on the MCT. It is convenient to think of these processes in the context of five tectonothermal stages in the history of the central Himalaya. These stages are depicted in figure 12 as a series of schematic cross sections through the central Himalaya, accompanied by generalized *P–T* paths for an arbitrary sample from the middle of the Greater Himalayan sequence.

### Early continental subduction: middle Eocene to early Oligocene

Studies of continent–continent collisional belts suggest that some fraction of the post-collisional shortening between the continents involves 'subduction' of portions of the continental lithosphere (Hodges *et al.* 1982). This may involve interplate ('B-type') subduction of the leading edge of one of the continental masses, or intraplate ('A-type') subduction in one or both masses (terminology after Bally 1980). Geologic and geophysical data indicate that the leading edge of India became involved in both of these processes during Himalayan orogenesis (Roecker 1982; Mattauer 1986). Although many studies have emphasized the importance of the Main Central and Main Boundary Thrusts in the development of the Himalaya, recent mapping in southern Tibet (Burg & Chen 1984) and northwest India (Searle 1986) has demonstrated the existence of numerous compressional structures north of and structurally above the Greater Himalayan sequence. Some of these structures appear to have accommodated several tens of kilometres of crustal shortening before development of the MCT (Burg & Chen 1984).

It seems likely that the Greater Himalayan sequence experienced high-pressure, high-tempreature metamorphism as a consequence of early Tertiary A-type and B-type subduction of the Indian Plate margin, and we infer that this was the early metamorphic event recognized in Nepal and Garhwal (figure 12a). Thermobarometric data from several transects indicate

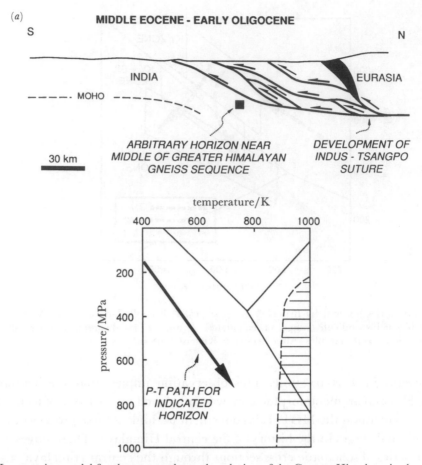

FIGURE 12. Interpretive model for the tectonothermal evolution of the Greater Himalaya in the central Himalaya. Half arrows on the schematic cross sections indicate active structures during the time intervals shown. Solid arrows indicate $P–T$ trajectory during each time interval for the arbitrary horizon marked by a solid box in the cross sections. Field with horizontal lines in the $P–T$ diagrams indicates approximate conditions of water-saturated anatexis of pelites. See text for further explanation.

that the basal portions of the metamorphic sequence were tectonically buried to depths in excess of 30 km.

### Early uplift: early to late Oligocene

The Garhwal and eastern Nepal transects yield data which indicate that the Greater Himalayan sequence experienced a minimum of 10 km of uplift after high-pressure, high-temperature metamorphism and before intermediate-pressure, high-temperature metamorphism (figure 9b). The Garhwal $P–T$ paths are similar to those predicted by theoretical models of 'erosion-controlled' uplift (England & Richardson 1977), and do not clearly suggest that tectonic denudation played a role in their unroofing. None of the available data constrain uplift rates during this interval.

### Late heating and burial: late Oligocene

The top of the Greater Himalayan sequence is characterized by the occurrence of leucogranite sills, dikes and plutons that commonly yield late Oligocene–middle Miocene crystallization ages. Metamorphic studies in Garhwal, central Nepal and eastern Nepal

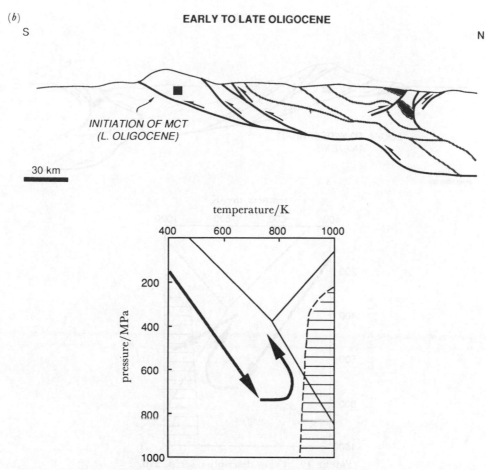

FIGURE 12. (*b*) For description see opposite.

demonstrate that these granites were intimately associated with sillimanite ± cordierite grade metamorphic assemblages in the surrounding country rocks. An abundance of geochemical data (Le Fort *et al.* 1987) indicates that the granites are anatectic melts of portions of the Greater Himalayan Metamorphic Sequence. In some sections (e.g. Burhi Gandaki–Darondi), the structurally high granite plutons appear to have been derived from the presently exposed, structurally lower portions of the gneiss sequence. In other sections (e.g. Alaknanda–Dhauli), granites are conspicuously rare in the basal part of the sequence, and the granites must be *in situ* melts or, alternatively, they must have an unexposed provenance. Where granites are common throughout the Greater Himalayan sequence, anatexis may have buffered the tempreature of the sequence for a significant period of time (Hodges *et al.* 1988).

The relation between the second metamorphic event, the Greater Himalayan leucogranites, and movement on the MCT remains unclear. Textural relations and thermobarometric data suggest that the intermediate-pressure, high-temperature metamorphic event in central Nepal and in the Dudh Kosi–Hongu–Hinku transect of eastern Nepal was synchronous with early movement along the MCT. However, most transects show some evidence of post-metamorphic movement along the MCT and in some areas the MCT appears to be exclusively post-metamorphic. In practice, the MCT is mapped at the boundary between the physiographic Greater and Lesser Himalaya. It seems clear that this topographic break does not everywhere

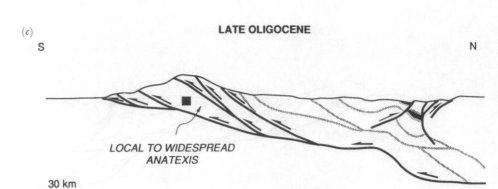

(c)                                LATE OLIGOCENE

S                                                                              N

LOCAL TO WIDESPREAD
ANATEXIS

30 km

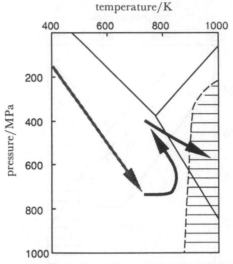

FIGURE 12. (c) For description see p. 270.

correspond to the same fault, and this diachroneity is a major contributing factor to the
controversy concerning the thermal significance of the 'MCT'.

Some of the most thought-provoking results of the Garhwal and east–central Nepal
petrologic research were $P-T$ paths indicative of several kilometres of tectonic burial during
intrusion of the leucogranites and second-stage metamorphism. The similarity of these paths for
samples from widely separated areas indicates that the burial event affected a large portion of
the metamorphic core of the Himalaya (figure 12c). Many compressional structures in
southern Tibet and northwest India appear to be the right age to account for this burial (Burg
& Chen 1984; Searle 1986), but the general lack of detailed mapping in much of the Tibetan
Sedimentary Zone limits our understanding of the process.

The ultimate heat source for anatectic melting of the Greater Himalayan gneisses remains
the greater unanswered question concerning the thermal evolution of the Himalaya. A variety
of explanations have been proposed.

Le Fort (1981) suggested that melting was triggered by the influx of metamorphic fluids
from the footwall of the MCT, and that the hanging wall had retained enough residual heat
from the first metamorphic event to permit melting. Unfortunately, if the preliminary
hornblende Ar–Ar data from Garhwal are to be trusted, then the Greater Himalaya (in that
region, at least) had cooled to below 775–825 K (the nominal range of closure temperatures
for Ar in hornblende; Harrison 1981) well before the anatectic event.

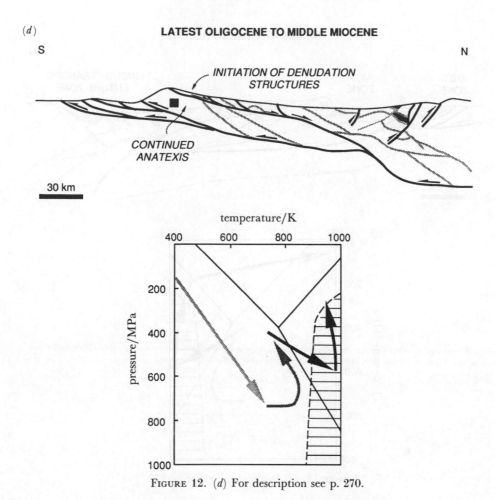

FIGURE 12. (d) For description see p. 270.

Some workers (Le Fort 1975; Scholtz 1980) have attributed the melting to frictional heating along the MCT. Although this phenomenon may have contributed some heat to the system, it is unlikely to have been the dominant cause of melting (Molnar *et al.* 1983).

Bird (1978) proposed that the Greater Himalaya could have been heated as a consequence of delamination of the Indian lithosphere, subduction of the lower lithosphere, and consequence upwelling of the asthenosphere. Stern *et al.* (1988) have suggested that this process may have triggered anatectic melting. This hypothesis is virtually impossible to test, but it seems unusual that it would have produced melting in such a restricted area.

Jaupart & Provost (1985) attributed the melting to 'heat focusing' near the top of the Greater Himalayan sequence. In effect, they suggested that the contrast in thermal conductivity across a thrust contact between the Greater Himalayan gneisses and the Tibetan sedimentary sequence acted as a thermal barrier that led to unusually high temperatures at the top of the gneiss sequence. There are two problems with this model: (1) the predicted temperature discordance is not sufficiently abrupt to produce the distinct metamorphic break observed at this contact; and (2) where the Greater Himalayan sequence – Tibetan sedimentary sequence contact has been mapped as a fault, sense of shear indicators suggest normal rather than reverse movement, and the structure is demonstrably post-metamorphic (Burg & Chen 1984; Burchfiel *et al.* 1986; Herren 1987).

Another possibility is that the limited distribution of anatectic melts reflects local

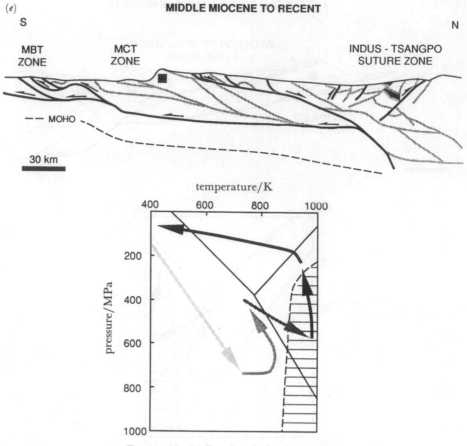

FIGURE 12. (*e*) For description see p. 270.

concentrations of heat-producing elements in the Greater Himalayan sequence (Pinet & Jaupart 1987). We have been impressed by the fact that many of the pelitic samples that we have studied include up to 1 modal % zircon, monazite, apatite and xenotime. Vidal *et al.* (1982) determined U, Th, K concentrations in a few Greater Himalayan gneiss samples from central Nepal that imply heat production rates of up to 3 mW m$^{-3}$, roughly three times greater than 'nominal' heat production rates in the Earth's crust (Jaupart & Provost 1985). Heat production of this magnitude would have a profound influence on the thermal structure of the gneiss sequence. If the heat-producing elements were unevenly distributed, then it seems plausible that more radioactive portions of the sequence might have melted during late Oligocene burial whereas less radioactive portions did not. Although Pinet & Jaupart (1987) argue for the importance of this process based on the trace-element chemistry of the leucogranites, we know too little as yet about heat-production rates in the gneisses to realistically evaluate the affects of internal heat production on the thermal evolution of the Greater Himalaya.

*Tectonic denudation: latest Oligocene to middle Miocene*

An increasing body of structural and geochronological data indicate that large-scale low-angle normal faulting occurred along the Greater Himalayan–Tibetan series contact in latest Oligocene–middle Miocene time (Burg *et al.* 1984; Burchfiel *et al.* 1986; Herren 1987).

Although we have no precise estimates, we feel that the amount of tectonostratigraphic throw on these structures must have been large. In the Rongbuk Valley of southern Tibet, for example, the principle normal fault zone places essentially unmetamorphosed Ordovician strata onto Greater Himalayan lithologies, which were apparently metamorphosed to sillimanite grade in late Oligocene–early Miocene time.

The structural characteristics of these fault zones are analogous with the detachment systems that separate metamorphic core complexes of the North American Cordillera from their unmetamorphosed structural cover (Coney 1980). Numerical models of the thermal consequences of detachment development (Furlong & Londe 1986; England & Jackson 1987; Ruppel *et al.* 1988) demonstrate that footwall metamorphic rocks experience $P-T$ trajectories characterized by substantial decompression with little cooling followed by rapid cooling at shallow levels. Three lines of evidence, all circumstantial, suggest to us that tectonic denudation may have had an important thermal effect on the Greater Himalayan sequence (figure 12 *d*). First, we have seen little indication in the M2 assemblages from Garhwal or Nepal for substantial re-equilibration during cooling from peak temperatures. Such re-equilibration is ubiquitous in high-grade metamorphic rocks from most compressional belts (Tracy *et al.* 1976; Hodges & Royden 1984), and its virtual absence in the central Himalaya may indicate rapid cooling from near-peak temperatures. Second, the apparent preservation of a lithostatic pressure gradient in the Burhi Gandaki–Darondi section, as well as inverted temperature gradients immediately above the MCT in the Kali Gandaki and Everest transects, suggest 'quenching' of the Greater Himalaya rather than slow uplift.

Our third argument in favour of tectonic denudation involves the observation that leucogranite plutons in the Greater Himalaya yield a range of high-precision radiometric ages: in the Manaslu and Everest leucogranite suites, Deniel *et al.* (1987) and Schärer *et al.* (1986) have documented multiphase intrusive events that occurred over an interval of several million years. Tectonic denudation of the Greater Himalaya after late Oligocene burial could result in a $P-T$ path that remained within the region of water-saturated granite melting over much of the latest Oligocene–middle Miocene interval (figure 12 *d*).

### Final uplift: middle Miocene to Recent

In the central Himalaya, final uplift of the metamorphic core was accommodated by: (1) simple isostatic readjustment of a thickened crust; (2) movement over ramps in major faults which are structurally lower (e.g. the Main Boundary Fault); and (3) continued tectonic denudation (figure 12 *e*). Pressure estimates from the petrologic studies cited above indicate that the Greater Himalaya have experienced roughly 15–20 km of uplift since late Oligocene–early Miocene time, for an average uplift rate of 0.6–0.8 mm a$^{-1}$. Exactly how much post-Oligocene uplift occurred before the middle Miocene and how much occurred after remains poorly constrained. Based on the arguments presented above, we believe that several kilometres of tectonic denudation occurred before middle Miocene time at rates significantly greater than 0.6–0.8 mm a$^{-1}$.

### Conclusions

Reconstructions of the $P-T$ evolution of the Greater Himalayan sequence in five widely spaced areas in the central Himalaya suggest a complicated Tertiary thermal history. Several of the areas exhibit clear evidence for an early intermediate-to high-pressure, intermediate-temperature thermal event, followed by a later low-to intermediate-pressure, high-temperature event. We believe that the early event can be attributed to middle Eocene–early Oligocene loading of the sequence as a consequence of intracontinental subduction. The later event appears to be intimately associated with high-temperature movements of the MCT and the generation and intrusion of leucogranites in latest Oligocene–middle Miocene time. Calculated $P$–$T$ paths uniformly indicate that the second event involved several kilometres of tectonic burial, presumably as a consequence of structurally higher thrust imbrications. The lack of substantial high-temperature retrogression of the assemblages produced in the second event, the preservation of inverted temperature gradients and normal lithostatic pressure gradients above the MCT, and the multi-episodic nature of leucogranite production in the Greater Himalaya imply that tectonic denudation by movement on north-dipping normal fault systems resulted in rapid uplift of the sequence over the latest Oligocene–middle Miocene interval.

We thank B. C. Burchfiel, P. Le Fort, P. Molnar, R. Parrish, A. Pêcher, L. H. Royden and K. S. Valdiya for fruitful discussions on Himalayan tectonics. Our research has been funded primarily through a U.S. National Science Foundation grant to K. V. H. and P. Molnar, with additional grants to M. S. H. from the American Alpine Club, the Explorers Club, and the Geological Society of America, and to D. S. S. from Sigma Xi and the MIT EAPS Student Research Fund.

## References

Bally, A. W. *et al.* 1980 U.S. Geol. Surv. Open File Report 80–501. (100 pages.)
Bird, P. 1978 *J. geophys. Res.* **83**, 4975–4987.
Brookfield, M. E. & Reynolds, P. H. 1981 *Earth planet. Sci. Lett.* **55**, 157–162.
Brunel, M. 1983 Ph.D. thesis, University of Paris. (381 pages.)
Brunel, M. 1986 *Tectonics* **5**, 247–265.
Brunel, M. & Kienast, J. R. 1986 *Can. J. Earth Sci.* **23**, 1117–1137.
Burchfiel, B. C. & Royden, L. H. 1985 *Geology* **13**, 679–682.
Burchfiel, B. C., Hodges, K. V. & Royden, L. H. 1986 *Geol. Soc. Am. Abs. Prog.* **18**, 553.
Burg, J. P. & Chen, G. M. 1984 *Nature, Lond.* **311**, 219–223.
Burg, J. P., Brunel, M., Gapais, D., Chem, G. M. & Liu, G. H. 1984 *J. struct. Geol.* **6**, 535–542.
Caby, R., Pecher, A. & Le Fort, P. 1983 *Rev. Geol. dyn. Geog. phys.* **24**, 89–100.
Coney, P. J. 1980 In *Cordilleran metamorphic core complexes* (ed. M. D. Crittenden *et al.*), *Geol. Soc. Am. memoir* no. 157, pp. 4–34.
Deniel, C. *et al.* 1987 *Contr. Miner. Petr.* **96**, 78–92.
England, P. & Jackson, J. 1987 *Geology* **15**, 291–294.
England, P. C. & Richardson, S. W. 1977 *J. geol. Soc. Lond.* **134**, 301–313.
England, P. C. & Thompson, A. B. 1984 *J. Petr.* **25**, 894–928.
Ferry, J. M. & Spear, F. S. 1978 *Contr. Miner. Petr.* **66**, 113–117.
Frank, W. & Fuchs, G. R. 1970 *Geol. Rdsch.* **59**, 552.
Furlong, K. P. & Londe, M. D. 1986 In *Extensional tectonics of the southwestern United States: a perspective on processes and kinematics*, (ed. L. D. Mayer). Geol. Soc. Am. Special Paper no. 208, pp. 23–30.
Gansser, A. 1964 *Geology of the Himalayas.* (289 pages.) London: Wiley Interscience.
Ghent, E. D. & Stout, M. Z. 1981 *Contr. Miner. Petr.* **76**, 92–97.
Harrison, T. M. 1981 *Contr. Miner. Petr.* **78**, 324–331.
Heim, A. & Gansser, A. 1939 *Mém Soc. helv. Sci. Nat.* **73**, 1–245.
Herren, E. 1987 *Geology* **15**, 409–413.
Hodges, K. V., Bartley, J. M. & Burchfiel, B. C. 1982 *Tectonics* **1**, 441–462.
Hodges, K. V. & Crowley, P. 1985 *Am. Miner.* **70**, 702–709.
Hodges, K. V. & McKenna, L. W. 1987 *Am. Miner.* **72**, 671–680.
Hodges, K. V. & Royden, L. H. 1984 *J. geophys. Res.* **89**, 7077–7090.
Hodges, K. V. & Silverberg, D. S. 1988 *Tectonics* (In the press.)
Hodges, K. V., Le Fort, P. & Pêcher, A. 1988 *Geology.* (In the press.)

Honneger, K., Dietrich, V., Frank, W., Gansser, A., Thoni, M. & Trommsdorff, V. 1982 *Earth planet. Sci. Lett.* **60**, 253–292.

Hubbard, M. 1988 *J. metamorph. Geol.* (In the press.)

Jaupart, C. & Provost, A. 1985 *Earth planet. Sci. Lett.* **73**, 385–397.

Johnson, G. D., Johnson, N. M., Opdyke, N. D. & Tahirkheli, R. A. K. 1979 In *Geodynamics of Pakistan*, (ed. A. Farah & K. A. De Jong), pp. 149–166. Quelta: Geol. Surv. Pakistan.

Le Fort, P. 1975 *Am. J. Sci.* **275** A, 1–44.

Le Fort, P. 1981 *J. geophys. Res.* **86**, 10545–10568.

Le Fort, P. 1986 In *Collision tectonics* (ed. M. P. Coward & A. C. Ries), Geological Society of London Special Publication no. 19, pp. 159–172 London: Blackwell.

Le Fort, P., Cuney, M., Deniel, C., France-Lanord, C., Sheppard, S. M. F., Upreti, B. N. & Vidal, P. 1987 *Tectonophysics* **134**, 29–57.

Le Fort, P., Pêcher, A. & Upreti, B. N. 1986 In *Sciences de la Terre Memoire 47* (ed. P. Le Fort, C. Colchen & C. Montenat), pp. 211–228. Foundation Scientifique de la Geologie et de ses Applications.

Lyon-Caen, H. & Molnar, P. 1983 *J. geophys. Res.* **88**, 8171–8191.

Maluski, M., Proust, F. & Xiao, X. C. 1982 *Nature, Lond.* **298**, 152–154.

Mattauer, M. 1986 In *Collisional Tectonics* (ed. M. P. Coward & A. C. Ries), Geol. Soc. Lond. Special Publication no. 19, pp. 37–50.

Milnes, A. G. 1978 *Tectonophysics* **47**, 369–392.

Molnar, P. 1984 *A. Rev. Earth planet. Sci.* **12**, 489–518.

Molnar, P., Chen, W.-P. & Padovani, E. 1983 *J. geophys. Res.* **88**, 6415–6429.

Newton, R. C. & Haselton, H. T. 1981 In *Thermodynamics of minerals and melts* (ed. R. C. Newton *et al.*), pp. 131–147. New York: Springer-Verlag.

Oxburgh, E. R. & Turcotte, D. 1974 *Schweiz. miner. petrogr. Mitt.* **54**, 641–662.

Pêcher, A. 1978 Doctorates–Sciences, University of Grenoble. (354 pages.)

Pêcher, A. & Le Fort, P. 1986 In *Sciences de la Terre Memoire 47* (ed. P. Le Fort, M. Colchen & C. Montenat), pp. 285–309. Foundation Scientifique de la Geologie et de ses Applications.

Pinet, C. & Jaupart, C. 1987 *Earth planet. Sci. Lett.* **84**, 87–99.

Powell, C. M. & Conaghan, P. 1973 *Earth planet. Sci. Lett.* **20**, 1–12.

Roecker, S. W. 1982 *J. geophys. Res.* **87**, 945–959.

Ruppel, C., Royden, L. & Hodges, K. V. 1988 (In preparation.)

Schärer, U. 1984 *Earth planet. Sci. Lett.* **67**, 191–204.

Schärer, U., Allègre, C. J. & Xu, R. H. 1986 *Earth planet. Sci. Lett.* **77**, 35–48.

Scholtz, C. H. 1980 *J. geophys. Res.* **85**, 6174–6184.

Searle, M. P. 1983 *Episodes* **4**, 21–26.

Searle, M. P. 1986 *J. struct. Geol.* **8**, 923–936.

Seitz, J. F., Tewari, A. P. & Obradovich, J. 1976 *Geol. Surv. India Misc. Publ.* **24**(2), 332–337.

Spear, F. S. & Selverstone, J. 1983 *Contr. Miner. Petr.* **83**, 348–357.

St-Onge, M. R. 1987 *J. Petr.* **28**, 1–27.

Stern, C. R., Kligfield, R., Schelling, D., Virdi, N. S., Futa, K., Peterman, Z. E. & Amini, H. 1988 (In preparation.)

Stöcklin, J. 1980 *J. geol. Soc. Lond.* **137**, 1–34.

Thakur, V. C. 1976 *Eclog. geol. Himalaya* **268**, 433–442.

Thakur, V. C. 1980 In *Thrust and nappe tectonics* (ed. K. R. McClay & N. J. Price), pp. 381–392. Oxford: Blackwell.

Thompson, A. B. & England, P. C. 1984 *J. Petr.* **25**, 929–955.

Tracy, R. J., Robinson, P. & Thompson, A. B. 1976 *Am. Miner.* **61**, 762–765.

Valdiya, K. S. 1979 *J. geol. Soc. India* **20**, 145–157.

Valdiya, K. S. 1980 *Geology of the Kumaun Lesser Himalaya*. Dehra Dun: Wadia Institute of Himalayan Geology. (291 pages.)

Valdiya, K. S. 1981 In *Zagros-Hindu Kush-Himalaya, Geodynamic Evolution* (ed. H. K. Gupta & F. M. Delaney), pp. 87–110. Washington, D.C.: American Geophysics Union.

Valdiya, K. S. 1986 *Geol. Soc. Am. Abs. Prog.* **18**, 778.

Vidal, P., Cocherie, A. & Le Fort, P. 1982 *Geochim. cosmochim. Acta* **64**, 2274–2292.

Windley, B. F. 1983 *J. geol. Soc. Lond.* **140**, 849–865.

## Discussion

P. J. TRELOAR (*Department of Geology, Imperial College, London, U.K.*). Professor Hodges equates an 'M1' metamorphism with deformation associated with the MCT, and an 'M2' metamorphism with post-MCT crustal thickening related to breakback thrusting in the MCT hanging wall. 'M1' followed an earlier Barrovian-type regional metamorphism, which I shall call 'M0', which was presumably related to pre-MCT post-collisional crustal thickening. 'M0' may be the same age, tectonically if not necessarily temporally, as the regional metamorphism described in Zanskar where the isograds were subsequently folded (Searle *et al.* this symposium) and in N Pakistan where they were subsequently imbricated during late-stage thrusting (Coward *et al.* this symposium). Is there evidence in Professor Hodges's area that, as in Zanskar and N Pakistan, the 'M0' metamorphism accompanied a normal, as opposed to inverted, thermal gradient? The thrusting along the MCT may be part of the same broad deformation event as both the recumbent folding of isograds in the MCT hanging wall in Zanskar, and the thrusting and imbrication of isograds in N Pakistan along structures that appear to be analogous to the MCT. The continuation of metamorphism post-MCT (the 'M2' event) in Nepal, suggests that metamorphism there continued later into the deformation sequence than in the areas farther west where the MCT-age deformation was not followed by a major late phase of regional metamorphism. Is there a simple regional or thermal explanation for this?

The status of the MCT-related inverted metamorphism seems problematic. Is it: (*a*) a tectonic overfolding of earlier isograds, the original Le Fort model; (*b*) a tectonic imbrication due to post-metamorphic thrusting within the MCT zone; (*c*) an inverted metamorphism driven by downward heating from an overlying slab; or a combination of all three? Searle (this symposium) favours (*a*) for Zanskar and (unpublished) for Darjeeling; Professor Hodges appears to favour (*b*) for the Garhwal schuppen zone; whereas Hubbard (cited by Professor Hodges) appears to favour (*c*) although acknowledging a substantial syn- to post-metamorphic modification of the inverted profile. Is it possible to combine all of these disparate interpretations into a single model, possibly involving a syn-MCT nappe (locally a fold nappe, locally a thrust nappe) developed in the MCT hanging wall. This would have recumbently folded earlier isograds, in places telescoping them or even cutting them out altogether along thrust surfaces along the inverted limb within the MCT zone. The complete cutting out of parts of the metamorphic sequence would create the right local conditions for the Hubbard model of downward heat transfer. In such an overall model the MCT, a ductile shear zone of variable (up to 10 km) thickness, could be viewed as a ductile detachment zone under the nappe within which an earlier metamorphic sequence could be inverted, telescoped and imbricated along a number of shear surfaces with a local, and maybe rare, direct imposition of hot slabs on top of cold ones driving a second stage syn-MCT inverted metamorphism. In such a complicated zone, and the use of the term 'zone' rather than 'thrust' is deliberate, it may not always be possible to separate which of the possible causes of 'inverted metamorphism' is dominant at any particular locality.

K. V. HODGES. Albert Einstein has been quoted as saying that explanations of physical phenomena should be as simple as possible, but no simpler. Inverted metamorphic field gradients are common in a variety of tectonic settings, and it is safe to assume that different mechanisms could be responsible for this phenomenon in different places. In the Himalaya specifically, I am not convinced that a single model satisfies all of the data gathered throughout the orogen. In some areas (such as the Kali Gandaki and Dudh Kosi drainages) the simple Le Fort (1975, 1981) model of conductive cooling of an overthrust MCT nappe appears to be consistent with the vast majority of the petrologic data. In other areas, models involving post-metamorphic imbrications or recumbent folding of pre-existing isograds seem to provide better fits to the data. Dr Treloar's suggestion of a 'unified' hypothesis to explain the general relations between inverted metamorphism and the MCT is certainly worthy of further scrutiny, but I think that we have to be careful not to over-generalize a complicated process. We must recognize that 'Main Central Thrust' has become an unfortunate generic term for any fault zone that separates the Greater and Lesser Himalaya. It is quite likely that the MCT as mapped in Garhwal is totally unrelated the MCT as mapped in Darjeeling. In part, I think that much of the confusion about the relations between the MCT and metamorphism arises because the 'MCT' developed at different times with different thermal significance in different places.

Moreover, polymetamorphism has greatly complicated the issue, at least in the central Himalaya. I am convinced that the occurrence of high-temperature (sillimanite + cordierite ± potassium feldspar) assemblages above lower-temperature (kyanite + staurolite) assemblages in this sector is the consequence of two distinctive metamorphic events. The early (essentially 'Barrovian') event affected the entire Greater Himalayan sequence, whereas the late (essentially 'Buchan') event affected the entire sequence in some areas (e.g. the Burhi Gandaki section) but was restricted to the uppermost portions of the sequence in most other areas. (To specifically answer Dr Treloar's question, there is no convincing evidence from the central Himalaya of inverted thermal gradients during the Barrovian event.) The second event seems intimately related to the generation of the Greater Himalayan leucogranites, and those areas which were most strongly affected by the second event correspond to zones of intense migmatization. In our paper we argue that the second event was associated with several kilometres of tectonic burial, but the dominant heat source for this metamorphism appears to have been within the Greater Himalayan sequence. Several workers have speculated on the possible cause of this metamorphic/anatectic event, and we have tried to summarize their arguments in our paper. I do not think we have sufficient data to critically evaluate the models proposed, but I suspect that the sporadic development of second-event assemblages was caused by locally high concentrations of heat-producing elements within the Greater Himalayan sequence (Pinet & Jaupart 1987).

A. MOHAN (*Department of Geology, Banaras Hindu University, India*). My point is regarding Professor Hodges's approach of conventional rim temperature estimates. Garnet profiles from different zones of inverted metamorphic sequences generally reveal prograde zoning from core to near rim. But this trend is often reversed when rim composition is taken into account. Therefore, the near-rim compositions should be used in estimating the temperature.

Professor Hodges shows in his diagrams of $P/T$ against distance from MCT that there are situations when both $P$ and $T$ increase upwards above the MCT, in addition to increasing $T$–decreasing $P$ points. Why has he dropped those earlier points reflecting increasing $P/T$ and only considered those data with increasing $T$–decreasing $P$? That would explain the cause of inversion of metamorphic isogrades through the model proposed by Le Fort (1975). I have thermobarometric data from Sikkim–Darjeeling region for the garnet to sillimanite zones where $P$ increases upwards with increasing $T$. These suggest that none of the models proposed for the cause of inverted isograds is fully relevant.

K. V. HODGES. I would disagree that the garnet profiles from Himalayan inverted metamorphic sequences are 'generally' indicative of prograde (i.e. increasing temperature growth). In my experience, a variety of zoning patterns indicative of prograde, retrograde, and perhaps even constant temperature growth occur in both M1 and M2 garnets. It is true that many garnets from the Himalaya show a pronounced inversion in their zoning patterns over their outermost 20–150 μm. This inversion is often characteristic of retrograde reequilibration during uplift. I consistently use the outermost rim compositions of phases in mutual contact for thermobarometry because these compositions are the most likely to reflect equilibrium. Rim equilibrium may have been attained substantially after peak temperature during uplift, but it is a fairly straightforward matter to access the higher-temperature portions of the $P$–$T$ path for a sample by inclusion thermobarometry or thermodynamic modelling. It is tempting to try to reconstruct peak temperatures by using near-rim garnet compositions (rather that outermost rim compositions) as Dr Mohan suggests, but this technique if very dangerous: it is virtually impossible to know with any degree of certainty which part of the garnet profile represents 'peak temperature', and it is even more difficult to establish the composition of other phases that were in equilibrium with the chosen garnet composition without finding inclusions in the garnet. The common practice of calculating a 'peak temperature' by using the core composition of a garnet and the composition of a distant matrix biotite is completely inappropriate because there is no guarantee that the core of the garnet was ever in equilibrium with the biotite.

I believe that the second part of Dr Mohan's discussion refers to Mary Hubbard's data from eastern Nepal. There is certainly some scatter in the data, but we must remember that all measurements are subject to uncertainty. The uncertainties in thermobarometry are rather large (see Hodges & McKenna 1987), and it's not prudent to try to interpret each and every inversion in the apparent $P$–$T$ gradient when those inversions occur over pressure and temperature ranges smaller than the analytical uncertainties. I think that the preponderance of Hubbard's data favour the Le Fort model, but late- to post-metamorphic imbrication within the MCT zone is likely to have occurred. Dr Mohan's data from Sikkim–Darjeeling may be inconsistent with the Le Fort model, just as our data from Garhwal seem inconsistent. Again, the thermal and mechanical processes that produced inverted metamorphism in the Himalaya are complex, and no single model seems universally applicable.

*Phil. Trans. R. Soc. Lond.* A **326**, 281–299 (1988)    [ 281 ]

*Printed in Great Britain*

# Granites in the tectonic evolution of the Himalaya, Karakoram and southern Tibet

By P. Le Fort†

*Centre de Recherches Pétrographiques et Géochimiques, B.P. 20, 54501 Vandoeuvre-lès-Nancy, France*

Four major plutonic belts are related to the Meso-Cainozoic orogenic evolution of the Himalaya–Transhimalaya–Karakoram realm: the Transhimalaya belt and its satellite Kohistan arc, the Karakoram batholith, the High Himalaya belt and the North Himalaya belt. A fifth one results from the lower Palaeozoic epirogenic events: the 'Lesser Himalaya' belt. The tectonic settings of their production and emplacement are successively reviewed. Among the first four, two result from oceanic subduction along an Andean margin locally branching into an island arc and two result from intracontinental subduction after closure of the oceanic realm. Both Andean belts are made up of very large quantities of highly diversified granitoids produced more or less continuously during 70 Ma at least, whereas the intracontinental ones are limited to a small volume of very uniform anatectic granite produced during a 10–15 Ma period. The production and emplacement in the Andean belts is partly controlled by the obliquity of the convergence between India and Eurasia. The emplacement of the intracontinental belts is even more dependent on the regional tectonic setting. These contrasting belts are case studies probing the depths and mechanisms of their production and giving adequate models for older geodynamic frames.

## 0. Introduction

Five large plutonic belts occur in the Karakoram–Transhimalaya–Himalaya region (Debon *et al.* 1981) (see figure 1).

1. The Karakoram belt extending for some 800 km to the north of the western syntaxis of the Himalaya and made up of a large variety of calc-alkaline to sub-alkaline quartz diorite to granite.

2. The Transhimalaya belt lying just north of the Indus–Tsangpo Suture Zone (ITS) and forming a nearly continuous 2500 km long batholith made up of gabbros to granites. To the west lies the large Kohistan Zone, comprising a wide variety of mainly mafic plutonic and volcanic rocks.

3. The North Himalaya belt running some 50 km south of the suture and forming a series of some 20 domes of two-mica adamellite.

4. The High Himalaya belt intruding the High Himalayan sedimentary series, between 100 and 150 km south of the suture and forming a series of a dozen or so main lenticular slabs and innumerable dykes made up of muscovite, biotite and/or tourmaline leucoadamellite (Le Fort *et al.* 1987).

5. The 'Lesser Himalaya' belt lying mostly in the southern part of the Himalaya but extending widely over the entire Himalaya and south Tibet realms, forming discontinous masses of both gneissic and non-gneissic granites (Le Fort *et al.* 1986).

All first four belts are related to the Mesozoic–Cainozoic evolution of the India and Eurasia

† Present address: Institut Dolomieu, 15 rue Maurice Gignoux, 38031, Grenoble, France.

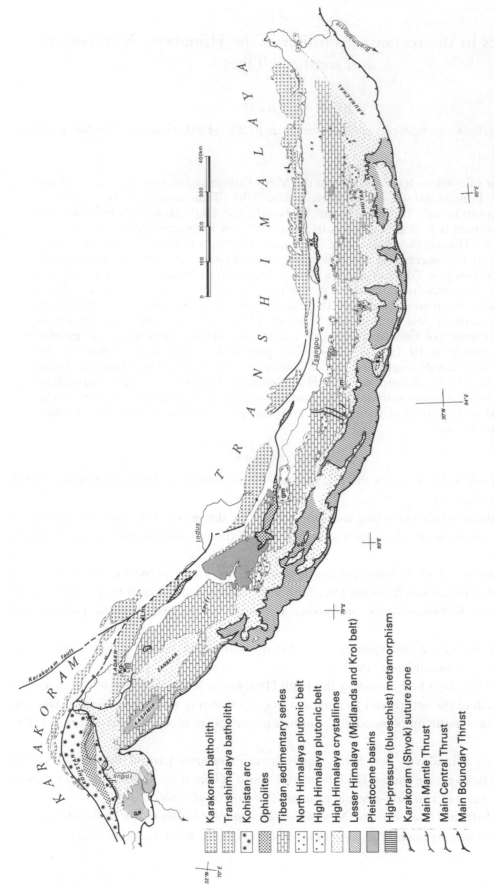

FIGURE 1. Main structural divisions and Meso–Cainozoic plutonic belts of the Himalaya–Transhimalaya–Karakoram. Based on Desio (1964), Gansser (1977), Gamerith (1979), Bard et al. (1980), Stöcklin & Bhattarai (1980), Academia Sinica (1981), Bordet et al. (1981), Fuchs (1981), Polino (1981), Valdiya (1981), Gansser (1981), Colchen et al. (1986) and Le Fort (1986b). The zone shown as ophiolitic in Kohistan corresponds to the lowermost part of the island arc. 'Lesser Himalaya' granites and Subhimalaya series are not represented; e = Everest, gm = Gurla Mandata, K = Kathmandu, Kg = Kargil, L = Lhasa, Lh = Leh, m = Manaslu, P = Peshawar, T = Thimpu, X = Xigaze.

Karakoram batholith
Transhimalaya batholith
Kohistan arc
Ophiolites
Tibetan sedimentary series
North Himalaya plutonic belt
High Himalaya plutonic belt
High Himalaya crystallines
Lesser Himalaya (Midlands and Krol belt)
Pleistocene basins
High-pressure (blueschist) metamorphism
Karakoram (Shyok) suture zone
Main Mantle Thrust
Main Central Thrust
Main Boundary Thrust

Plates. Only the fifth one of Lower Palaeozoic age is not directly connected to the system as its granites had long ago crystallized and cooled down. However, their presence modifies the rheological properties of the Himalayan deforming crust and in some cases they are probably responsible for the shape of the Himalayan structures.

Let us review the major characteristics of the five belts before trying to relate more precisely their generation and emplacement to the tectonic framework, and to evaluate their bearing on the evolution of the Himalayan Orogeny *s.l.*

## 1. GEOLOGICAL SETTING AND DESCRIPTION
### *Transhimalaya batholithic belt*

The Transhimalaya batholith, nearly continuous for some 2500 km, with a width of about 50 km, has been studied mainly along two segments: some 200 km in Ladakh (Honegger *et al.* 1982; Sharma & Choubey 1983), and another 200 km in southern Xizang south of Lhasa (Academia Sinica 1980; Tu *et al.* 1981; Debon *et al.* 1982, 1986a), where it is also known as the Gangdese (or Kangdese) batholith. Characteristics are broadly similar in both segments. As summarized by Debon *et al.* (1986a), the batholith is composite, made up of numerous plutonic bodies often continuous with gradational contacts. Postmagmatic cleavage is often superimposed on flow structures. Compositions range from noritic gabbro to adamellite through quartz monzonite and granodiorite. Contrary to rather common statements, true tonalite has not been described. Most rocks are typically metaluminous and subalkaline but not calcalkaline as mostly considered (figure 2). It has been known to intrude Mesozoic rocks since Hayden (1905) described it intruding Jurassic and Cretaceous series.

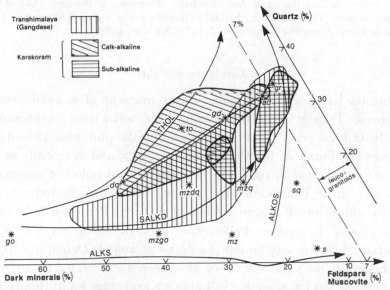

FIGURE 2. Distribution of the Transhimalaya and Karakoram plutonic belts in the triangular quartz–dark minerals–feldspars + muscovite diagram ('Q–B–F' diagram) (from Debon *et al.* 1986a, 1987). The parameters in percentage (by mass) are directly calculated from chemical analysis (La Roche 1964; Debon & Le Fort 1982). Different subtypes among the three main types of cafemic and alumino–cafemic associations are distinguished: THOL (tholeiitic), CALK (calc-alkaline), SALKD (dark coloured sub-alkaline, i.e. monzonitic), ALKS and ALKOS (alkaline saturated and oversaturated, respectively). Representative points of the different petrographic types are also shown. The 120 analyses of the Transhimalaya (Gangdese region) fall in a typical sub-alkaline field. For the 60 analyses from Karakoram, part of them are calc-alkaline while others fall in the same sub-alkaline field.

Age determinations from various methods mostly range from 120 to 40 Ma (Lower Cretaceous to Upper Eocene). There is no obvious quiescence in the magmatic activity at the scale of the entire range although, locally, ages generally cluster around one or two values.

Initial ratios of $^{87}Sr/^{86}Sr$ are moderately low and present a general tendency to increase from west to east (from 0.704 to 0.707) and when available, from south to north (figure 3). This can be related to an increasing contribution of continental crust toward the east during the subduction of the Tethys ocean floor.

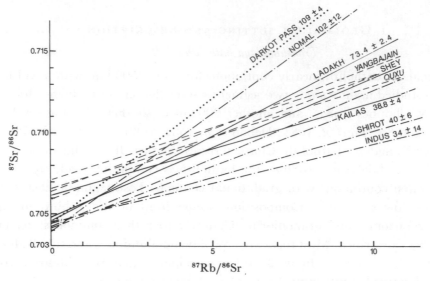

FIGURE 3. Whole-rock Rb–Sr isochrons for some plutonic (and volcanic) bodies of the Karakoram, Kohistan and Transhimalaya showing a general increase in $^{87}Sr/^{86}Sr$ initial ratio from west to east. Lines: dots, Karakoram (Debon *et al.* 1987); broken lines and dots, Kohistan (Petterson & Windley 1985; Debon *et al.* 1987); continuous lines, western Transhimalaya (Ladakh) (Honegger *et al.* 1982; Schärer *et al.* 1984); broken lines, eastern Transhimalaya (Gangdese) (Debon *et al.* 1982; Xu *et al.* 1985).

### Karakoram batholith

In Karakoram, the backbone of the main range is made up of an axial composite batholith intruding Palaeozoic–Triassic (Desio 1964) sedimentary series lying on the northern side of the range. During the last few years it has been shown that the plutonism extended as far back as middle Cretaceous (Le Fort *et al.* 1983 *b*; Debon *et al.* 1987) and as recently as Upper Miocene (Debon *et al.* 1986 *b*; Searle *et al.* 1988). In between, several pulses of magmatism have been dated, particularly during Palaeocene and Eocene, with light-coloured sub-alkaline plutons that extend to the south into Kohistan (Debon *et al.* 1987; Petterson & Windley 1985).

The chemistry and Cretaceous–Palaeogene ages obtained on the Karakoram and Transhimalaya batholiths are very similar (Le Fort *et al.* 1983 *b*; Debon *et al.* 1987) and suggest that they both belong to the same belt. They have been disconnected by a right-lateral strike-slip movement of some 300 km along the still-active Karakoram Fault (figure 4). Accordingly, the Shyok ophiolitic zone (or northern suture zone, NSZ, of some authors) is only a branch of the Indus–Tsangpo Suture (Stöcklin 1977; Sengör 1984, 1985), the back-arc suture.

To the southwest, the Karakoram Zone corresponds with the Central Mountains of Afghanistan through the eastern Hindu Kush (Le Fort *et al.* 1983 *b*) (figure 4). From the Lhasa Block to the Central Mountains, one can thus follow the same Tethyside continental strip

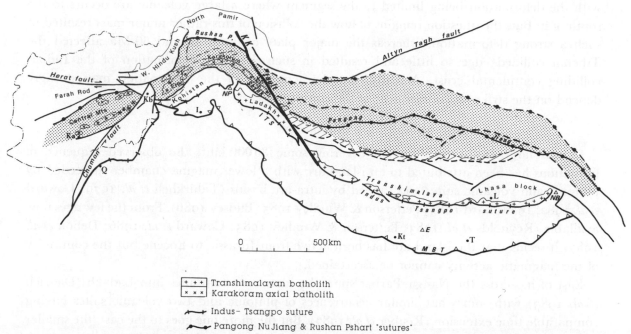

FIGURE 4. Global structural map of Afghanistan, Hindu-Kush, Karakoram, Himalaya and Tibet after various sources. A right-lateral strike-slip movement of some 300 km along the Karakoram Fault has been restituted (broken lines) to illustate the previous geodynamic setting where the Karakoram prolongs the Transhimalaya, the Rushan Pshart prolongs the Pangong NuJiang Sutre Zone and the Shyok Zone is only a branch of the Indus Tsangpo Suture (ITS). The Tethyside south Tibetan (Lhasa) Block, stippled, extends from the Central Mountains of Afghanistan, through Karakoram, to eastern Transhimalaya. To the SE, the Karakoram Fault seems to merge into the ITS. E = Everest, EHK = Eastern Hindu-Kush, I = Islamabad, ITS = Indus Tsangpo Suture, K = K2, Ka = Kandahar, Kb = Kabul, KK = Karakoram, Kt = Kathmandu, L = Lhasa, NB = Namche Barwa, NP = Nanga Parbat, Q = Quetta, T = Thimpu.

through Karakoram. The simple Transhimalayan arc of southern Xizang prolongs westward into the double-arc system of Kohistan–Ladakh to the south and Karakoram to the north.

Magmatically and tectonically, two major differences appear between Karakoram and Transhimalaya belts. In Karakoram the plutonic activity outlasts the Eocene timing of India–Eurasia collision, and continues at least up to Miocene, when the Baltoro pluton and numerous granitic dykes and pods are emplaced (Debon *et al.* 1986*b*; Searle *et al.* 1988), whereas, in Transhimalaya, granitoid production seems to stop rather abruptly some 10 Ma after collision. Actually, upper Miocene volcanic activity has been recently dated in the Kangdese region (Maqiang andesites and acid ignimbrites, 15–10 Ma; Coulon *et al.* 1986), and the presence of magma is now suspected at shallow depth in the same region (Pham *et al.* 1986). Thus the difference between the two batholiths may be more in the rates of uplift and erosion, otherwise known to be very fast in the Karakoram region (Zeitler 1985).

The other difference appears in the deformation pattern of Karakoram when compared to that of the Transhimalaya. In Karakoram, Cretaceous plutons such as the Hunza have been intensely sheared toward the south under high-grade regional metamorphic conditions before the middle Eocene (Le Fort *et al.* 1983*b*; Bertrand & Debon 1986; Coward *et al.* 1986; Debon *et al.* 1987), whereas Transhimalayan plutons remain almost unaffected. Such a deformation may result from an Upper Cretaceous–Palaeocene collision of the Kohistan arc after subduction of the northern Neo-Tethys oceanic crust along the so-called NSZ. This explanation agrees with the deformation being limited to the segment where a large volcanic arc occurs to the south of it. But, the question remains of how the collision of this rather minor mass resulted in such a strong deformation, whereas the major plate collision around 50 Ma affected the Tibetan collided edge so little and resulted in such dramatic deformation of the Indian colliding continental crust. Actually, the characteristics of the deformation may strongly depend on the strike-slip component of the convergence.

### Kohistan arc

In this large oval-shaped region covering some $36000\ km^2$, the observed sequence of formations has been attributed to an island arc with a lower magma chamber succeeded by plutonic and volcanic suites and topped by intra-arc basins (Tahirkheli *et al.* 1979; Coward *et al.* 1982, 1986; Bard 1983; Petterson & Windley 1985; Pudsey 1986). From the few ages now available (Reynolds *et al.* 1983; Petterson & Windley 1985; Coward *et al.* 1986; Debon *et al.* 1987) it seems that the island arc has been active from Jurassic to Eocene but the continuity of the magmatic activity cannot be ascertained.

East of it, across the Nanga–Parbat spur, the island arc continues into Ladakh (Dietrich *et al.* 1983) with somewhat similar occurrences of plutonic and two volcanic suites having comparable time extension (Reuber *et al.* 1987). But the more one goes to the east, the smaller abundance of volcanics one meets. The Dras volcanics, for example, laterally pass into a flysch formation (Nindam flysch of Bassoullet *et al.* 1978 and Colchen *et al.* 1986 *a*). Concurrently, as noted above, the $^{87}Sr/^{86}Sr$ initial ratios present a general tendency to increase (figure 3). These two facts can be related to the increasing contribution of continental crust material in the magmas and to the screen effect of this material and of the water that it contains against the ascent of the magma to the surface.

### High Himalaya belt

Numerous studies have been devoted to the High Himalaya granitoid belt during the past 15 years (see Le Fort *et al.* 1987 for an extensive set of references) although they cover only a few percent of the Himalaya and account for not more than 0.5‰ of its volume. The dozen or so main plutonic bodies have a lenticular shape. They are accompanied by a dense network of aplopegmatitic dykes that extends on a much larger area than the plutons themselves and

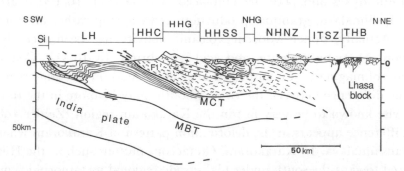

FIGURE 5. Schematic cross section giving the present structure of the Himalaya in Central Nepal, the disposition of the different tectonic units and the localization of three of the plutonic belts discussed here (from France-Lanord & Le Fort 1988). HHC = High Himalaya Crystalline (Tibetan Slab), HHG = High Himalaya leucogranite (Manaslu), HHSS = High Himalaya sedimentary series, ITSZ = Indus-Tsangpo Suture Zone, LH = Lesser Himalaya, MBT = Main Boundary Thrust, MCT = Main Central Thrust, NHG = North Himalaya granite, NHNZ = North Himalaya Nappe Zone, Si = Siwaliks, THB = Transhimalaya batholith.

that often connects them. The granitic rocks occur close to the limit between the crystallines (HHC) and the sedimentary series (HHSS) of the High Himalaya (figure 5). The plutons may intrude the HHSS up to the Cretaceous.

The High Himalaya granites have a very homogeneous composition of muscovite, biotite and/or tourmaline-bearing leucoadamellite reflected in their major element chemical composition. This composition is close to minimum melt composition in the haplogranitic system under addition of B and F, and variable water saturation of the magma as revealed by the slight variations of the Na/K ratio (Le Fort 1981, 1986; Le Fort et al. 1987; France-Lanord & Le Fort 1988). Trace element and isotopic characteristics are for most of them inherited from the source rock in the HHC, shown to be represented in Central Nepal by the quartzo-pelitic Formation 1 of the Tibetan Slab (Le Fort 1981; Vidal et al. 1982; Cuney et al. 1984; Deniel et al. 1987; France-Lanord et al. 1988, France-Lanord & Le Fort 1988).

From field, petrographical and geochemical studies I have proposed a model of generation of the High Himalaya granites (Le Fort 1975, 1981, 1986; Vidal et al. 1982). It associates the major zone of thrusting at continental scale along the Main Central Thrust (MCT), the generation of inverted metamorphism in the underthrusted Midlands formations, the liberation of large quantities of fluids from these formations, and the begetting of anatectic melts in the hot gneisses of the overthrusting HHC. The degree of migmatization lies around 10–12 % of the total volume of underlying gneisses as estimated from the projection of their field extension. Thus collection of the melts is a slow process that proceeds in a discontinuous way, and releases multiple successive batches of leucogranitic magmas rising through the overlying crystallines.

The different batches emplace at the limit between the HHC and the HHSS, which corresponds to the disharmonic boundary between the infrastructure and the superstructure. This emplacement occurs at syn- to late-metamorphic and deformation time as evidenced by the variable intensity of deformation taken up by the granite.

Isotopic (cooling) ages by various methods span from 25 Ma by U–Pb on monazite from the Manaslu (Deniel et al. 1987) to around 10 Ma by K–Ar on micas from plutons such as in Bhutan (Dietrich & Gansser 1981). The magmatic part of the evolution may have lasted for more than 10 Ma.

### The North Himalaya belt

Mentioned as such for the first time by Academia Sinica (1980), the North Himalaya belt groups, in a series of dome, Lower Palaeozoic porphyritic granite and Cainozoic two-mica adamellite (Debon et al. 1981, 1985, 1986a; Burg et al. 1984b; Le Fort 1986). The two types of rock may be associated or independently present in the above-mentioned plutonic domes. I have suggested (Le Fort 1986) that these plutons were diapirically emplaced and had carried away in their ascent portions of the cap of porphyritic granite lying above their migmatitic zone of production.

Compared to the HHG, the diapirically emplaced NHG necessitates a much higher degree of melting in the migmatitic production zone, probably several tens percents. Actually, the two-mica adamellite rock is similar to the two-mica leucoadamellite of the High Himalaya in its major and trace element characteristics, heterogeneous Sr isotopic ratios, high Sr initial ratios, Pb isotope compositions and high $\delta^{18}O$ values. Its age seems to be slightly younger than the average age of the High Himalaya plutons, which agrees with its production being linked to the same mechanism but its localization being more to the north in the overthrusted slab.

'LESSER HIMALAYA' BELT

The 'Lesser Himalaya' granitic belt (Le Fort *et al.* 1980, 1983*a*, 1986; Debon *et al.* 1986*a*) is made up of some fifteen independent plutons that appear in the HHC but at a short distance to the north of the Main Boundary Thrust (figure 6). The plutons are generally composed of porphyritic peraluminous quartz-rich granite, rich in dark igneous and metasedimentary inclusions. Isotopic ages have cofirmed the identity of the belt emplaced around the Cambro-Ordovician boundary and showing a high $^{87}Sr/^{86}Sr$ initial ratio.

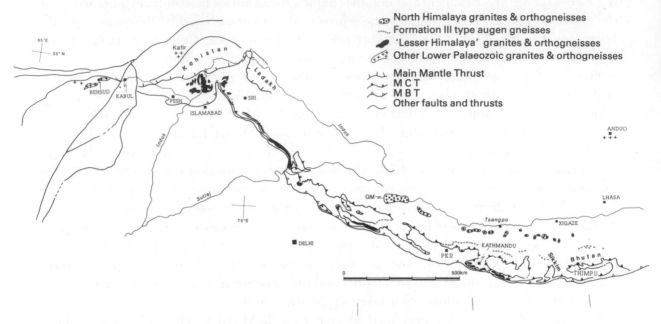

FIGURE 6. Geological sketch map showing the location of the 'Lesser Himalaya' granitic belt and the extension of the magmatic formations with an isotopic age around 500 Ma (from Le Fort *et al.* 1986). GM = Gurla Mandata, PESH = Peshawar, PKR = Pokhra, SRI = Srinagar.

This magmatism that has still preserved its plutonic characteristics along the Lesser Himalaya belt, is part of a much larger ensemble extending in a region exceeding $10^6 \, km^2$ (figure 6). In most of this region it has been affected by the Mesozoic–Cainozoic deformations resulting in the production of large augen gneiss formations such as the Formation III of the HHC of central Nepal. In western Himalaya it may be associated with some peralkaline plutonism (Rafiq 1987). It also forms the masses of porphyritic granite and gneisses of the North Himalaya belt.

These granites are the last plutonic generation before the Himalayan times. As such they have conferred fundamental rheological and geochemical characteristics to the Himalayan basement. In particular, I have suggested (Le Fort *et al.* 1983*b*, 1986) that their southern limit has induced the position of the Main Central Thrust (MCT) of the Himalaya.

2. TECTONIC SETTING OF THEIR PRODUCTION

In the overall convergent framework of India and Eurasia, the different belts of granitoid relate to two main types of subduction: oceanic and continental. The youngest plutons are more ambiguous.

*Oceanic subduction: Transhimalaya and Karakoram belts*

From Jurassic to Eocene, the drift of the Indian Plate has progressively closed the Neo-Tethys, its oceanic floor being subducted to the north under the southern Eurasian margin and producing the Transhimalaya and Karakoram belts. The cafemic to alumino-cafemic character, the calc-alkaline to sub-alkaline nature and some rather low $^{87}Sr/^{86}Sr$ initial ratios of the first stage of magmatism support the inference that magmas were generated from oceanic subduction of oceanic material along an Andean-type margin involving some sialic crust (Debon *et al.* 1981, 1987). This margin shows lateral variations. From east to west, the Transhimalaya Andean margin progressively passes into the Kohistan island arc, branching off the south Tibet continental crust in the Ladakh region (figure 7). According to Debon *et al.* (1986a), the island arc has extended, during part of its existence, for 1400 km further eastward, at least as far as Quxu where the Transhimalaya batholith has emplaced simultaneously in the continental margin and anphibolites of the arc, originally built onto its side. To the northwest, the early evolution of the Karakoram prolongs that of the Transhimalaya.

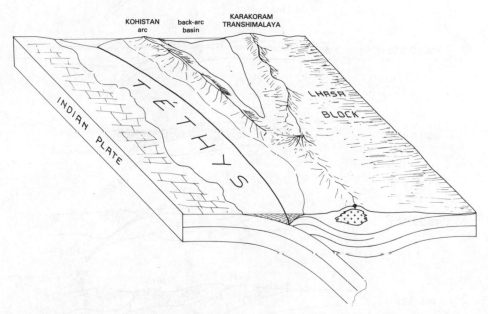

FIGURE 7. Tentative block diagram showing the relation between Transhimalaya Karakoram, Kohistan and India around Upper Cretaceous time. The Kohistan island arc branches off the Transhimalaya belt in the Ladakh region.

Thus the oceanic subduction of the Tethys occurs either below continental crust or below oceanic crust. We have seen that the increasing contribution of the continental crust in the production of the Transhimalayan granitoids was already visible in the $^{87}Sr/^{86}Sr$ initial ratio variations (figure 3). The results from Pb isotope geochemistry (Gariépy *et al.* 1985) also suggest the contribution of lead derived from the crust of the Lhasa Block in the granitoids of the Kangdese region, the contribution being higher in the more acid types. In Karakoram, inherited Precambrian components are found in the zircons of the only analysed pluton, the Baltoro one (Searle *et al.* 1988; Rex *et al.*, this symposium), and the $\delta^{18}O$ values for eastern Karakoram increase with time and evolution of the granitoids (Srimal *et al.* 1987).

We have already noted the subalkaline frequent character of the Transhimalaya and

Karakoram plutons. Debon *et al.* (1986*a*) have suggested that such trends could be related to a strike-slip component along the subduction zone. In the case of the Karakoram-Transhimalaya batholiths, the character is, in fact, met along their entire length.

*Intracontinental subduction: High and North Himalaya*

Once the India–Eurasia collision occurred, the building of the Himalaya proceeded in two main steps (Mascle 1985; Le Fort 1988). During the first, a system of nappes developed from the suture zone, thickening the crust by at least 15 km (figure 8*b*, *c*). The second step succeeded with the birth of intracontinental thrusting along a thick ductile zone known as the Main Central Thrust Zone (MCT) (figure 8*d*). In between these steps, carbonatite sheets have been emplaced around 31 Ma in the SW part of the Himalayan realm, SW of the northwestern syntaxis (Le Bas *et al.* 1987).

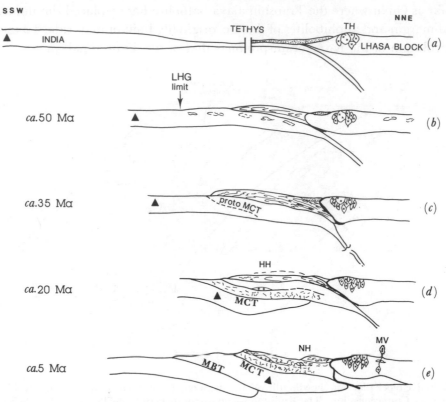

FIGURE 8. Schematic evolution of the Transhimalaya and Himalaya regions from Mesozoic to Recent. (*a*) Mesozoic. The Tethys oceanic crust is subducted under the Lhasa Block Andean margin; sedimentation occurs on both sides of the ocean: dominantly carbonated on India passive margin, of flysch-type south of the active margin. (*b*) Eocene. Collision of India and Eurasia (Lhasa Block) Plates induces the nappe thrusting on the northern margin of India Plate; these nappes include ophiolitic material mélanges, fore-arc sediments and flyschs; the extent of the 500 Ma plutons is shown in both continental plates. (*c*) Eocene–Oligocene. The North Himalaya nappe system had its largest extension; subduction of Indian crust becomes impossible because of buoyancy forces; MCT is about to break out, probably on a former limit of the 500 Ma magmatism; Transhimalaya becomes quiescent. (*d*) Lower Miocene. Underthrusting of the Lesser Himalaya sediments and Indian crust along the MCT releases large quantities of fluids, induces anatectic melting in the High Himalaya crystallines, produces the High Himalaya granites that emplace at the limit between the HHC and the sedimentary series, and slightly later the North Himalaya granites that emplace diapirically. (*e*) Mio–Pliocene. MCT has been relayed by MBT; High and North Himalaya plutonic belts are cooling; to the north of the Transhimalaya, Indian crustal material and or sediments trapped below the Lhasa Block has produced the Maqiang volcanics; erosion in the Himalaya continues at a rate of about 1 mm a$^{-1}$. HH = High Himalaya, LH = Lesser Himalaya, MBT = Main Boundary Thrust, MCT = Main Central Thrust, MV = Maqiang volcanics, NH = North Himalaya, TH = Transhimalaya.

The first step, of Alpine type, occurred shortly after collision, around 50 Ma and lasted for some 20 Ma. The second step, purely Himalayan, probably started around 30 Ma and lasted for another 20 Ma. It was relayed by another thrusting movement more to the south along the Main Boundary Thrust (MBT) some 10 Ma ago.

The time elapsed between collision and the initiation of the MCT movement enabled the crust already thickened by the nappes, to heat up. Thus the future HHC reached a level of temperature where anatexis was only prevented by the dryness of this basement already granitized during early Palaeozoic time.

The production of High and North Himalayan melts accompanied the movement along the MCT. During this movement, the hotter High Himalaya crystallines (HHC) were thrusted over the weakly metamorphosed Lesser Himalaya. The latter underwent dehydration and released large quantities of fluids that rose into the HHC, and thereby in turn, induced partial melting of them. This kind of process resembles that of dehydration and melting in oceanic subduction such as that of the Transhimalaya. But in the present case, the intracontinental subduction generates an entirely crustal magma with a very narrow range of composition obtained by close to minimum melt anatectic conditions.

In addition, what we know from field and geochemical studies shows that the source rocks of the granite were rather homogeneous along the entire Himalaya and suffered similar conditions of metamorphism and melting.

The mineralogical and geochemical differences between the High Himalaya and the Cainozoic North Himalaya plutons seem to lie mostly in the degree of anatexis; higher anorthite contents of the plagioclase and higher dark-mineral content of the North Himalaya adamellites being related to the higher temperature and therefore the higher percentage of melting attained in the source migmatites.

Dating of the Himalayan granites has proved to be difficult and hazardous. Actually, the magmatism lasts over a period of 10 Ma at least, as shown by U–Pb determinations from 25 to 14 Ma on monazites and zircons (Schärer 1984; Schärer et al. 1986; Deniel et al. 1987). Such determinations need to be multiplied on the High Himalaya and even more on the North Himalaya plutons to ascertain the slight difference in age (a few million years) that seems to exist globally between the two belts. Taking an average velocity of around 2 cm a$^{-1}$ on the MCT and a distance of some 40–60 km between the two belts, the North Himalaya should be younger by 2–3 Ma, everything remaining similar otherwise.

### The late magmatism

In Karakoram (Baltoro granite) as in the Lhasa region of the Transhimalaya (Maqiang volcanites), Miocene magmatism has occurred at a time when the Himalayan magmatism was already active and the MCT was about to be relaid by the MBT. Searching for the cause of this magmatism, one may look some 20–30 Ma back, the time necessary to thermally re-equilibrate a portion of crust (England & Thompson 1984), at the epoch when the system of nappes was replaced by the intracontinental thrusting. The Alpine system probably became stuck by the buoyancy of the marginal terranes underplating the Lhasa Block s.l. These terranes, composed of portions of the volcanic arc and back-arc fillings to the west and more flyschoid to the east, dumped below the Karakoram–Transhimalaya range, around Oligo-Miocene time would have undergone melting during Upper Miocene (figure 8e).

The production of this late magmatism illustrates rather well the time gap existing between the tectonic impulse and the magmatic response. This gap represents the time necessary for the crust to re-equilibrate the thermal disorders induced by the tectonics. It varies with the physical properties of the crust and the thermal characteristics of the surrounding media but remains in the order of 20–30 Ma. Similarly, one may predict that responding to the blocking of movement along the MCT, around 10–15 Ma ago, crustal melting of the tip of the underplated material (the equivalents of the Lesser Himalaya) should occur within a few million years from now. The present-day activity observed in southern Xizang with abundance of hot springs with high chemistry (Zhang *et al.* 1981), high heat flow measurements (Francheteau *et al.* 1983) and the possible presence of isotropic material at shallow depth shown by magnetotelluric methods (Pham *et al.* 1986) may be precursors of this coming magmatic activity.

## 3. Tectonic setting of their emplacement

Our knowledge of the emplacement of the intracontinental subduction related plutonism is quite different from that of the oceanic related one. The latter is less well known because the characteristics of this enormous quantity of granitoid rocks as well as of their surrounding formations has, in very few regions, been studied in detail. The former, the High Himalaya and North Himalaya belts, although more inaccessible in a way, are less extensive and have been paid more attention.

### Emplacement of the Transhimalaya and Karakoram belts

In both belts, emplacement of granitoid plutons has been going on for several tens of million years between around 100 Ma and 40 Ma. Whether the production and emplacement of magmas has been a more or less continuous process or whether they occurred as discontinuous pulses, remains conjectural. But the more isotopic ages obtained, the smaller the time gap between the different plutons seems to be. This is especially true if one considers the entire belt and not only a small portion of it. At this scale, there seems to be no systemic migration of the magmatic activity with time (Debon *et al.* 1986 *a*).

In the Transhimalaya, but probably also in the Karakoram, it seems that the rate of granitoid emplacement and volcanic production is triggered by the collision with the Indian continent (and Koshistan island arc) (Debon *et al.* 1986 *a*; Le Fort 1986). Two factors may have contributed to this increase; the dragging of water-rich sediments along the subduction zone during the final stages of India–Eurasia convergence (Debon *et al.* 1986 *a*) and the deformation of the collided margin facilitating the ascent of magma in a weakened crust (Debon *et al.* 1985; Le Fort 1988).

Few plutons have been studied with enough detail to appreciate the characteristics of the deformation of the rocks in which they have been emplaced. To our knowledge, only four such plutons, in the Lhasa–Quxu region, have been studied (Brun *et al.* 1983). From the shape and internal asymmetry of the microgranular mafic inclusions, these authors have deduced that the plutons have been emplaced in a portion of the crust undergoing strong left-lateral wrenching. A similar conclusion is reached on geochemical grounds (Afzali *et al.* 1979; Debon *et al.* 1986 *a*) to explain the subalkaline nature of the magmatism (cf. figure 2). Strike-slip movement tends to generate tension fractures down to very deep levels in the lithosphere and thus promote alkaline to subalkaline magmatism.

In the Karakoram belt, such strike-slip movement has only been documented at post-magmatic stage (Coward *et al.* 1986) when the mid-Cretaceous plutons were entirely crystallized. In the southern part of the Hunza pluton, for example, a right-lateral strike-slip movement has been documented (Coward *et al.* 1986).

In the Kohistan arc, a first stage of plutonic intrusions has been folded together with the surrounding volcanic and sedimentary rocks. This deformation is related by Petterson & Windley (1985) to the formation of the northern suture between Kohistan and Karakoram, around Upper Cretaceous.

In the best exposed sections of the two belts, the visible thickness of granitoid material frequently exceeds 3000 m as in the Hunza Valley, the Ladakh batholith along the Indus and the Kangdese along some north–south valleys. The floor of the plutons never seems to be exposed. Inclusions of the surrounding rocks in which the plutons are intruded are particularly abundant towards the borders of the more mafic plutons such as Hunza (Debon *et al.* 1987), Ladakh (Sharma & Choubey 1983) and Kangdese (Debon *et al.* 1984). There are more sediments in the surrounding rocks of the Karakoram batholith but more volcanics for the Transhimalaya batholith (see Rai 1983). The inclusion of such metamorphosed shales, limestones, quartzites and volcanics indicates a somewhat high level of emplacement.

The crust of the Transhimalaya was likely to be thick and elevated at the time of India–Eurasia collision. In fact, the Cainozoic compressive deformation of the Gangdese range of southern Tibet has been reported to be rather limited (Tapponnier *et al.* 1981; Burg 1983). According to England & Searle (1986), a precollision elevation of 3000 m in the Lhasa Block would have limited its post-collisional thickening and favoured the deformation and thickening of the suture zone and northern margin of the Indian Plate. A smaller elevation contrast could have prevailed in Karakoram during Upper Cretaceous and would partly explain the high thickening strain of the Karakoram active margin when it collided with the Kohistan arc around Cretaceous–Tertiary boundary.

The level of intrusion being already quite high in the crust, the mean rate of denudation for the Transhimalaya must have been relatively small, in the order of a few kilometres for a minimum of 40 Ma (around 0.1 mm a$^{-1}$). The same rate estimated for the Karakoram is much higher as some of the plutons are much more recent. However, this rate does not have to be continuous. In fact, a recent study (Copeland *et al.* 1988) along a section southwest of Lhasa, with the dating techniques of $^{39}$Ar/$^{40}$Ar, has shown a strong acceleration in the rate of uplift in the interval 20–17 Ma, from 0.07 to around 3.7 mm a$^{-1}$.

### Emplacement of the High and North Himalaya plutons

Resulting from the same process of production, the two belts differ in their mode of emplacement.

### The High Himalaya granites (HHG)

Mapping of the High Himalaya plutons clearly shows that they start to emplace along the main boundary between the isoclinally folded High Himalaya crystallines, or Tibetan Slab, and the openly folded High Himalaya sedimentary series (Le Fort 1973, 1981, 1986) (figure 5). This is particularly conspicuous when the pluton is thin as for the 300 m thick and 50 km long 'arm of Chhokang', stretching eastward of the Manaslu pluton (Colchen *et al.* 1986 b). There, the slab of granite corresponds to the disharmonic boundary between the infrastructure and the superstructure, it also corresponds to a regional metamorphic gap, the sillimanite grade below the granite giving way to a mild epizonal metamorphic assemblage only a few hundred metres above the slab in many cases (Manaslu, Shisha Pangma, Everest, etc.), this boundary remains the location of the floor of the main pluton that extends higher up, intruding the sedimentary series from Lower Palaeozoic up to Triassic, Jurassic and even Cretaceous.

Geochemical studies have shown that emplacement of the HHG is a slow process that operates by incremental gathering of small batches of magma (Deniel *et al.* 1987). It probably spreads on more than 10 Ma, the average volume of magma emplaced per year being less than 0.1 km³ for the entire Himalayan range.

During this very slow process, the HHG do not rise diapirically as they never form a large mass of molten magma. We will see that this is different from the North Himalaya granites (see below). The room necessary for the magma may be produced by two main mechanisms: tectonic movements and fluid convection around the pluton.

The relation of the HHG to regional tectonics may be considered at several scales. On the outcrop, the granite often presents a magmatic layering underlined by micas and tourmaline, superimposed by a tectonic cleavage especially well developed towards the edges of the pluton or in smaller lenses and sheets where it can lead to the formation of an augen gneiss (Le Fort *et al.* 1987). The emplacement of the granite is syn-kinematic to late-kinematic (Le Fort 1975, 1981, 1986). When the pluton is not so big that the entire shape cannot be appreciated (as for the Manaslu and Everest), it appears that it develops a pinch and swell kilometric structure (for the Garhwal plutons; Scaillet *et al.* 1988). Although this structure had developed when the granite was already much more competent than the surrounding sedimentary rocks, it may have offered part of the room necessary for the additional magma to coalesce. In fact the flow and deformation structures vary in intensity, sometimes within very short distances. Thus it is not unlikely that, during deformation, the stretching of the competent granite creates the room for the ascending magma to emplace within it. But this remains difficult to observe in the field as the different batches of leucogranite are generally very much alike in their mineralogy, colour and grain size, and boundaries are extremely hard to follow.

Another way of providing some room for the granite is by arching the roof of the pluton. Actually there is a definite ballooning effect at the top of the granite. Longitudinal sections show this very broad anticlinal doming (France-Lanord *et al.* 1988), which is probably not entirely caused by the boudinage effect mentioned above, but, cutting across the general sedimentary bedding and warping it, may be ascribed to the buoyancy of the granitic magma.

Finally and especially in the case of large plutons, Le Fort (1981) has suggested that the 'caving out' of the sedimentary country rocks, when dominantly carbonated, by fluid convection and dissolution around the pluton, could help to emplace significant quantities of magma and make the granite pluton grow. A frozen image of this process is given by the very dense network of granitic and aplo-pegmatitic dykes that crosscut the country rocks of the plutons for several kilometres around them (Le Fort 1981).

Thermobarometric data on the contact aureole towards the roof of major plutons such as the Manaslu (Roy-Barman & Le Fort 1988) indicate that the pressure at the time of the maximum temperature was not less than 270 MPa and could have reached 400 MPa at somewhat deeper levels. Such rather high values indicate that the granite did not reach a much higher level than 10–15 km below the surface. This is a major difference with some of the other leucogranites to which the Himalayan ones are often compared, such as the Hercynian leucogranites of Western Europe, for which high level of emplacement have been obtained especially for the most specialized ones such as Echassières (Cuney & Autran 1987), although they are around 300 Ma old.

For the HHG, such deep level of emplacement and young ages, between 25 and 14 Ma (cf. Le Fort *et al.* 1987), suggest very rapid denudation, in the order of 1 mm a$^{-1}$. Some values calculated from closure temperatures of different minerals from the Manaslu and Everest plutons give the same order of magnitude: between 0.6 and 0.8 mm a$^{-1}$ for 4–10 Ma (Krummenacher *et al.* 1978; Kai 1981; Le Fort 1988). As suggested by Caby *et al.* (1983), such rapid denudation is probably not only by erosion but implies tectonic denudation. Burg (1983) and more recently Burg *et al.* (1984a), Gapais *et al.* (1984) and Herren (1987), for example, have described large longitudinal normal faults lying generally north of the HHG but deforming by northward shear some of them and having a down dip movement of several kilometres, up to 25 km. Burg (1983) mapped several of these north of Everest, one of which is cut by a two-mica leucogranite dated by U–Pb on two fractions of monazite at 25–22 Ma (Copeland *et al.* 1987). Such a relatively old age suggests that tectonic denudation has been active since almost the beginning of the granite emplacement in the High Himalaya.

### The North Himalaya granites (NHG)

Satellite pictures have shown, before mapping, the elliptical shape of the North Himalaya plutons (Gansser 1977; Academia Sinica 1980; Burg 1983). Their emplacement is not controlled by the HHC–HHSS disharmony as for the HHG. I have suggested (Le Fort 1986) that the NHG belt could be best explained by diapiric emplacement of magma. The NHG plutons are probably intermediate between the two extreme cases of Brun (1981), the gneiss domes and the diapiric plutons, and have risen for more than 10 km through the crust.

The generally good alignment of the belt some 50 km south of the trace of the suture probably corresponds to a broad anticline of the entire continental crust that can be compared with the outer rise observed in oceanic domain at a constant distance from the subduction trench.

To the west, the NHG seems to disappear, although a number of domes such as the Gurla Mandata (figures 1 and 6) and Rupshu are more or less aligned with it and could prolong it westwards all the way to Zanskar and southern Ladakh. However, the interruption of the regularly spaced plutons corresponds with a sharp bend of the MBT trace and its merging with the MCT; one may relate it to the former configuration of India and Eurasia Plate margins (Le Fort 1986).

The presence of chlorite, chloritoid, garnet and staurolite ± andalousite and cordierite (Burg *et al.* 1987) suggests that the NHG emplaced at a roughly similar level to the HHG and that the rate of uplift was of the same order of magnitude (around 1.5 mm a$^{-1}$). There also, discontinuous normal faulting may be responsible for a good part of this fast denudation. But more-thorough field, petrographical and geochemical work must be collected to constrain better the model of emplacement of the NHG.

### CONCLUSION

The four belts of the Himalayan Orogeny *s.l.* divide in two very contrasting groups (Blattner *et al.* 1983; Pitcher 1983), related to the Tethys oceanic and the India intracontinental subductions respectively. They contrast notably by the duration of their production (more than

70 Ma against about 10 Ma), by the volume of granitoid rock produced (several $10^5$ km$^3$ against a few $10^4$ km$^3$), by the origin of the melt (dominantly oceanic against almost entirely continental), by the petrographic and geochemical nature of the granitoids (diverse against very homogeneous) and by their relation to tectonics (generally little affected against syntectonic).

Their bearing on the tectonic evolution of the region is also quite contrasted. In Karakoram and Transhimalaya, the tectonic activity seems to be low during the emplacement of the plutons and, in turn, these do not seem to foster the deformation or to weaken the crust of the active margin (but they probably form no more than 10% of the crust). On the contrary, in the High Himalaya there is an intricate link between the granite and the migmatitic zone of production as well as in the disharmonic zone of emplacement. It is likely that the periods of normal faulting that occur during the overall compressive régime are made possible by the presence of leucogranitic mushes not yet entirely crystallized. The granites of the Himalaya are an intrinsic part of the tectonic evolution.

All the characteristics of the different belts are in a way the signature of the tectonic situation that has generated them. Time passing and erosion getting to deeper levels, the contrast between the different signatures will even increase. The patterns observed in this region help to interpret similar situations from the past, specially from Precambrian shields where granitic belts are so profuse and other evidences so limited.

These belts are among the best to work on. In fact, deep sections enable us to study and sample at crustal scale, and their relatively recent age has preserved many details and optimizes the geochronological studies. In addition, the absence of orogenic activity in the region for most of the Palaeozoic and Mesozoic, that is for about 400 Ma, clarifies the situation and greatly enhances the legibility of the tectonic and petrologic evolution. But not even half of these belts have yet been geologically walked across and studied on a reconnaissance basis. Thus the hope is great that, with an average amount of work, we will gain an unrivalled knowledge of the most important orogenic processes.

Field work has been supported mostly by the GRECO Himalaya–Karakoram (CNRS). Laboratory work has been supported by the Centre de Recherches Pétrographiques et Géochimiques (Nancy). Fruitful discussions, suggestions and comments from many colleagues have been greatly appreciated. Among them J. M. Bertrand, F. Debon, C. France-Lanord, G. Mascle, P. Molnar, A. Pêcher and S. M. F. Sheppard are specially thanked. Technical assistance from J. Gerbaut and A. Legros is gratefully acknowledged. This is CRPG contribution no. 747.

## REFERENCES

Academia Sinica 1980 *Symp. on Qinghai-Xizang (Tibet) plateau.* (104 pages.)

Afzali, H., Debon, F., Le Fort, P. & Sonet, J. 1979 *C. r. hebd. Séanc. Acad. Sci., Paris* D, 287–290.

Bard, J. P. 1983 *Earth planet. Sci. Lett.* **65**, 133–144.

Bard, J. P., Maluski, H., Matte, Ph. & Proust, F. 1980 Proc. Int. Comm. Geodyn., spec. issue, *Geol. Bull. Univ. Peshawar (Pakistan)* **13**, 87–94.

Bassoullet, J. P., Colchen, M., Marcoux, J. & Mascle, G. 1978 *C. r. hebd. Séanc Acad. Sci., Paris* **286**, 563–566.

Bertrand, J. M. & Debon, F. 1986 *C. r. hebd. Séanc. Acad. Sci., Paris* (II) **303**, 1611–1614.

Blattner, P., Dietrich, V. & Gansser, A. 1983 *Earth planet. Sci. Lett.* **65**, 276–286.

Bordet, P., Colchen, M., Le Fort, P. & Pêcher, A. 1981 In *Zagros–Hindu Kush–Himalaya geodynamic evolution* (ed. H. K. Gupta & F. M. Delaney), *Am. Geophys. Union, Geodyn, ser.* **3**, 1149–168. Washington, D.C.: American Geophysics Union.

Brun, J. P. 1981 Thèse Sci., University of Rennes, France. (197 pages.)

Brun, J. P., Pons, J. & Wang, F. P. 1983 Terra Cogn. 3, 264.

Burg, J. P. 1983 Thèse Doct. ès-Sci., University of Montpellier. (368 pages).

Burg, J. P., Brunel, M., Gapais, D., Chen, G. M. & Liu, G. H. 1984a J. struct. Geol. 6, 535–542.

Burg, J. P., Guiraud, M., Chen, G. M., & Li, G. C. 1984b Earth planet. Sci. Lett. 69, 391–400.

Burg, J. P., Leyreloup, A., Girardeau, J. & Chen, G. M. 1987 Phil. Trans. R. Soc. Lond. A 321, 67–86.

Caby, R., Pêcher, A. & Le Fort, P. 1983 Rev. Géogr. phys. Géol. dyn. 24, 89–100.

Colchen, M., Mascle, G. & Van Haver, T. 1986a In Collision tectonics (ed. M. P. Coward & A. C. Ries), Geol. Soc. London, spec. publ. no. 19, pp. 173–184.

Colchen, M., Le Fort, P. & Pêcher, A. 1986b In Recherches géologiques dans l'Himalaya du Népal, éd. Cent. Natl. Rech. Sci., Paris. (138 pages.)

Copeland, P., Harrison, T. M., Parrish, R., Burchfield, B. C., Hodges, K. & Kidd, W. S. F. 1987 Eos, Wash. 68, 1444.

Copeland, P., Harrison, T. M., Kidd, W. S. F. & Xu, R. H. 1988 Earth planet. Sci. Lett. 86, 240–252.

Coulon, C., Maluski, H., Bollinger, C. & Wang, S. 1986 Earth planet. Sci. Lett. 79, 281–302.

Coward, M. P., Jan, M. Q., Rex, D., Tarney, J., Thirlwall, M. & Windley, B. F. 1982 Nature, Lond, 295, 22–24.

Coward, M. P., Windley, B. F., Broughton, R. D., Luff, I. W., Petterson, M. G., Pudsey, C. J. & Rex, D. C. 1986 In Collision tectonics (ed. M. P. Coward, & A. C. Ries), Geol. Soc. London, spec. publ. no. 19, 203–219.

Cuney, M., Le Fort, P. & Wang, Z. X. 1984 In Proc. Symp. on Geology of granites and their metallogenetic relations, Nanjing University, China, 1982 (ed. K. Xu & G. Tu), pp. 853–873. Beijing: Science Press.

Cuney, M. & Autran, A. 1987 Géol. Fr. 2–3, 7–24.

Debon, F. & Le Fort, P. 1982 Trans. R. Soc. Edinb. 73, 135–149.

Debon, F., Le Fort, P. & Sonet, J. 1981 Symp. Geol. Ecol. Stud. of Qinghai Xizang plateau, vol. 1, Beijing, May 1980, pp. 395–405. Beijing: Science Press.

Debon, F., Sonet, J., Liu, G. H., Jin, C. W. & Xu, R. H. 1982 C. r. hebd. Séanc. Acad. Sci., Paris 295, 213–218.

Debon, F., Sonet, J., Liu, G. H., Jin, C. W. & Xu, R. H. 1984 In Etude géologique et géophysique de la croûte terrestre et du manteau supérieur du Tibet et de l'Himalaya (éd. J. L. Mercier & L. Li), CNRS-Acad. Sci. Géol. Chine rédaction, pp. 309–317.

Debon, F., Zimmermann, J. L., Liu, G. H., Jin, C. W. & Xu, R. H. 1985 Geol. Rdsch. 74, 229–236.

Debon, F., Le Fort, P., Sheppard, S. M. F. & Sonet, J. 1986a J. Petr. 27, 219–250.

Debon, F., Zimmermann, J. L. & Bertrand, J. M. 1986b C. r. hebd. Séanc. Acad. Sci., Paris (II) 303, 463–468.

Debon, F., Le Fort, P., Dautel, D., Sonet, J. & Zimmermann, J. L. 1987 Lithos 20, 19–40.

Deniel, C., Vidal, Ph., Fernandez, A., Le Fort, P. & Peucat, J. J. 1987 Contr. Miner. Petr. 96, 78–92.

Desio, A. 1964 Geological tentative map of the Western Karakoram. Scale 1/500.000. University of Milan: Institute of Geology.

Dietrich, V. J. & Gansser, A. 1981 Schweiz. miner. petrogr. Mitt. 61, 177–202.

Dietrich, V. J., Frank, W. & Honegger, K. 1983 J. volcan. geotherm. Res. 18, 405–433.

England, P. C. & Thompson, A. B. 1984 J. Petr. 25, 894–928.

England, P. C. & Searle, M. 1986 Tectonics 5, 1–14.

France-Lanord, C. & Le Fort, P. 1988 Trans. R. Soc. Edinb. 79. (In the press.)

France-Lanord, C., Sheppard, S. M. F. & Le Fort, P. 1988 Geochim. cosmochim. Acta. 52, 513–526.

Francheteau, J., Jaupart, C., Kang, W. H. & Shen, H. C. 1983 Terra Cogn. 3, 266.

Fuchs, G. 1981 Mitt. öst. geol. Ges. 74/75, 101–127.

Gamerith, H. 1979 Geologische Karte von Gilgit–Chitral–Wakhan (Nord Pakistan und Ost Afghanistan). Austromineral, Vienna (Austria).

Gansser, A. 1977 Structural map of Himalaya and South Tibet. Scale 1:3.500.000. Colloq. int. 268. Ecologie et géologie de l'Himalaya, Paris, (éd. Cent. Natl. Rech. Sci., vol. Sci. de la Terre), pp. 181–191.

Gansser, A. 1983 Denkschr. Schweiz. naturf. Ges. 96, 181.

Gapais, D., Gilbert, E. & Pêcher, A. 1984 C. r. hebd. Séanc. Acad. Sci., Paris (II) 299, 179–182.

Gariépy, C., Allègre, C. J. & Xu, R. H. 1985 Earth planet. Sci. Lett. 74, 220–234.

Hayden, H. H. 1905 Rec. geol. Surv. India 32, 160–174.

Herren, E. 1987 Geology, 15, 409–413.

Honegger, K., Dietrich, V., Frank, W., Gansser, A., Thöni, M. & Trommsdorff, V. 1982 Earth planet. Sci. Lett, 60, 253–292.

Kai, K. 1981 Geochem. J., 15, 63–68.

Krummenacher, D., Basett, A. M., Kingery, F. A. & Layne, H. F. 1978 In Tectonic geology of the Himalaya (ed. P. S. Saklani), pp. 151–166. New Delhi; Today & Tomorrow Publishers.

La Roche, H. de 1964 Sciences Terre 9, 293–337.

Le Bas, M. J., Mian, I. & Rex, D. C. 1987 Geol. Rdsch. 76, 317–323.

Le Fort, P. 1973 Bull. Soc. géol. Fr. 7, 555–561.

Le Fort, P. 1975 Am. J. Sci, 275 A, 1–44.

Le Fort, P. 1981 J. Geophys. Res. (Red Ser.) 86 B (11), 10545–10568.

298 P. LE FORT

Le Fort, P. 1986 In *Collision tectonics* (ed. M. P. Coward & A. C. Ries), Geol. Soc. London, spec. publ. no 19, 159–172.

Le Fort, P. 1988 In *Tectonic evolution of the Tethyan regions* (ed. A.M.C. Sengör) *Proc. NATO ASI meet., Istanbul, Oct. 1985*, Dordrecht: Reidel. (In the press.)

Le Fort, P., Debon, F. & Sonet, J. 1980 In Proc. of the Int. Geodyn. conf., Peshawar, Nov.–Dec. 1979, *Geol. Bull. Univ. Peshawar (Pakistan)* **13**, 51–61.

Le Fort, P., Debon, F. & Sonet, J. 1983a *Granites of Himalayas, Karakorum and Hindu Kush* (ed. F. A. Shams), pp. 235–255. Lahore, Pakistan: Institute of Geology, Punjab University.

Le Fort, P., Michard, A., Sonet, J. & Zimmermann, J. L. 1983b *Granites of Himalayas, Karakorum and Hindu Kush* (ed. F. A. Shams), pp. 377–387. Lahore, Pakistan: Institute of Geology, Punjab University.

Le Fort, P., Debon, F., Pêcher, A., Sonet, J. & Vidal, Ph. 1986 In *Evolution des domaines orogéniques d'Asie méridionale (de la Turquie à l'Indonésie)* (ed. P. Le Fort, M. Colchen & C. Montenat) *Mém. Sci. Terre*, **47**, 191–209.

Le Fort, P., Cuney, M., Deniel, C., France-Lanord, C., Sheppard, S. M. F., Upreti, B. N. & Vidal, Ph. 1987 *Tectonophysics* **134**, 39–57.

Mascle, G. H. 1985 *Bull. Soc. géol. Fr.* **8**, 289–304.

Petterson, M. G. & Windley, B. F. 1985 *Earth planet. Sci. Lett.* **74**, 45–57.

Pham, V. N., Boyer, D., Therme, P., Xue, C. Y., Li, L., & Guo, Y. J. 1986 *Nature, Lond.* **319**, 311–314.

Pitcher, W. S. 1983 *Mountain building processes* (ed. K. Hsü), pp. 19–40. London: Academic Press.

Polino, R. 1981 *Geological map of the upper Imja Khola (Mount Everest region, eastern Nepal)*. Scale 1/25000. S.E.L.C.A., Firenze.

Pudsey, C. J. 1986 *Geol. Mag.* **123**, 405–423.

Rafiq, M. 1987 Ph.D. dissertation, Centre of excellence in geology, University of Peshawar (273 pages.)

Rai, H. 1983 In *Geology of Indus Suture Zone of Ladakh* (ed. V. C. Thakur & K. K. Sharma), pp. 79–97. Dehra Dun: Wadia Institute of Himalayan Geology.

Reuber, I. 1988 *Tectonophysics*. (In the press.)

Reuber, I., Colchen, M. & Mevel, C. 1987 *Geodin. Acta* **1**, 283–296.

Reynolds, P. H., Brookfield, M. E. & Mcnutt, R. H. 1983 *Geol. Rdsch.* **72**, 981–1004.

Roy-Barman, X. & Le Fort, P. 1988 (In preparation.)

Scaillet, B., Dardel, J., Pêcher, A. & Le Fort, P. 1988 In *12ème réunion Sci. Terre, Lille*, p. 120. Soc. géol. France.

Schärer, U. 1984 *Earth and planet. Sci. Lett.* **67**, 191–204.

Schärer, U., Hamet, J. & Allègre, C. J. 1984 *Earth planet. Sci. Lett.* **67**, 327–339.

Schärer, U., Xu, R. H. & Allègre, C. J. 1986 *Earth planet. Sci. Lett.* **77**, 35–48.

Searle, M. P., Rex, A. J., Tirrul, R., Rex, D. C. & Barnicoat, A. 1988 *Bull. geol. Soc. Am.* (In the press.)

Sengör, A. M. C. 1984 *Geol. Soc. Am. spec. paper*, no. 195. (82 pages.)

Sengör, A. M. C. 1985 *Episodes*, **8**, 3–12.

Sharma, K. K. & Choubey, V. M. 1983 In *Geology of Indus Suture Zone of Ladakh* (ed. V. C. Thakur & K. K. Sharma), pp. 41–60. Dehra Dun: Wadia Institute of Himalayan Geology.

Srimal, N., Basu, A. R. & Kyser, T. K. 1987 *Tectonics*, **6**, 261–273.

Stöcklin, J. 1977 *Mém. Soc. géol. Fr.* **8**, 333–353.

Stöcklin, J. & Bhattarai, K. D. 1980 *Geological map of Kathmandu area and central Mahabharat range*. Scale 1/250000. H. M. Govt of Nepal, Dep. min. and geol.

Tahirkheli, R. A. K., Mattauer, M., Proust, F. & Tapponnier, P. 1979 In *Geodynamics of Pakistan* (ed. A. Farah & K. A. De Jong), pp. 125–130. Quetta: Geol. Survey Pakistan.

Tapponnier, P. *et al.* 1981 *Nature, Lond.* **294**, 405–410.

Tu G. C., Zhang, Y. Q., Zhao, Z. H. & Wang, Z. G. 1981 In *Geol. Ecol. Stud. of Qinghai–Xizang plateau*, vol. 1, pp. 353–361. Beijing: Science Press.

Valdiya, K. S. 1981 In *Zagros–Hindu Kush–Himalaya geodynamic evolution* (ed. H. K. Gupta & F. M. Delany), *Am Geophys. Union, Geodyn. Ser.* **3**, 87–110. Washington, D.C.: American Geophysics Union.

Vidal, Ph., Cocherie, A. & Le Fort, P. 1982 *Geochim. cosmochim. Acta*, **46**, 2279–2292.

Xu, R. H., Schärer, U. & Allègre, C. J. 1985 *J. Geol.* **93**, 41–57.

Zeitler, P. K. 1985 *Tectonics*, **4**, 127–151.

Zhang, Z. F., Zhu, M. X. & Liu, S. B. 1981 *Geol. Ecol. Stud. of Qinghai-Xizang Plateau, Beijing, May 1980*, vol. 1, pp. 895–904. Beijing: Science Press.

## Discussion

V. S. CRONIN (*Department of Geology, and Center for Tectonophysics, Texas A&M University, U.S.A.*). The exposures of the Transhimalayan batholith along the Indus–Tsangpo Suture Zone (ITSZ) seem to cluster in roughly half a dozen large masses that are separated from one another by valleys (see Gansser 1981, figure 2). Analysis of satellite imagery suggests that the valleys may be fault bounded. It is tempting to speculate that the large granitic masses are separated from one another by normal faulting and/or that they are separated along northwest-trending, right-lateral strike-slip faults. Both strike-slip and normal faults can accommodate east–west extension along the ITSZ at the southern edge of Tibet. Hirn *et al.* (1984) have inferred a major strike-slip fault along the ITSZ, based on deep seismic profiling. Armijo *et al.* (1986) describe right-lateral strike-slip faulting along the Karakoram–Jiali Fault Zone, *ca.* 200 km north of the ITSZ.

Does Dr Le Fort have any field data that might indicate whether the Transhimalayan batholith has been segmented by normal or strike-slip faulting? In a similar vein, does he know of any field data to support the hypothesis of strike-slip faulting along the ITSZ?

*References*

Armijo, R., Tapponnier, P., Mercier, J. L. & Han T. L. 1986 *J. Geophys. Res.* **91**, 13803–13872.
Gansser, A. 1981 In *Zagros–Hindu Kush–Himalaya geodynamic evolution* (ed. H. K. Gupta & F. M. Delany) *Am. Geophys. Union, Geodyn. Ser.* vol 3, pp. 111–121. Washington, D.C.: American Geophysics Union.
Hirn, A., Nercessian, A., Sapin, M., Jobert, G., Xu Zhong Xin, Gao, En Yuan, Lu De Yuan & Teng Ji Wen 1984 *Nature, Lond.* **307**, 25–27.

P. LE FORT. I thank Dr Cronin for stressing the segmented appearance of the Transhimalaya batholith (THB). I am also tempted to interpret this as a result of faulting. However, if we know a number of NS grabens such as the Thakkhola–Mustang one (Colchen *et al.* 1986; Tapponnier *et al.* 1986) that cut through the Indus–Tsangpo Suture Zone (ITSZ) and THB, they do not correspond to the major interruptions of the THB otherwise oriented NE–SW. Right-lateral strike-slip movement would better fit with these major gaps. However, we have very few indications that may support this interpretation and the relative movement should be rather limited at the level of the THB.

From the interpretative simplified maps of Tibet by Taponnier *et al.* (1986, figure 2) and Armijo *et al.* (1986, figure 33), there is a band of right-lateral decoupling along the Karakoram–Jiali Fault Zone (KJFZ), that presently separates northern Tibet from southern Tibet and India. It is likely that this zone was previously active some 150 km more to the south along the Transhimalaya zone when it was in the position of the KJFZ.

As for the ITSZ, strike-slip movement along it has been documented by Burg (1983) in the ophiolitic formations including the late Liuqu conglomerates in the region S and SW of Lhasa. Inferences of older right-lateral strike-slip movements have been put forward for the generation of the small episutural Eocene to Miocene, continental basins of the Ladakh region by Mascle *et al.* (1986). It should also be noted that after suturing, relative movement was transferred towrds the south where strike-slip has been documented in the High Himalaya crystallines (see Burg 1983; Pêcher *et al.* 1984; Gapais *et al.* 1984), this zone becoming the actual limit between India and Tibet during the Himalayan building.

*References*

Mascle, G., Hérail, G., Van Haver, T. & Delcaillau, B. 1986 *Bull. Cent. Rech. Explor.–Prod. Elf–Aquitaine*, Pau, **10**, 181–203.
Pêcher, A., Bouchez, J. L., Cuney, M., Deniel, C., France-Lanord, C. & Le Fort, P. 1984 In *10éme Réunion sci. Terre, Bordeaux*, p. 436. Soc. géol. France.
Tapponnier, P., Peltzer, G. & Armijo, R. 1986 In *Collision tectonics* (ed. M. P. Coward & A. C. Ries), Geol. Soc. London, spec. publ. no. 19, 115–157.

*Phil. Trans. R. Soc. Lond.* A **326**, 301–320 (1988)    [ 301 ]

*Printed in Great Britain*

# The mechanics of the Tibetan Plateau

By P. C. England[1] and G. A. Houseman[2]†

[1] *Department of Earth Sciences, University of Oxford, Parks Road,*
*Oxford OX1 3PR, U.K.*
[2] *Research School of Earth Sciences, Australian National University,*
*Canberra, ACT 2601, Australia*

Continental convergence results in compressional deformation over a distance, perpendicular to strike, that is comparable to the length of the convergent boundary. The compressional forces generated by the convergence are resisted, to some extent, by the extensional deviatoric stresses arising from isostatically balanced increases in crustal thickness; as a result a plateau may form, in front of a compressional boundary, whose elevation is limited by the strength of the continental lithosphere. However, the extensional stresses do not exceed the compressional stresses that generate the crustal-thickness contrasts unless there is a major change, either in the convergent velocity or in the potential energy of the elevated region. For the collision of India with Asia, it appears that there has not been a change in the convergent boundary condition sufficient to cause the late-Tertiary to present extension in the region. It is suggested that thermal evolution of the region, involving a delayed convective instability of the base of the thickened lithosphere, could have raised the surface elevation and the potential energy of the Tibetan Plateau, leading to the observed extension there.

## 1. Introduction

Continental lithosphere is subjected to stresses at its boundaries, owing to relative motion of the plates, and in its interior, arising from isostatically compensated elevation contrasts (Artyushkov 1973; Dalmayrac & Molnar 1981; Molnar & Lyon-Caen 1988). The fact that continental portions of the plates undergo large strain, whereas oceanic portions usually do not, presumably results from the lesser strength and greater buoyancy of continental lithosphere (McKenzie 1972; Tapponnier & Molnar 1976). To say this is to state only the broad outline of the process; explanation of even large-scale phenomena such as the presence of plateaux of uniform elevation, the plan form of deforming zones, or the variation of strain rates within these zones, requires a more detailed understanding of the forces acting on and the rheology of the lithosphere. The purpose of this paper is to review briefly work that has been carried out on these problems, with particular reference to the evolution of the Tibetan Plateau.

## 2. Mechanics of continental deformation

### 2.1. *The continuum approximation*

Most investigations of the mechanics of continental deformation have treated the lithosphere as a continuous medium (see, for example, Bird & Piper 1980; England & McKenzie 1982, 1983; Vilotte *et al.* 1982, 1984, 1986; Bird & Baumgardner 1984; England & Houseman 1985,

† Present address: Department of Earth Sciences, Monash University, Clayton, Victoria 3168, Australia.

1986; Houseman & England 1986). A superficial glance at any actively deforming region shows the importance of faulting in lithospheric deformation; it appears that earthquakes in regions of continental deformation account, in most cases, for over 30 % and, in some cases, for close to 100 % of the strain of the upper crust (Jackson & McKenzie 1988; Ekström & England 1988). It might, therefore, seem inappropriate to treat continental deformation as a continuous phenomenon.

No large strain of a solid is continuous, so the question of interest is whether the deformation of the continents may be *approximated* by that of a continuous medium. The continuum hypothesis identifies a length that is great compared with the length scale of unpredictable fluctuations in the properties of the system; if this length is also short compared with the scale of the system itself, then there is some hope of treating the system quantitatively by using continuum mechanics.

The scale of continental deformation is large: active regions have horizontal dimensions that are several times the thickness of the lithosphere, and many times the crustal thickness. Using continuum mechanics to investigate the behaviour of deforming continental lithosphere involves assuming that the discontinuities represented by faults and shear zones are at spacings short compared with the thickness of the lithosphere; if this assumption is correct, then the continuum approach to the mechanics of the continental lithosphere may be a reasonable one.

The occurrence of earthquakes shows that, at some scale, the deformation must be treated as discontinuous, but there is little understanding of the way in which the continents deform over scales between that of the individual faults that produce earthquakes and that of the deforming zones as a whole (a few hundred to a couple of thousand kilometres). Characterizing the distribution of strain within this range of scales will be a problem for some time to come. An approach that treats the continents as continuous on all observable scales will probably not provide all the answers; on the other hand, the simple physical arguments that follow from treating the lithosphere as being continuous on the large scale have given some useful insights to the mechanics of the continents. We concentrate here on those arguments, but the reader should bear in mind that they are not expected to hold at all scales; the area of a fault that breaks in a large continental earthquake may exceed $10^4$ km$^2$, so it is possible that the continuum approximation is not valid even at the largest scales on which it is customary to carry out geological mapping.

### 2.2. *The thin-sheet approximation*

Once the assumption of continuity is made, the next step is to specify the configuration of the deforming system; most of the investigations cited above make use of the thin-sheet approximation (first used in this context by Bird & Piper 1980). In this approximation, stresses acting on the top and base of the lithosphere are assumed to be negligible, as are the surface slopes (see figure 1), so shear stresses on horizontal planes within the lithosphere are negligible compared with the other components of the stress tensor. Under these conditions, one principal stress is vertical, and is equal to the weight per unit area of overlying rock, a condition that is often assumed to hold in geological processes:

$$\sigma_{zz} = g \int_{\text{land surface}}^{z} \rho(z') \, dz', \tag{1}$$

where $\rho$ is the density. Neglecting shear stresses on horizontal planes is equivalent to assuming that the top of the lithosphere is not displaced, in the horizontal plane, relative to its base; the behaviour of the system then depends upon vertical averages of the stresses acting within the lithosphere.

The assumption that the top of the lithosphere is not displaced appreciably with respect to its base may seem unwarranted when considered in the light of field observation; all thrusts and normal faults do contain such components of shear, and there is abundant evidence in the fabrics of rocks deformed in the ductile régime that they have undergone sub-horizontal shear. However, provided that the continuum hypothesis is tenable for a given region (see above), a distributed set of thrust or normal shear zones produces horizontal displacement of the top of the lithosphere with respect to its base that is small compared with the total shortening or extension. The assumptions made here are clearly inappropriate to regions, such as accretionary prisms, in which there is a rigid substrate that applies shear stresses to the base of the deforming region. For instance, the southernmost 200 km or so of the India–Asia collision zone is underlain at present by Indian Shield, and a detailed analysis of the mechanics of this region would need to take account of this; throughout most of the collision zone, however, the Asian lithosphere appears to be underlain by upper mantle whose temperature is normal, or even higher than normal, for its depth and is presumably much weaker than the lithosphere above it (see Molnar, this symposium). The virtual absence of regional surface slopes within the Tibetan Plateau supports the assumption that there are no organized shear stresses acting on its base.

There is a separate, rheological, issue concerning this assumption about the configuration of strain: it appears to neglect the observation that, even in the ductile régime, much of the deformation of the continental lithosphere is concentrated into narrow shear zones that separate regions of less intensely deformed rock. The rheology of such a piece of lithosphere is rather different from one deforming homogeneously. Consider, as a simple example, two pieces of continental lithosphere each increasing in thickness by a factor of two; one piece thickens by homogeneous pure shear, and the other by motion on a set of faults and shear zones, of arbitrary orientation. In each case, the thickened lithosphere has changed its gravitational potential energy, and work has been done (against or by gravity, and in deforming the lithosphere to achieve this state. To a first approximation, provided that the two pieces of lithosphere had similar density structures at the start of the deformation, the amount of work done against or by gravity during the deformation does not depend on the configuration of the strain. However, the amount of work done to deform the lithosphere may well depend on whether the strain is distributed uniformly throughout the lithosphere or concentrated in narrow zones.

In any real situation, there will be considerable uncertainty as to the rheology of the continental lithosphere. Equally, the density distribution for any given region will be ill-determined, so that the potential-energy change and the work done to deform the lithosphere will, separately, be uncertain by quite large amounts. As we shall see below, the behaviour of the system is governed by the *ratio* of gravitational forces to the forces required to deform the medium, not by the absolute value of either. The practical approach, therefore, is to determine the form of the function relating the gravitational forces to changes in lithospheric thickness and of the function relating the vertically averaged forces acting on the lithosphere to the average strain rates it undergoes. The ratio of the magnitudes of the two forces, being so uncertain, is then left as a parameter that we may hope to constrain by comparison of the results of calculations with the observations.

The work done by or against gravity in thickening the continental lithosphere has been discussed extensively, and it is generally assumed that elevated regions that are isostatically compensated have greater potential energy than their lower-lying surroundings (see, for example, Evison 1960; Frank 1972; Artyushkov 1973; Tapponnier & Molnar 1976; Dalmayrac & Molnar 1981). However, if the elevated region is produced by thickening the continental lithosphere, the buoyant thick crust may be underlain by a negatively buoyant root of thickened lithospheric mantle. For most of the likely density distributions within the continental lithosphere, the potential energy per unit area of the thickened lithosphere is likely to be greater than that of an equivalent column beneath unthickened lithosphere. There is, though, a range of conditions in which the reverse would be true; one of these is considered by Fleitout & Froidevaux (1982).

The choice of an average rheology has usually been made by one of two approaches. The first specifies purely plastic behaviour in the upper lithosphere (representing friction on faults) with a power-law rheology for the lower lithosphere (representing the steady-state creep of silicates); a vertically averaged rheology is calculated from point to point within the medium depending upon the state of stress and strain (Bird & Piper 1980; Vilotte *et al.* 1982, 1984, 1986); a special case of this is the assumption of purely plastic behaviour throughout the lithosphere, as was made by Tapponnier & Molnar (1976). A second approach, on which we shall concentrate, is to specify a single power-law rheology for the whole lithosphere and to leave the exponent in the power-law rheology as a free parameter (England & McKenzie 1982; Houseman & England 1986); variations in this exponent reflect variations in the relative contributions of creep and slip on faults to the vertically averaged rheology of the lithosphere (Sonder & England 1986).

Figure 1 illustrates the assumed configuration of deformation of the continental lithosphere as it might apply to the India–Asia collision zone. A piece of lithosphere that behaves in a plate-like fashion (that is, which deforms elastically) moves relative to a weaker region of continental lithosphere. This relative motion is accommodated by large strains in the weaker lithosphere (figure 1 *a*). The stronger lithosphere may be either oceanic, or continental material that has, perhaps, a lower geothermal gradient (see Molnar & Tapponnier 1981). The deformation of the weaker lithosphere involves work against gravity, because $\int \sigma_{zz}$ (proportional to the gravitational potential energy of the lithosphere), is generally greater beneath thicker crust than beneath thinner crust (figure 1 *b*). In many cases it is a good approximation to treat the average of this stress difference as depending linearly on the square of the vertical strain (England & McKenzie 1982; England & Houseman 1988); this is shown in figure 1 *c*.

The vertical integral of the difference in $\sigma_{zz}$ between any two columns of continental lithosphere in isostatic equilibrium (figure 1 *c*) is identical to the difference in gravitational potential energy between them, provided that equation (1) holds (for an accessible reference, see Molnar & Lyon-Caen 1987). This quantity, which has the dimensions of force per unit length, will be referred to as either a buoyancy force or a potential-energy contrast in what follows.

Work is also done against frictional and viscous forces within the deforming lithosphere (figure 1 *d*); the vertical average of these stresses can often be treated as though the lithosphere obeyed a power-law rheology (Sonder & England 1986); this is illustrated by figure 1 *e*.

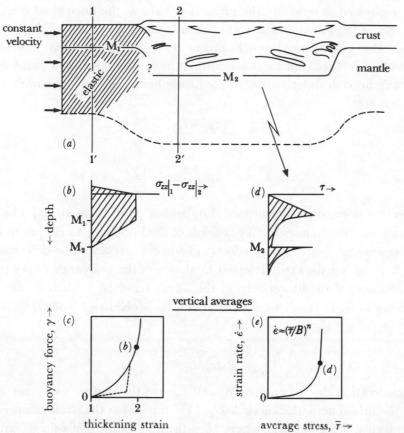

FIGURE 1. Sketch of the distributions of topography, buoyancy force and deviatoric stress assumed for continental deformation. (*a*) A plate that behaves elastically is moving towards a region of continental lithosphere that undergoes large compressional strain, by thrust faulting in its upper portions and by ductile deformation at greater depths. In the context of this paper, the elastic plate may be thought of as the Indian Shield and the region of distributed deformation would then correspond to the Eurasian Plate to the north of the Himalaya. The location of the transition between the two plates is not well known in Asia, and is deliberately ill-defined in this sketch, but the elastic plate is not considered to penetrate far beneath the Asian lithosphere (see Molnar, this symposium). There is about tenfold vertical exaggeration; regional surface slopes are a few degrees on the edges of the Tibetan Plateau, and much less in the interior.

(*b*) If equation (1) holds, the vertical stress, $\sigma_{zz}$, is equal to the weight per unit area of overlying rock; the figure shows, as a function of depth, the difference in vertical stress between two columns: 1–1′ in the unthickened lithosphere to the left of the figure and 2–2′ through the thickened lithosphere. The integral of this difference with respect to depth is the buoyancy force, $\gamma$, arising from isostatically compensated elevation differences (see Dalmayrac & Molnar 1981). The buoyancy force depends on the thickening strain (*c*); it may rise smoothly, as is shown by the solid line or, as is suggested in §3 of this paper, it may increase sharply when the base of the lithosphere is removed convectively. For most cases this buoyancy force acts so that work must be done against gravity to increase the thickness of the lithosphere, so regions of increased surface height have a tendency to spread. (*d*) Work must also be done to overcome frictional and viscous resistance to deformation; the figure, following Brace & Kohlstedt (1980), sketches the deviatoric stress as a function of depth in continental lithosphere deforming at geological strain rates. Uncertainties as to the physical mechanisms operating make it pointless to give absolute values of stress; Sonder & England (1986) show that the vertical average of such a rheological profile approximates to a power-law rheology over a wide range of conditions (*e*).

## 2.3. *Mathematical formulation*

The mathematical formulation of this problem is discussed by Bird & Piper (1980) and England & McKenzie (1983), and will not be repeated here. We reproduce the governing equations to allow us to relate the results discussed below to the underlying physics.

As the vertical stress is assumed to be determined by local isostatic balance (above) the problem may be expressed in terms of quantities that vary in the horizontal directions only: the vertical averages of the gravitational and viscous forces acting on the lithosphere. We use the symbol $\gamma$ for the gravitational potential energy contrast between a given piece of lithosphere and some reference column (for example the mid-ocean ridges), and the symbol $\bar{\tau}$ for the vertically averaged deviatoric stress in the lithosphere. The two horizontal components of the force balance are

$$\frac{\partial \bar{\tau}_{xx}}{\partial x} + \frac{\partial \bar{\tau}_{xy}}{\partial y} + \frac{\partial (\bar{\tau}_{xx} + \bar{\tau}_{yy})}{\partial x} = \frac{\partial \gamma}{\partial x}, \tag{2}$$

$$\frac{\partial \bar{\tau}_{yx}}{\partial x} + \frac{\partial \bar{\tau}_{yy}}{\partial y} + \frac{\partial (\bar{\tau}_{xx} + \bar{\tau}_{yy})}{\partial y} = \frac{\partial \gamma}{\partial y}, \tag{3}$$

where $x$ and $y$ are the horizontal coordinates (England & McKenzie 1983). The right-hand sides of these equations contain horizontal gradients of the buoyancy forces per unit area, and the left-hand sides contain horizontal gradients of viscous stresses. The deformation of the lithosphere then depends on the ratio, referred to above, of the buoyancy forces produced by changes in the thickness of the lithosphere to the forces required to deform the continental lithosphere. This ratio is referred to, by England & McKenzie (1982), as the Argand number, $Ar$:

$$Ar = \frac{g\rho_{\mathrm{c}}(1 - \rho_{\mathrm{c}}/\rho_{\mathrm{m}})\,L}{B(U_0/L)^{1/n}}, \tag{4}$$

where $g$ is the acceleration due to gravity, $\rho_{\mathrm{c}}$ and $\rho_{\mathrm{m}}$ are the densities of crust and mantle, respectively, $L$ is the lithospheric thickness. $B(U_0/L)^{1/n}$ represents the stress required to deform the lithosphere at a strain rate of $U_0/L$, where $U_0$ is the magnitude of a velocity applied to the boundary of the lithosphere. The parameter $n$ is the exponent in the power-law rheology that is assumed for the lithosphere (figure 1).

As defined above, the Argand number refers to specific conditions of rheology and density structure of the lithosphere, but for most of the discussion below a more general statement suffices:

$$Ar \approx \frac{\text{buoyancy stress that would result from a contrast, } L, \text{ in crustal thickness}}{\text{viscous stress required to deform lithosphere at reference strain rate}}. \tag{5}$$

The stresses referred to in both the numerator and the denominator above are averaged vertically through the lithosphere. If the strength of the lithosphere were very large $Ar$ would be small, and the right-hand sides of equations (2) and (3) would be negligible; deformation would then depend only on the boundary conditions that were applied to the lithosphere by the relative motion of the plates. On the other hand, when $Ar$ is very large, the stresses due to elevation contrasts dominate the deformation and the lithosphere would not be able to support appreciable topographic contrasts.

### 2.4. *Length scales of continental deformation*

When buoyancy forces are negligible (as might be the case at the onset of deformation, when variations in crustal thickness are small, or if the lithosphere is very strong), approximate solutions to the equations can be found that have a simple form (England *et al.* 1985). For example, if a compressional or extensional boundary condition of characteristic length $D$ is applied to the edge of a piece of lithosphere whose behaviour is described by equations (2) and

(3), the resulting deformation dies out approximately exponentially with distance from that boundary with a scale length that depends only upon the value of $D$, and of the exponent, $n$, in the power-law rheology chosen to describe the lithosphere (figure 1). Thus if a relative velocity $U_0$ is applied perpendicular to the edge of a continent, the velocity in the interior at a distance $x$ from the boundary will be given approximately by

$$u(x) = U_0\, e^{-(x\sqrt{n\pi/2D})} \qquad (6)$$

(England *et al.* 1985, equation 29; Houseman & England 1986, equation 7). At distances greater than about $4D/\sqrt{n\pi}$ from the boundary, this velocity is small compared with the boundary velocity, so the width of a deforming compressional or extensional zone would be approximately equal to its length if $n$ were equal to 3, or about half that if $n$ were equal to 10. (A power law exponent of 10 has no significance in terms of a single deformation mechanism operating within the lithosphere, but could represent the vertical average of creep at depth and sliding on faults near the surface (Sonder & England 1986).)

One of the many interesting questions concerning the India–Asia collision zone is whether there is any interaction between deformation produced by the convergence of India and the extensional deformation on the eastern boundary of Asia. The discussion above shows that, if Asia acts as a thin continuous sheet, extensional and compressional stresses on different boundaries may be accommodated independently by deformation of the appropriate kind close to the respective boundaries.

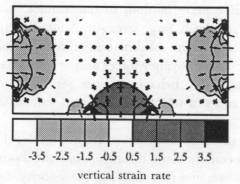

-3.5  -2.5  -1.5  -0.5  0.5  1.5  2.5  3.5

vertical strain rate

FIGURE 2. Results of a thin-sheet calculation superimposing deformation due to extensional and compressional boundary conditions. In this, and subsequent figures, a map view is presented of the results of thin sheet calculations of the kind described in detail by Houseman & England (1986) and England & Houseman (1988). The solid lines show the magnitude and orientation of the calculated horizontal principal deviatoric stresses. Thinner lines correspond to extensional deviatoric stresses, and the thicker lines to compressional stresses. Vectors shorter than the thickness of the lines are not shown. The sum of the horizontal principal strain rates is the negative of the vertical strain rate, $\dot{e}_{zz}$, which is contoured in this figure. Both velocities and strain rates are dimensionless in this and subsequent figures (see Houseman & England 1986 for the non-dimensionalization); alternatively, for comparison with the India–Asia collision, the width of the region of calculation may be taken as 10000 km, and the maximum velocity as 50 mm a$^{-1}$, in which case one unit of strain rate corresponds to $3 \times 10^{-16}$ s$^{-1}$.

This is illustrated for one set of conditions in figure 2, which shows the results of a simple calculation of the deformation of a sheet of material with power law exponent 3.

A convergent velocity of unit magnitude is imposed over a length $D$ of one boundary of the sheet, and there is an axis of symmetry through the middle of this boundary. The boundaries at right-angles to the influx boundary have an outward velocity of unit magnitude imposed over a length $D$ of their central sections. The deformations associated with the two boundary conditions take place almost independently of one another. Although there are horizontal extensional strain rates over much of the region of calculation in figure 2 (because there are two extensional boundaries to one compressional) they are appreciable only close to the extensional boundaries. Note that the width of each zone over which the strain rate is large is approximately the length of the boundary over which the convergent or divergent velocity is applied – appropriate, as discussed above, for a fluid of power-law exponent 3. If a calculation with a higher power-law exponent had been illustrated, these widths would have been smaller, and the separation of styles of deformation would have been more pronounced (see equation (6)).

Note that the results obtained would have been completely different if one had assumed plane horizontal strain for the region. The condition of plane strain forces the net mass flux across the vertical boundaries to be zero; under such circumstances the convergent velocity over one boundary must be accommodated by outward motion wherever a boundary condition exists that would permit it, irrespective of the distance to that boundary; see, for example, the plane-strain experiments of Tapponnier *et al.* (1982) and Vilotte *et al.* (1982).

## 3. MECHANICS OF THE INDIA–ASIA COLLISION ZONE

### 3.1. *Plateau formation*

Continental lithosphere deforms in a distributed fashion under forces of the same order of magnitude as those that drive plate motion, so one cannot, in general, assume that the continental lithosphere has arbitrarily large strength. A full description of the behaviour of a collision zone should then consider the combination of deformation caused by the boundary conditions and that caused by the buoyancy forces arising from changes in the thickness of the lithosphere. Figure 3 outlines this behaviour in a set of sketch cross sections through hypothetical compressional zones in which the buoyancy forces vary between negligible and dominant.

In figure 3a it is assumed that buoyancy forces are negligible; then, as discussed above, the deformation dies out exponentially with distance from the boundary. As time progresses compressional strain accumulates at a rate that, like the velocity, decreases with distance from the boundary. This behaviour is illustrated more fully in the thin-sheet calculations of Houseman & England (1986, figure 3).

On the other hand, if the buoyancy forces are dominant, it may readily be seen from equations (2) and (3) that even the smallest gradients of buoyancy force will dominate the flow; in the limit that $Ar$ goes to $\infty$, no surface slopes will be supported. Presumably each such deforming continental region would be bounded somewhere, if only by surrounding oceanic lithosphere, so that cross sections through the system at successive times would show a region of no surface slope whose area decreased and whose surface height increased with time (figure 3b).

The intermediate case is illustrated in figure 3c; here the Argand number is greater than zero, but still small; thickening takes place in front of the compressional boundary until the compressional stresses are no longer great enough to increase appreciably the gravitational potential energy of the lithosphere there. At greater distances the lithosphere has lower surface elevation and is still under compression. These lower regions continue to thicken rapidly until they reach the thickness of the plateau behind them (Tapponnier & Molnar, 1976; Molnar &

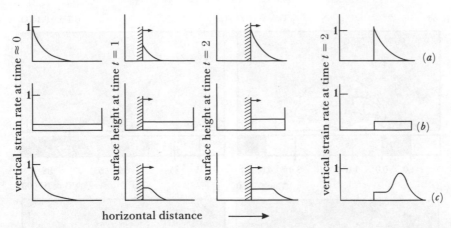

horizontal distance ⟶

FIGURE 3. Sketch of the evolution of vertical strain rates and surface heights in hypothetical portions of continental lithosphere having differing strengths. A cross section is shown through regions of lithosphere subjected to an indenting boundary condition that moves to the right, as depicted by the hatches and arrows in the two central columns. It is assumed that the lithosphere is always in isostatic equilibrium, so that changes in crustal thickness are reflected in changes in surface elevation. The left-hand column shows the distribution of vertical (thickening) strain rate shortly after deformation begins, the two central columns show the distribution of surface heights at two subsequent times, and the right-hand column shows the distribution of strain rate at the latest of these times. When the continental lithosphere is arbitrarily strong (a), the buoyancy forces do not influence the deformation and vertical strain accumulates in proportion to the vertical strain rate, which, in a frame of reference fixed to the convergent boundary, does not change with time. The width of the deforming zone depends only on the along-strike length of the boundary and the power-law exponent of the sheet (equation (6)). When the buoyancy forces dominate (b), even the smallest crustal thickness gradients cannot be supported by the strength of the lithosphere and both surface heights and vertical strain rates are laterally homogeneous. In intermediate cases (c) the lithosphere can support crustal thickness contrasts, and in the early phases of deformation the strain rate field resembles that of the strong lithosphere (a). Eventually the crust becomes sufficiently thick for the buoyancy forces to become comparable to the strength of the lithosphere, thickening slows near the indenting boundary, and the convergence is then accommodated largely around the edges of a plateau whose area increases, although its surface height does not appreciably increase.

Tapponnier 1978; Molnar & Lyon-Caen 1988). In this case, the deformation is characterized by a mountain belt of roughly constant surface height whose area increases with time.

These arguments are based on inspection of the governing equations, but are borne out by solutions obtained, for example, by England & McKenzie (1983), Vilotte et al. (1984) and Houseman & England (1986). A wide range of solutions exhibit the characteristics discussed above, many of which are illustrated by Houseman & England (1986); the reader is referred to this paper for details of the calculation procedures.

Figure 4 reproduces one of these calculations, in which the Argand number, $Ar$, is 3 and the power-law exponent, $n$, in the continental rheology is 10 (Houseman & England 1986, figures 4 and 5). The calculation shows the development with time of a region of thickened crust with very low gradients of crustal thickness, except near the corners of the indenter (figure 4b, d, f). Over the same interval, the locus of maximum thickening strain rate migrates away from the boundary, and is concentrated on the outer slopes of the plateau (figure 4a, c, e).

Contours are shown at three dimensionless times; by using the values for parameters given in the caption to figure 2, these may be converted to times by multiplying by 100 Ma.

The solutions obtained to the thin-sheet equations with an indenting boundary condition show that the lithosphere needs to be able to support vertically averaged deviatoric stresses of about 100 MPa to maintain surface elevation contrasts comparable to the Tibetan Plateau

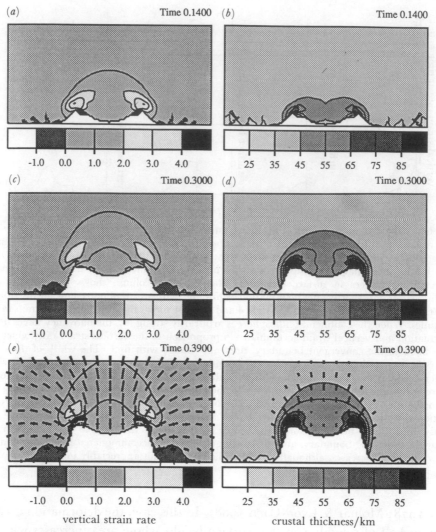

FIGURE 4. Contours of vertical strain rate and crustal thickness for a thin-sheet calculation with $n = 10$ and $Ar = 3$ (equations (1)–(5); see Houseman & England 1986 for details of calculation). Shaded regions correspond to the deforming lithosphere, the solid lines surrounding them correspond to the boundaries to the deforming material. One portion of the bottom boundary moves with time. Dimensionless times, shown above the parts of the figure, can be converted to times by multiplying by 100 Ma. As in figure 2, these are map views of the deformation. Figure 4$a$, $c$, $e$ show contours of vertical strain rate. Figure 4$b$, $d$, $e$ show contours of crustal thickness; before deformation starts the crustal thickness is 35 km throughout. Solid lines superimposed on ($e$) show the orientations and relative magnitudes of the calculated horizontal principal stresses; thicker lines correspond to compressional stress, the few thinner lines, around the corners of the indenter, to extensional stresses. The longest symbol corresponds to a compressional deviatoric stress of $1.0 \times 10^8$ Pa averaged vertically through the lithosphere. The solid lines on ($f$) show the calculated horizontal principal strain rates; these are parallel to the respective principal stresses, but depend on the $n$th power of the deviatoric stress ($e$). Note the concentration of compressional strain rates around the edges of the plateau. The maximum compressional strain rate is $1.4 \times 10^{-15}$ s$^{-1}$.

(England & Houseman 1986). If this condition holds, then most of the convergent motion applied by the indenting boundary is accommodated by thickening of the lithosphere in front of that boundary; this is so for several different assumptions about the state of stress on the lateral boundaries to the lithosphere (England & McKenzie 1983; Vilotte *et al.* 1982; Houseman & England 1986; England & Houseman 1986; Vilotte *et al.* 1986).

Note that even though the outward displacement of the maximum thickening strain rates (figure 4 *a*, *c*, *e*) shows that buoyancy forces are important in the deformation of the region, compressional stresses still dominate within the elevated region (figure 4*e*) and the width of the plateau is still very similar to what would be predicted by the simple arguments of §2.4: namely about half the length of the covergent boundary condition. Compare this with a width of 1000–1500 km for the Tibetan Plateau, and a length of 2500 km for the Himalayan front. (Other comparisons between calculation and observation are given by England & Houseman 1986.) If the relative importance of the buoyancy forces is greater than in figure 4, then the width of the deforming region becomes larger, but the maximum thickening strain is correspondingly diminished for an equivalent amount of convergence, and crustal thickness contrasts as great as that between the Tibetan Plateau and its surroundings are not formed (Houseman & England 1986).

### 3.2. *Extension of a thickened plateau*

The results of thin-sheet calculations have been compared with observations in the India–Asia collision zone by England & Houseman (1986) and Vilotte *et al.* (1986). These calculations appear to account satisfactorily for the major features of the Tertiary deformation in Asia: the present size and shape of the plateau (above), the northward displacement of the southern margin of the region by at least 1000–2000 km (Zhu *et al.* 1977, 1981; Molnar & Chen 1978; Achache *et al.* 1984; Lin & Watts, this symposium), the distributed thrusting early in the deformation of Tibet, that is followed by strike-slip deformation (Chang *et al.* 1986) and the concentration of thickening strain around the edges of the plateau at the present day (Molnar & Deng 1984; Molnar, this symposium). The presence of low-lying and relatively aseismic areas within the deforming region, of which the most prominent is the Tarim Basin, can be accounted for by quite modest lateral variations in the strength of the lithosphere (Molnar & Tapponnier 1981; Vilotte *et al.* 1984; England & Houseman 1985). (The calculation illustrated in figure 4 is only one of a set described by Houseman & England (1986), and the reader is referred to that paper, and to England & Houseman (1986) for illustrations of deformation with different values of the parameters *n* and *Ar*.)

The most striking feature of the active strain in the Tibetan Plateau is the extension that occurs on roughly north–south normal faults (Molnar & Tapponnier 1978; Tapponnier *et al.* 1981; Armijo *et al.* 1982; 1986). The origin of this extension is explained in principle by the buoyancy forces associated with crustal thickness contrasts (Tapponnier & Molnar 1976), but this argument is a purely static one that does not take into account the compressional stresses associated with the convergence of India with Asia. Calculations that do take these stresses into account (one of which is illustrated in figure 4) do not exhibit extension in the region of thickened crust at any stage: although some stretching does occur parallel to the indenting boundary, it is always smaller in magnitude than the horizontal shortening, so that there is no net thinning of the lithosphere.

The extension in Tibet began comparatively recently, probably in the past 5 Ma (Armijo *et al.* 1982, 1986). The discrepancy between calculation and observation can be resolved if the India–Asia collision zone underwent, in this time interval, a change in the balance between the horizontal deviatoric stresses applied on the boundaries of Asia, and the buoyancy stresses in the interior of the continent. Three possible changes that could lead to extension within the plateau are:

(1) a change in the strength of the lithosphere;

(2) a change in the boundary conditions acting on Asia: either a reduction in the north–south

compressional stress applied by the convergence of India with Asia, or an increase in east–west extensional stresses owing to a condition applied on the eastern boundary of Asia.

(3) an increase in the potential energy of the Tibetan Plateau.

### 3.2.1. *Changes in strength of the lithosphere*

Before discussing mechanisms that can produce extension, it is worth mentioning another mechanism that has been proposed several times to the authors, which would not. The Argand number relates the strength of the continental lithosphere to the stresses arising from crustal thickness contrasts (equations (4) and (5)). It might seem at first sight that, if some mechanism diminishes the strength of the thickened lithosphere with time, any plateau that forms would eventually become sufficiently weak to flow away under its own weight. The most likely cause of a reduction in strength of the continental lithosphere during deformation is the heating of rocks by thermal relaxation of thickened lithosphere. Presumably, these changes occur first in the thickened crust near the indenting boundary and, it might be thought, would lead to exactly the imbalance between viscous stresses and buoyancy required for extension of the elevated region.

Such a change, if it affected the entire sheet, would indeed lead to extension; as the governing equations show, this is equivalent to a reduction in the convergent velocity (see equation (4)), a case that is considered below. However, we cannot consider local changes in viscosity in the same way as changes that affect the Argand number of the whole system: there is a balance between the compressional forces applied to the plateau by its surroundings (both the convergent boundary and the deforming foreland) and the work that is done within the plateau (both against gravity and to deform the material of the plateau). If the strength of the plateau is decreased while the boundary stresses are maintained, the rate of thickening in the plateau must increase.

Figure 5 illustrates this point. The topography and velocity boundary conditions are the same in each portion of the figure, and are those calculated for the set of parameters used in figure 4 (figure 5$b$, $d$, $f$), but the strain rate fields are calculated under three different assumptions. In figure 5$c$, $d$ no modification to the original viscosity structure has occurred, and this calculation would carry on to yield the results shown in figure 4. In figure 5$a$, $b$ the factor $B$ in the Argand number (figure 1 and equation (4)) has been decreased by 20 % for the region of the lithosphere in which the crustal thickness exceeds 55 km; this change would result in a tenfold increase in strain rate for a constant stress. The change is imposed arbitrarily and instantaneously to illustrate the influence of what would presumably be a gradual warming and weakening of the material of the thickened lithosphere. Comparison of figures 5$a$ and $c$ shows that the result of this change in rheology is to increase the compressional strain rates in the plateau, particularly at its edges, where the gradients of buoyancy forces are greatest, and that there is no appreciable change in the extensional strain rates parallel to the boundary of the plateau.

This result may be understood by considering a simple two-dimensional calculation, as might be appropriate close to the axis of symmetry in figure 5, where deviatoric stresses and strain rates in the east–west direction are small.

Figure 6 gives a simplified sketch of such a region, which is treated as being in two parts having lengths $l_1$ and $l_2$, gravitational potential energies $\gamma_1$ and $\gamma_2$ and vertically averaged viscosities $\eta_1$ and $\eta_2$, respectively. For simplicity we consider newtonian viscous fluids, but the

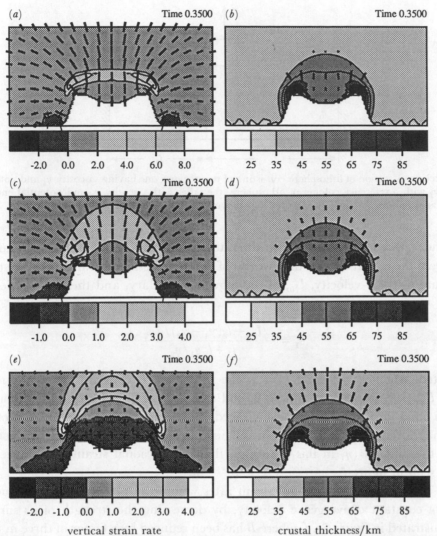

vertical strain rate           crustal thickness/km

FIGURE 5. Contours of vertical strain rate and crustal thickness for calculations illustrating the influence of local changes in strength of the lithosphere. Figures 5c, d are taken from the calculation shown in figure 4, at a dimensionless time of 0.35 (35 Ma after the start of deformation). Figures 5a, b are calculated for the same distribution of topography, but with a decrease in strength in the region where the crustal thickness exceeds 55 km; figures 5e, f are also calculated with this distribution of topography, but the region in which the crustal thickness is less than 55 km has its strength reduced by a factor of three. The band of high compressional strain rates in (a) coincides with the edge of the region whose crustal thickness exceeds 55 km, and the equivalent region in (f) is seen as an area of lithospheric thinning in front of the convergent boundary. Symbols as in figure 4. Vectors indicating principal horizontal deviatoric stresses are superimposed on figures 5a, c and e (maximum values $8.6 \times 10^7$, $1.0 \times 10^8$ and $6.3 \times 10^7$ Pa); principal horizontal strain rates are shown on figures 5b, d and e (maximum values 1.8, 1.3 and $1.2 \times 10^{-15}$ s$^{-1}$).

same arguments apply to a non-newtonian lithosphere. If deformation in the $x$ direction is neglected, the $y$ component of the force balance (equation (3)) reduces to

$$\frac{\partial(2\bar{\tau}_{yy} - \gamma)}{\partial y} = 0 \tag{7}$$

so that

$$4\eta_1\dot{\epsilon}_1 - \gamma_1 = 4\eta_2\dot{\epsilon}_2 - \gamma_2 = \text{constant},$$

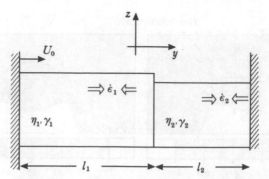

FIGURE 6. Sketch cross section of lithosphere consisting of two regions, one having viscosity $\eta_1$ and potential energy contrast $\gamma_1$ with a reference column of lithosphere, the other with viscosity $\eta_2$ and potential energy contrast $\gamma_2$. They are confined by rigid boundaries that move towards each other with velocity $U_0$.

where $\dot{\epsilon}_1$ and $\dot{\epsilon}_2$ are the horizontal strain rates in the two regions. The constant may be evaluated from the condition that the integral of the compressional strain across the two regions must be equal to the $y$-velocity, $U_0$, of the moving boundary, and these expressions may be re-arranged to yield:

$$\dot{\epsilon}_1 = -\frac{U_0 - l_2(\gamma_1 - \gamma_2)/4\eta_2}{l_1 + l_2\eta_1/\eta_2}. \tag{8}$$

As expected, when $\gamma_1 = \gamma_2$ and $\eta_1 = \eta_2$ the strain rate is constant across the region, and $\dot{\epsilon}_1 = \dot{\epsilon}_2 = -U_0/(l_1+l_2)$; as $\eta_1 \to \infty$, $\dot{\epsilon}_1 \to 0$, and as $\eta_2 \to \infty$, $\dot{\epsilon}_1 \to -U_0/l_1$. The important result displayed in equation (8) is that if, as envisaged above, some initial combination of viscosities and buoyancy forces that support a plateau in compression is perturbed by decreasing the viscosity of the plateau ($\eta_1$ in this example), the compressional strain rate in the plateau is *increased*, as seen in figure 5.

Equation 8 shows that extensional strain rates within the plateau can be induced, under conditions of constant convergence velocity, by decreasing the strength of its surroundings, $\eta_2$; this is illustrated in figure 5*e, f*, where *B* has been reduced by a factor of three in the portion of the lithosphere where the crustal thickness is less than 55 km. *Ad hoc* ways might be imagined of decreasing rapidly the strength of the Asian lithosphere relative to that of the Tibetan Plateau, but they are not pursued here.

### 3.2.2. *Changes in boundary condition*

As can be seen from the discussion of the governing equations and the Argand number in §2.3, and from the simple case described by equation (8), changes in the convergent velocity would influence the balance between compressional stresses and buoyancy forces within the plateau. Complete cessation of the convergence of India with Asia would remove the compressional stress from the plateau, and presumably lead to widespread extension. The definition of the Argand number (equation (4)) shows that dividing the convergent velocity $U_0$ by a given amount is equivalent to increasing the Argand number by the $(1/n)$th power of that amount. England & Houseman (1988) show that a reduction in compressive stress by a factor of three is required before a change in boundary condition in the thin-sheet calculations leads to extension within the plateau. For a lithosphere with a power-law exponent $n = 3$, this would correspond to a decrease in India's velocity by a factor of around 30 (equation (4)) before the extension began in the Pliocene. Such a decrease can be ruled out from present-day observations of plate motion (see, for example, Minster & Jordan 1978) and from the moment release of large earthquakes in Asia (Molnar & Deng 1984), both of which suggest that India's rate of convergence with Asia is still about 50 mm $a^{-1}$.

The arguments of §2.4 suggest that extensional stresses on eastern or western boundaries to the deforming region (such as appear to exist in Eastern China) should die out exponentially with distance from the boundary and would, therefore need to exceed considerably the compressional stresses applied by India to the southern boundary of Asia if they were to be responsible for extension within the Tibetan Plateau. Regional stresses are difficult to determine; the extensional strain rates in Eastern China determined by Molnar & Deng (1984) are similar to those determined for the Tibetan Plateau. This, together with the observation that extensional faulting, although covering most of the plateau, is confined within the 4 km contour (Chen & Molnar 1983) suggests that conditions on the eastern boundary of Asia do not provide the explanation for extension within the plateau.

### 3.2.3. *Increase in potential energy of the plateau*

The removal of compressional stresses by cessation of convergence, or the weakening of the lithosphere surrounding an elevated region, are acceptable mechanisms for producing extension in elevated regions, and may have been responsible for the onset of extension in places such as the Basin and Range or the Andes. However, the discussion of the previous sections suggest that these mechanisms probably did not operate in the case of the Tibetan Plateau.

Equation (8) shows that there is a third perturbation to the force balance that could produce extension in the elevated region: increasing the potential energy of the plateau ($\gamma_1$, equation (8)) may cause extension, even under conditions of constant convergence rate.

Conductive relaxation of the lithosphere can lead to increases in its potential energy, but this takes place over tens of millions of years. An alternative mechanism is suggested by the numerical experiments of Houseman *et al.* (1981) who investigate the stability of the continental lithosphere after thickening. Parsons & McKenzie (1978) point out that the plates should be thought of as consisting of two distinct layers: each of them transfers heat predominantly by conduction but the upper layer, consisting of the crust and uppermost mantle, is mechanically strong whereas the lower layer is fluid, and represents the upper thermal boundary layer to the convecting mantle. Houseman *et al.* (1981) showed that thickening such a rheologically layered lithosphere resulted in gravitational instability of the lower of these layers and its rapid descent into the convecting mantle. Replacement of the thermal boundary layer with hot asthenosphere results in increases to the surface elevation and gravitational potential energy of the mechanical boundary layer. Figure 7 illustrates the influence of this process on the mechanical evolution of thickened lithosphere; a fuller discussion is given by England & Houseman (1988).

For discussion of the mechanics of the Tibetan Plateau, the important result of the experiments of Houseman *et al.* is that the descent of the thickened thermal boundary layer into the convecting mantle follows some time after the thickening. The duration of the delay depends strongly upon the Rayleigh number of the convecting mantle and upon the degree of thickening of the thermal boundary layer, and may be from several million years to several tens of millions of years. If this mechanism operates the conditions exist, in a thickened orogenic belt, for relatively rapid increase in potential energy of the belt at some time after the belt

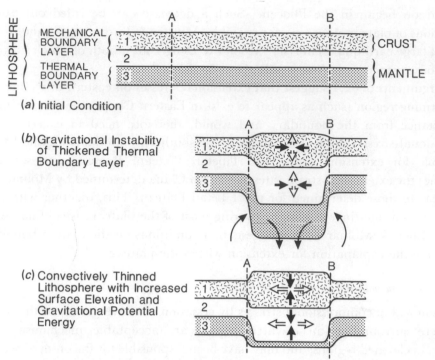

FIGURE 7. Sketch of the mechanical evolution of thickened continental lithosphere, following Houseman *et al.*
(1981). (*a*) In its undisturbed state the continental lithosphere consists of (1) crust, (2) upper mantle that is
sufficiently cold to be much stronger than the convecting mantle and (3) mantle that is hot and weak enough
to take part in mantle convection; this layer forms the upper thermal boundary layer to the convecting upper
mantle. (*b*) A portion of the continental lithosphere is thickened; horizontal thermal gradients are generated
and amplified within the thermal boundary layer and, after a time that depends on the viscosity of the
convecting mantle and the strain in the boundary layer, the thermal boundary layer drops off, to be replaced
by hot material from below. This process increases the surface height and potential energy of the lithosphere
(*c*). As sketched here, the process occurs in a uniform fashion across the deformed region; there is no reason to
believe that this would be the case in reality.

forms; if the increased potential energy exceeds that supportable by the boundary conditions,
extension of the orogen will occur.

Treating the full three-dimensional problem of mantle convection coupled to a deforming
lithosphere is prohibitively expensive of computer time, and a simplified approach to the
convective removal of the thermal boundary layer is required. England & Houseman (1988)
investigate the influence of this process by treating the removal of the thermal boundary layer
as an instantaneous event that occurs at a given value of the thickening strain experienced by
the continental lithosphere. They also show that, for a lithosphere that does not lose its thermal
boundary layer, increases in potential energy due to crustal thickening are largely offset by
decreases due to thickening of dense lithospheric mantle, so that overall changes in potential
energy under these conditions are small compared with those that take place once the thermal
boundary layer leaves the thickened lithosphere.

England & Houseman (1988) consider a range of conditions for the relative magnitudes of
buoyancy forces before and after loss of the thermal boundary layer, but we illustrate their
results with a single calculation (figure 8) that neglects buoyancy forces until the loss of the
thermal boundary layer. In figure 4, the Argand number (equation (4)) is kept at 3
throughout; the conditions of figure 8 are those of figure 4, except that the Argand number for

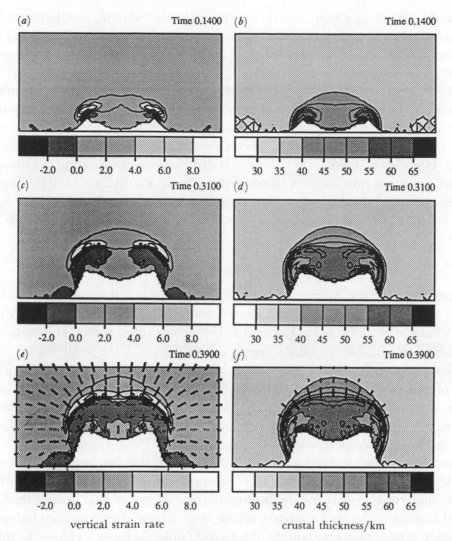

FIGURE 8. As figure 4, except that the potential energy of any piece of lithosphere increases abruptly if its thickening strain exceeds 70% (see text). Note the predominance of extensional horizontal strain rates in the regions where this has occurred (crustal thickness greater than 60 km), and that the principal extensional strain rate is perpendicular to the convergent velocity in front of the indenter. The strength of the lithosphere is assumed here to be unaffected by the convective removal of the boundary layer. Maximum deviatoric stress in $(e)$ is $3.0 \times 10^7$ Pa and maximum principal strain rate in $(f)$ is $1.9 \times 10^{-15}$ s$^{-1}$.

a particular piece of lithosphere is kept at zero until that region has experienced 70% thickening strain, when it is set to 10. The change in Argand number corresponds to a change in the potential energy of the lithosphere, without any change in its strength; it is analagous to changing the quantity $(\gamma_1 - \gamma_2)$ in equation (8), leaving the other parameters unchanged.

The differences between the two cases may be clearly seen in the contours of strain rate (figures 4$a$, $c$, $e$ and 8$a$, $c$, $e$) and crustal thickness (figures 4$b$, $d$, $f$ and 8$b$, $d$, $f$): once crustal thickness exceeds 60 km in figure 8, the buoyancy forces dominate the deformation, and there is no further thickening of the lithosphere, indeed there is net thinning of the lithosphere throughout the region where this condition is achieved. Figure 8$e$ shows that this region can be as extensive, relative to the indenting boundary condition, as the Tibetan Plateau is in relation to the Himalayan front. Furthermore, in this calculation, one horizontal principal stress is extensional and is perpendicular to the direction of motion of the indenting boundary; the focal mechanisms of earthquakes (Molnar & Chen 1983) indicate that this is also the case within the Tibetan Plateau.

There are four parameters to the problem discussed in this section: the value of the power-law exponent, $n$, the values of the Argand number before and after the thermal boundary layer leaves the lithosphere, and the value of the compressional strain at which this transition is assumed to occur. There is not space to discuss the full range of solutions here; this is done by England & Houseman (1988). They show that distributions of topography, crustal thickness and present day strain rate similar to those of the Tibetan Plateau are exhibited by a range of calculations, provided that the vertically averaged deviatoric stresses within the viscous sheet are a few times $10^7$ Pa and its power-law exponent, $n$, is greater than or equal to 3.

## 4. CONCLUSIONS

Calculations that treat the Asian continental lithosphere as a thin continuous sheet appear to be able to account adequately for the distribution of Tertiary strain within Asia (see, for example, Vilotte *et al.* 1984, 1986; England & Houseman 1985, 1986). The horizontal extent of the region of appreciable crustal thickening is consistent with the predicted length scales for the deformation of such a sheet (England *et al.* 1985, and §2.4 of the present paper). The elevation of this plateau, and the concentration of present-day compressional strain rates around its margins (Molnar & Deng 1984) are consistent with calculations that treat the lithosphere as a sheet whose strength is of order 100–200 MPa at geological strain rates (England & Houseman 1986; Vilotte *et al.* 1986).

However, those calculations do not exhibit the thinning that is observed as the most recent phase of deformation on Tibetan Plateau. The arguments of §3 suggest that syn-convergent thinning of thickened crust is not the inevitable consequence of that crustal thickening: the extensional horizontal deviatoric stresses arising from crustal thickening are balanced by the compressional stresses that generate the thick crust; unless there is a change in this balance, extension will not occur in the thickened crust. One simple way in which the balance may change is by the removal of the convergent boundary stresses; this may be the explanation for the extension in the Basin and Range province of western North America. We argue in §3 that this does not seem acceptable for the present extension in the Tibetan Plateau, as convergence between India and Asia is still active, and we suggest that the balance there has been perturbed by the loss of the lower portion of the continental lithosphere, which forms the upper thermal boundary layer to the convecting mantle (Parsons & McKenzie 1978; Houseman *et al.* 1981).

The loss of the thermal boundary layer at the base of the lithosphere would have the effect of raising the surface height of the region of thickened crust by between 1 and 3 km (depending upon the density distribution within the lithosphere, and the thickness of material lost); the consequent change in the potential energy per unit surface area of lithosphere would be equivalent to an extensional driving force of several times $10^{12}$ N per metre length of the elevated region (England & Houseman 1988). The calculation illustrated in figure 8 shows that such a change in the potential energy of the thickened lithosphere is sufficient to effect a transition from horizontal shortening to horizontal extension within the region where the change occurs.

This work was supported by National Science Foundation grant EAR84-08352, and by a visiting fellowship at the Research School of Earth Sciences, Australian National University. We are grateful to John Platt and Peter Molnar for many helpful comments.

### REFERENCES

Achache, J., Courtillot, V. & Zhou, Y. X. 1984 *J. geophys. Res.* **89**, 10311–10339.
Armijo, R., Tapponnier, P., Mercier, J. L. & Han, T. 1982 *Eos, Wash.* **63**, 1093.
Armijo, R., Tapponnier, P., Mercier, J. L. & Han, T. 1986 *J. geophys. Res.* **91**, 13803–13872.
Artyushkov, E. V. 1973 *J. geophys. Res.* **78**, 7675–7708.
Bird, P. & Baumgardner, J. 1984 *J. geophys. Res.* **89**, 1932–1944.
Bird, P. & Piper, K. 1980 *Phys. Earth planet. Inter.* **21**, 158–175.
Brace, W. F. & Kohlstedt, D. L. 1980 *J. geophys. Res.* **85**, 6248–6252.
Chang, C. *et al.* 1986 *Nature, Lond.* **323**, 501–507.
Chen, W.-P. & Molnar, P. 1983 *J. geophys. Res.* **88**, 4183.
Dalmayrac, B. & Molnar, P. 1981 *Earth planet. Sci. Lett.* **55**, 473.
Ekström, G. & England, P. 1988 *J. geophys. Res.* (Submitted.)
England, P. C. & Houseman, G. A. 1985 *Nature, Lond.* **315**, 297–301.
England, P. C. & Houseman, G. A. 1986 *J. geophys. Res.* **91**, 3664–3676.
England, P. C., Houseman, G. A. & Sonder, L. S. 1985 *J. geophys. Res.* **90**, 3551–3557.
England, P. C. & Houseman, G. A. 1988 *J. geophys. Res.* (In the press.)
England, P. C. & McKenzie, D. P. 1982 *Geophys. Jl R. astr. Soc.* **70**, 292–321.
England, P. C. & McKenzie, D. P. 1983 *Geophys. Jl R. astr. Soc.* **73**, 523–532.
Evison, F. F. 1960 *Geophys. Jl R. astr. Soc.* **3**, 155–190.
Fleitout, L. & Froidevaux, C. 1982 *Tectonics* **1**, 21–56.
Frank, F. C. 1972 In *Flow and fracture of rocks* (Geophys. Monogr. no. 16) (ed. H. C. Heard, I. Y. Burg, N. L. Carter & C. B. Rayleigh), pp. 285–292. Washington, D.C.: American Geophysical Union.
Houseman, G. A. & England, P. C. 1986 *J. Geophys. Res.* **91**, 3651–3663.
Houseman, G. A., McKenzie, D. P. & Molnar, P. 1981 *J. geophys. Res.* **86**, 6115–6132.
Jackson, J. & McKenzie, D. 1988 *J. R. astr. Soc.* (In the press.)
McKenzie, D. P. 1972 *Geophys. Jl R. astr. Soc.* **30**, 109–185.
Minster, B. & Jordan, T. 1978 *J. geophys Res.* **83**, 5331–5354.
Molnar, P. & Chen, W. P. 1978 *Nature, Lond.* **273**, 218–220.
Molnar, P. & Chen, W. P. 1983 *J. geophys. Res.* **88**, 1180–1196.
Molnar, P. & Deng, Q. 1984 *J. geophys. Res.* **89**, 6203–6228.
Molnar, P. & Lyon-Caen, H. 1988 *Geol. Soc. Am. Spec. Publ.* (In the press.)
Molnar, P. & Tapponnier, P. 1978 *J. geophys. Res.* **83**, 5361–5375.
Molnar, P. & Tapponnier, P. 1981 *Earth planet. Sci. Lett.* **52**, 107–114.
Molnar, P., W.-P. Chen, & Tapponnier, P. 1981 In *Geological and ecological studies of Qinghai-Xizang Plateau* vol. 1, pp. 757–762.
Molnar, P. & Tapponnier, P. 1975 *Science, Wash.* **189**, 419–428.
Molnar, P. & Tapponnier, P. 1978 *J. geophys. Res.* **83**, 5361–5375.
Parsons, B. & McKenzie, D. 1978 *J. geophys Res.* **83**, 4485–4496.
Sonder, L. J. & England, P. C. 1986 *Earth planet. Sci. Lett.* **77**, 81–90.
Tapponnier, P., & Molnar, P. 1976 *Nature, Lond.* **264**, 319–324.
Tapponnier, P. *et al.* 1981 *Nature, Lond.* **294**, 405–410.
Tapponnier, P., Pelzer, G., Le Dain, A. Y., Armijo, R. & Cobbold, P. 1982 *Geology* **10**, 611–616.
Vilotte, J. P., Daignieres, M. & Madaraiga, R. 1982 *J. geophys. Res.* **87**, 10709–10728.
Vilotte, J. P., Daignieres, M., Madaraiga, R. & Zienckiewicz, O. C. 1984 *Phys. Earth planet. Inter.* **36**, 236–259.
Vilotte, J. P., Madariaga, R., Daignieres, M. & Zienckiewicz, O. 1986 *Geophys. Jl R. astr. Soc.* **84**, 279–310.
Zhu, X., Liu, C., Ye, S. & Liu, J. 1977 *Sci. Geol. Sin.* **1**, 44–51.
Zhu, Z., Zhu, X. & Zhang, Y. 1981 In *Geological and ecological studies of Qinghai-Xizang Plateau* vol. 1, pp. 931–939. Beijing: Science Press.

## Discussion

M. F. OSMASTON (*The White Cottage, Woking, Surrey, U.K.*). I am concerned about the force balance across the High Himalaya that Dr England's kind of modelling would seem to require to achieve and maintain (if that is the case) its extreme elevation. From the north, how can the spreading of Tibet, which is significantly lower, provide enough force? From the south, ocean-ridge push, as customarily calculated at present, is quite inadequate. Work that I am doing on the shape-evolution histories of subduction interfaces makes me increasingly doubtful that slab pull is the major force it has been made out to be. I therefore suggest that the High Himalayan elevation may be being maintained thermally, by crustal reheating, at a rate that sufficiently balances its rheological tendency to spread. The migration of upthrust activity to areas of lower elevation seems consistent with this.

P. C. ENGLAND. There is no appreciable difference in *average* elevation between the High Himalaya and the Tibetan Plateau (see, for example, Bird 1978).

*Reference*

Bird, P. 1978 *J. geophys. Res.* **83**, 4975–4987.

S. GHOSH (*Grant Institute of Geology, University of Edinburgh, U.K.*). It appears that although we know with a fair amount of confidence much about the early history of the tectonic evolution of the Himalayas and Tibet, the present-day movement scenario around the Himalayas including its crustal–subcrustal parts is still not very clear. There is hardly any doubt that the Himalayas is still an active mountain chain. Would it not be worthwhile for us now to explore the possibilities of identifying thrust areas for Himalayan research so that the important gaps in our knowledge about the geology and tectonics of the Himalayas are filled up within a comparatively short time?

P. C. ENGLAND. It would perhaps be more appropriate if this reply were left to Dr Molnar, who succinctly summarizes elsewhere in this volume the major problem areas, from the Himalaya to the Qilian Shan (with many references), but this comment raises an important point. It should be emphasized that the mechanism that is proposed in this paper for the evolution of the Tibetan Plateau, although consistent with our understanding of the behaviour of very viscous fluids, and with the observations of bathymetry and heat flow in the oceans is not established as occurring within the Earth. Observations of the history of surface elevation and extensional faulting and of present-day rates of movement in the Himalaya and the Tibetan Plateau will yield important insights into the dynamics of the lithosphere and mantle in zones of continental convergence. The technology will exist within the next few years for the precise determination of strain rates on these scales, and the successful pursuit of this very important data set will depend strongly both upon international cooperation, and upon cooperation between Earth scientists using a wide range of tools, from field observations to space geodetic techniques.

*Phil. Trans. R. Soc. Lond.* A **326**, 321–325 (1988)     [ 321 ]
*Printed in Great Britain*

# General discussion

J. G. RAMSAY, F.R.S. (*ETH-Zentrum, Zürich, Switzerland*). Several of the cross sections and profiles through various parts of the Himalayas that have been presented at the Meeting were based on geometric techniques appropriate to the frontal thrust belt of the Rocky Mountains of Canada and the U.S.A., or to the soft sediment deformation in and around Taiwan. In these reconstructions, faults, especially thrusts, are considered to exert the dominant control on the forms of structures. Fold forms appear only as fault bend folds developed as a consequence of movement of thrust sheets over irregular step-like thrust plane topography. In these models the effect of rock competence only seems to be considered as a characteristic rock property controlling the ramp-flat geometry of the fault planes, and the rock properties seem to exert little or no influence on the fold style. I would suggest that this current fashion of making constructions to depth is not only mechanically unsound, but it does not accord with the observations of structural geometry. First, the folds we see often show rounded hinges, tight interlimb angles, and changes of layer thickness that are not those of the constant thickness open kink-like folds arising from nappe transport over ramps and flats. The observed geometric variations of bed thicknesses in folds also often imply quite marked difference of ductility of the rock layers: the beds do not behave in a passive way to retain a constant layer length. In all folded rocks found in the crystalline units and in many in the non-metamorphic terranes, cleavage and schistosity are characteristically present. When deformation markers are seen it is very clear that rock deformation has been ductile and that the strains can be very high ($X \cdot Z = 50 \cdot 1$), quite outside the values predicted from the development of fold bend folds, and that cleavage also implies moderate to high rock strains. I fail to understand why some workers persist in using a model that is so much at odds with the undisputed observations and accept geometric constraints on their constructions that are inappropriate to the behaviour of deformation styles in a ductile crust.

R. W. H. BUTLER (*Department of Earth Sciences, Open University, U.K.*). Professor Ramsay criticizes the use of thrust tectonic models to explore the geological structure of mountain belts. He allies these models to a rather narrow and out-moded idea of balanced section construction, namely that for sections to balance they must conform to a simple 'Rocky Mountain' style (see Ramsay & Huber 1987) where line lengths are preserved and all deformation is contained on very narrow faults. If the aim of such section construction in the metamorphic parts of mountain belts was to completely predict structure to depth as in some oil and gas fields (Suppe 1983), his criticisms would be valid. However, that has never been the aim nor indeed the method adopted. So what have been the aims of regional section construction and structural restorations across broad tracts of mountain belts?

(i) Section balancing can be used to test gross crustal structure. For example, in northern Pakistan we predict that where cover rocks alone are stacked up, the crystalline basement from which these sheets have detached must continue further north, ultimately under the Kohistan arc terrane (Coward *et al.* this symposium, and references therein). In contrast, the Everest transect can accommodate crustal shortening beneath the surface expression of thrust structures with only limited detachment (Burg & Chen 1984; Butler & Coward 1988). Models such as these are important in planning the location of vastly more expensive seismic experiments and in the interpretation of resulting reflection profiles.

(ii) Structural restorations can indicate where plate convergence is being accommodated within the continental lithosphere. The crude restorations in the western Himalayas suggest that approximately one third of the bulk convergence between India and stable Asia has been accommodated south of the suture since collision. It may well be more but India certainly can be shown not to behave as a rigid indenter as once believed (Butler & Coward 1988).

(iii) Structural restorations can provide time-slice views of mountain belts from which it is possible to predict the evolution of the main orogenic loads that drive subsidence and sedimentation on the Indian foreland (Lyon-Caen & Molnar 1985). Coupled with gravity, stratigraphic and sedimentological data it is possible then to model the bulk thermo-mechanical behaviour of the lithosphere.

(iv) Combining structural restorations with pressure–temperature estimates within mountain belts, it is possible to reconstruct the interplay between horizontal and vertical movements that, coupled with erosion, control the morphology of the Himalayan ranges.

So what of the method? Clearly to gain truly representative estimates of orogenic shortening or to predict subsurface structure on a fold-by-fold, thrust-by-thrust basis we require the complete incremental restoration of all ductile strains, rotations and displacements in three dimensions together with estimates of volume changes associated with fluid flow and metamorphism. This has yet to be achieved even in the most well-studied and structurally simple parts of mountain belts. But in areas approximating to plane strain it is relatively straightforward to restore the complex catalogue of non-fault-bend straining folds outlined by Ramsay (see Butler 1988 for discussion) by using formational area balancing with local bed-thickness restoration. It is possible to simplify the structure for restoration purposes so that extending bed-length segments are not included in the final shortening value, an approximation which considers displacements to be localized onto infinitesimally narrow faults. The simplifications have been outlined at length elsewhere; they produce minimum estimates of orogenic shortening. They are not dependent on the narrow range of structural styles as advocated by Professor Ramsay.

Despite being minimum values, shortening in the western Himalayas determined from balanced sections is so large (over 470 km together with at least 150 km on the MMT) that it is possible to test simple models of India–Asia collision and begin to tackle the aims outlined above. At present the preferred model, as outlined by Coward *et al.* (this symposium; see also Butler 1986), of varying levels of detachment within the Indian crust developed during thrusting is consistent with the mappable surface geology (as interpreted by section balancing), the generation of foredeep basins, available geophysical data (gravity, earthquake epicentres and fault-plane solutions, seismic refraction and shallow reflection profiles) and uplift patterns in the mountain belts. It is interesting that the same broad conclusion was reached by Argand (1924), albeit with a far weaker data base. Professor Ramsay's 'undisputed observations' are entirely consistent with this model. The vast bulk (say 90%) of the net convergence between India and Asia that is located south of the suture appears to be accommodated by movements along relatively discrete thrust sense zones of simple shear rather than by pure shear buckling, upright cleavage or schistosity (say 10%). Although recognizing their limitations, certainly in markedly non-plane-strain settings, cross sections balanced for formational area and displacement provide a test of structural models and a regional framework to incorporate entirely separate geological information. Small-scale structures provide important information on the kinematics of large-scale detachment and ultimately can be used to determine strain rates etc., at least when combined with microstructural work. But the integration with 'non-structural' data is crucial if we are to understand the tectonics of mountain belts. It is unclear how myopic focusing on small-scale structures and strain measurements without the broader view offered by regional section construction will achieve our aims.

*References*

Argand, E. 1924 In *Proc. 13th. Int. Geol. Congr., Brussels.*
Burg, J. P. & Chen, C. M. 1984 *Nature Lond.* **311**, 219–23.
Butler, R. W. H. 1986 *J. geol. Soc. Lond.* **143**, 857–873.
Butler, R. W. H. 1988 In *Tectonic evolution of the Tethyan Regions* (ed. A. M. C. Sengör). Proc. NATO Advanced Study Inst.
Butler, R. W. H. & Coward, M. P. 1988 In *Tectonic evolution of the Tethyan Regions* (ed. A. M. C. Sengör). Proc. NATO Advanced Study Inst.
Lyon-Caen, H. & Molnar, P. 1985 *Tectonics* **4**, 513–538.
Suppe, J. 1983 *Am. J. Sci.* **283**, 684–721.
Ramsay, J. G. & Huber, M. I. 1987 *The techniques of modern structural geology. Volume 2. Folds and fractures.* London: Academic Press.

**M. P. Coward** (*Department of Geology, Imperial College, U.K.*). Professor Ramsay's discussion is claimed to be a general comment on several contributions. I shall, therefore, answer it in a general way, not specifically about any one contribution. He claims to fail to understand why some workers persist in using a model based on techniques appropriate to the frontal thrust belt of the Rocky Mountains, etc. I presume that even the most ardent critic of thrust geometry has to admit that thrusts do dominate the frontal structures of the Himalayas. They are clearly imaged on seismic profiles and are exposed in the frontal ranges from Nepal to Pakistan. However, few of these thrusts carry kink-band, fault-bend folds (Suppe 1983); many have folds or fold-trains developed by buckling of beds on fault hanging-walls due to variations in displacement on the fault. As Professor Ramsay claims, the fold inner arcs may show complex structures formed by the accumulation of weak material such as shale or salt. When constructing cross sections it is unwise to opt for one particular fold model or construction technique, e.g. the fault-bend fold model of Suppe (1983), constant or variable heave models (Verrall 1981; Williams & Vann 1987; Wheeler 1987), isogen models of Ramsay & Huber (1986). The construction should be compatible with the field and/or seismic observations. To follow this further: it is often very unwise to produce cross sections from simple map or seismic data alone, even though apparently there may be adequate orientation data, until clear observations have been made in the field on fold shape, fault sequence, strain, etc.

Professor Ramsay claims that the current fashion in interpreting sections to depth does not accord with observations on geometry. Viable cross sections must use all the available data, from the surface and from depth. These could include measurements of dip, bed thickness, strain and cleavage, incremental strains, compaction and diagenesis, fault kinematics plus seismic, gravity and magnetic data. If using a reasonable spread of seismic or field data, it is important to contour bed and fault surfaces and produce cut-off and branch-line maps (Coward 1984). In regions of consistent or only slightly variable plunge, it is possible to attempt down-plunge projections. However, as faults and their associated buckle or ramp-induced folds do vary in geometry along strike, such projections can lead to serious errors and this method must not be relied on. Section construction is not a quick slap-dash method, but a rigorous discipline, based on established structural principals. It certainly should encourage, not discourage, detailed field observations among its protagonists.

Further, it is important to determine the fault-slip direction. All the simple balancing techniques involve an assumption of plane strain, or a knowledge of the area change that has affected the particular section. Sections drawn oblique to the fault-slip direction always have material that has moved in or out of section and it is questionable as to whether such sections maintain constant area during deformation. In Pakistan, there are variable thrust movement directions in the eastern parts of the Salt Ranges and near the Hazara syntaxis (Coward *et al.*, this symposium) and hence here, simple two-dimensional (2D) balancing techniques should not be used. Three-dimensional (3D) balancing or sequential partial restoration should be attempted.

Apart from encouraging a rigorous approach to structural geology, balanced section construction has several uses. Predictions can be made and tested with extra surface or seismic data (or well data in exploration!). Estimates can be made of detachment depths and amounts of displacement. Indeed the concept of necessary basal detachment essentially came from early balanced section work (Chamberlain 1910). Before and often since then, many geologists have drawn sections with fold or fault geometry that clearly cannot work. If a section does not balance then that particular section or some of its assumptions are wrong. A balanced section need not be wrong. However, section-balancing techniques are iterative and rarely result in a unique solution, especially as they depend largely on the amount of input data. Details of the sections will change as more data are forthcoming. Thus, as claimed by Ramsay & Huber (1987) 'beware of the geologist who announces ... that his sections are "perfectly balanced" and who gives the impression that these sections are above reproach from any source. If you meet such a character you can be pretty sure that he is naïve, dishonest, or combines these traits.' The only sections that are 'beyond reproach' are those that can be photographed on hillsides or cliffs, and it is surprising how many geologists copy these incorrectly!

Certainly, as Professor Ramsay claims, the internal parts of orogenic belts have a complex history and it is difficult to produce reliable and reproduceable sections in these regions. Professor Ramsay seems concerned about fold shapes and 2D ductile strains in internal regions. It is possible to deal with some problems using area balancing methods. Kink-band methods are clearly not applicable to internal zones and line-length balancing methods need to take into account rock strain. Far more problematical are complex 3D strain histories involving rotations and thrust movements in more than one direction. These problems are certainly encountered in the internal zones of the western Himalayas (see Coward *et al.*, this symposium) and any sections to depth through these areas, or similar sections through, say, the internal zones of the Caledonides or western Alps, must be considered only as educated tectonic sketches. The sections can be balanced; there is no point in producing a section that looks glaringly ridiculous without discussing why, but the balancing is only as good as the input.

Section balancing is similar to many other techniques and concepts in structural geology, such as plate-tectonic modelling, fold analysis, finite and incremental strain measurement. It can be done well or badly. As with the other techniques, it can lead to tremendous increases in knowledge of 2D and 3D structural geology, but can also be abused. The structural geologist should attempt to understand and use all the techniques available.

*References*

Chamberlain, R. T. 1910 *J. Geol.* **18**, 228–251.
Coward, M. P. 1984 *Scott. J. Geol.* **20**, 87–106.
Ramsay, J. G. & Huber, M. 1986 *The techniques of modern structural geology, Volume 2. Folds and fractures.* London: Academic Press.
Suppe, J. 1983 *Am. J. Sci.* **283**, 684–721.
Verrall, P. 1981 *Structural interpretation with application to North Sea problems* (Joint Association for Petroleum Exploration Courses (U.K.), vol. 3.
Wheeler, J. 1987 *J. struct. Geol.* **9**, 1047–1049.
Williams, G. D. & Vann, I. 1987 *J. struct. Geol.* **9**, 789–795.